Pharmacovigilance for Herbal and Traditional Medicines

Joanne Barnes
Editor

Pharmacovigilance for Herbal and Traditional Medicines

Advances, Challenges and International Perspectives

Adis

Editor
Joanne Barnes
School of Pharmacy
Faculty of Medical and Health Sciences
University of Auckland
Auckland, New Zealand

ISBN 978-3-031-07273-4 ISBN 978-3-031-07275-8 (eBook)
https://doi.org/10.1007/978-3-031-07275-8

This Adis imprint is published by the registered company Springer Nature Switzerland AG
The registered company address is: Gewerbestrasse 11, 6330 Cham, Switzerland

For APB and JKB.
For Hazel, and for Pete.
Everlong.

Prologue

A large proportion of the world's population relies on traditional and herbal medicines as part of their normal healthcare. Many countries have their own systems of traditional medicine, including indigenous forms of treatment using plant species and other natural substances native to these regions, but while some countries have formalized systems—for example, Ayurveda or Traditional Chinese Medicine—in other places, herbal and traditional medicines are widely used without written or organised prescribing or regulatory systems. Thousands of herbal products are available worldwide and, even in regions with highly developed schemes for monitoring the safety of medicines, such as Europe or North America, many 'natural health' products (NHPs) or 'complementary and alternative' remedies are sold and administered outside regulatory or other healthcare frameworks. Importantly, patients and their families and carers may be oblivious that herbal and other NHPs purchased in a local pharmacy, supermarket or online, may have had little or no formal efficacy and safety testing during product development.

Patients worldwide are vulnerable to the adverse effects of herbal and traditional medicines, which may range from mild reactions, such as nausea or rash, to life-threatening or fatal events, such as hepatotoxicity or liver failure. Adverse drug reactions are often a surprise to those consuming products viewed as 'healthy' supplements or 'natural' remedies for common ailments. Some populations have increased susceptibility, for example, women using herbal medicines during pregnancy and labour, or elderly people with reduced renal function and coexisting medical conditions. Assessment of the risks of medicines is always challenging in these groups, and it is especially difficult if products do not have documented evidence of safety, including registered clinical trials with sufficient data from relevant populations.

In the past 15 years, use of herbal and traditional medicines worldwide has increased substantially: there are now many millions of people using a wide variety of NHPs and dietary supplements on a regular basis. Global sales have risen, with the herbal market estimated at hundreds of billions of dollars in the USA alone. In other countries, it is often more difficult to estimate the value of herbal pharmaceutical markets—or the potential danger to those who consume the formulations available on the shelves or websites of companies aiming to profit from these sales. In addition to safety concerns, uncertainties around the efficacy of herbal medicines remain. During the COVID-19 pandemic, there have been claims that some alternative and complementary remedies prevent or treat the novel coronavirus. These assertions have not been supported by evidence from robust clinical trials and promoting such products to fearful and vulnerable communities—often taking money from those who can least afford it—is both unacceptable marketing behaviour and potentially harmful to patients.

All of these issues, from the promotion of herbal medicines to the complex assessments undertaken in monitoring the safety of NHPs, underline the necessity and relevance of this important book. It has never been more crucial to examine the use of herbal and traditional medicines worldwide, to review known and potential safety signals and discuss uncertainties around evidence of efficacy and to consider how these evaluations may best be communicated to those taking these products. Such risk-benefit assessments are the daily work of the experts who have contributed to this fascinating book: within these pages you will find discussion of the challenges to those working in pharmacovigilance, and valuable perspectives from several different international regions. I encourage you to read with an open and interested mind, to share the knowledge gained with colleagues and, most importantly, to use the information this book provides to improve the care of patients across the globe.

International Society of Pharmacovigilance, ISoP Mira Harrison-Woolrych,
Dunedin, New Zealand

Preface

I conceived the idea for this book in 2004 when I began planning a dedicated conference (held in London in April 2006) on 'Pharmacovigilance of herbal medicines: current status and future directions' in conjunction with the (then) Royal Pharmaceutical Society of Great Britain and partners, including the International Society of Pharmacovigilance (ISoP), the Uppsala Monitoring Centre (UMC) and the European Scientific Co-operative on Phytotherapy, among others. This book has been a personal vision and goal since then, and several of the chapter authors are colleagues who delivered excellent presentations at that conference.

There is a pertinent need for an authoritative text focusing solely on the science and practice of monitoring harms associated with the use of herbal medicinal products, (other) traditional medicines and natural health products (HTMs/NHPs) in their broadest sense. The use of these types of products is ubiquitous globally and, in many low- and middle-income countries, and in traditional-medicine systems, their important place in healthcare has been recognised for very long periods of time. There is every indication that the use of herbal and traditional medicinal preparations and of manufactured, contemporary, natural health products will continue. It is to be expected that, if these preparations and products are pharmacologically active, then use of them, under certain circumstances and for some patients, will be associated with experiences of adverse reactions. This is not, however, universally recognized, including among some practitioners and users of HTMs/NHPs, and contributes to challenges in assessing and monitoring their safety profiles. This should not deter us from having safe use as the priority with respect to patient and consumer access to these products, underpinned by robust regulatory frameworks, including pharmacovigilance for HTMs/NHPs.

The concept of the book (and the conference in 2006) arose from a desire to draw together current knowledge and practices with respect to pharmacovigilance for herbal medicinal products, as well as to identify, showcase and celebrate advances and innovation in monitoring the safety of this unique and complex category of products and preparations. The book covers all fundamental aspects of pharmacovigilance for herbal medicines in particular and, by and large, also relates to the broader categories of (other) traditional medicines; these may, for example, include other ingredients of natural origin (such as insect and animal parts), and manufactured/finished natural health products (which covers a wide range of other types of products, including 'food/dietary supplements', probiotics and so forth), which present very similar challenges in pharmacovigilance and regulatory science in general. The general content and structure for the book arose, in part, from some of the key topics discussed at the 2006 conference and were further informed by my (the editor) interests and insights into this niche area in pharmacovigilance.

The book is in two parts. Part I covers the current status of pharmacovigilance for herbal medicines, including advances and challenges in the discipline. This part begins with a fascinating historical journey exploring the origins of safety and safety monitoring for herbal medicines, followed by a comprehensive account of the contemporary prevalence of use of HTMs/NHPs. Other chapters highlight the potential toxicity of certain herbal medicinal products with reference to specific groups of toxic herbal constituents and illustrate the importance of natural products chemistry to harms associated with herbal medicines, and its relevance in considering how pharmacovigilance for these products should be approached. Several other chapters discuss methodological approaches and ongoing challenges in pharmacovigilance for herbal medicines, including issues relating to nomenclature, coding and classification, and the idiosyncrasies and nuances involved in causality assessment for suspected adverse reactions associated with herbal medicines. These topics are discussed in detail in these chapters, along with possible solutions and considerations for moving forward.

It is, indeed, important to consider what advances have been made in pharmacovigilance for herbal medicines, and there have been several over the last decade or so. Several chapters, such as a purchase-event monitoring method piloted in New Zealand (Chap. 7), an active surveillance model developed in Canada (Chap. 19) and an ethnobotanical approach applied in Brazil (Chap. 21), were designed with the intention of improving pharmacovigilance specifically for HTMs/NHPs; others, such as the introduction in some countries of direct patient reporting of suspected adverse reactions to spontaneous reporting systems, have been implemented with a more general purpose. Beyond these techniques, there has been progress in regulatory pharmacovigilance for HTMs/NHPs in many countries, with 'light-touch' regulatory frameworks for HTMs/NHPs (or 'complementary medicines') mandating manufacturers to undertake pharmacovigilance activities for their products authorized under these regulations. Progress has also occurred in the professional space, with the launch, in 2017, of an International Society of Pharmacovigilance special interest group dedicated to Herbal and Traditional Medicines.

Still, spontaneous reporting remains the cornerstone of pharmacovigilance for HTMs/NHPs. Against this background, in Part II, the book dedicates several chapters to pharmacovigilance for HTMs/NHPs around the world, beginning with an overview and new analysis of international case safety reports held in VigiBase, the World Health Organization's (WHO) global database of individual case safety reports (ICSRs), maintained by the UMC. Ten other chapters from different countries, representing diverse historical, ethnic, cultural, social and political contexts, provide deeper insights and perspectives into pharmacovigilance, namely spontaneous reporting, for herbal and traditional medicines in those countries, and in the context of the local use, practice and regulatory landscape for these products.

Perhaps most of all, we should seek and applaud better understanding of consumers' and patients' use, access and beliefs around herbal and traditional medicines and, most particularly, how people experience, identify and behave in response to lived adverse events during or following the use of HTMs/NHPs. The book is peppered with references and insights to this incredibly important aspect—the human factors—concerning herbal and traditional medicines' use and their relevance for pharmacovigilance for these products and preparations.

My aspirations for this book at the outset were that it would be informative, interesting and inspiring. From my perspective, my vision for this book has been realized, and with such excellent contributions from an outstanding, diverse set of authors. I warmly and sincerely thank all the contributors for sharing their ideas, expertise and enthusiasm for the topics in this book and, of course, for their time and efforts in writing their respective chapters and pro-/epilogues. I also thank the authors, the publisher and Nitin Joshi (Editor of Drug Safety), for their patience and support while this book has evolved and, finally, come to fruition.

As always, my deepest thanks and appreciation are to my beautiful family, and my amazing friends, for all their love and support throughout the creation of this book and beyond.

Auckland, New Zealand Joanne Barnes

Acknowledgement

My grateful thanks go to E Lyn Lee for her administrative support and other brilliant assistance in helping to finalise the manuscript for this book.

Contents

Part I Advances and Challenges in Pharmacovigilance for Herbal Medicines

1 **The Historical Development of Pharmacovigilance for Herbal Medicines** 3
Jeffrey K. Aronson

2 **Prevalence of Use of Herbal and Traditional Medicines** 15
E Lyn Lee and Joanne Barnes

3 **Pyrrolizidine Alkaloids in Herbal Medicines and Food: A Public Health Issue** 27
Mina Kalantar zadeh and Elizabeth M. Williamson

4 **Pharmacovigilance for Herbal and Traditional Medicine-Induced Liver Injury** 41
Jia-bo Wang, Yuan Gao, Yu-ming Guo, and Xiao-he Xiao

5 **Herb-Drug Interactions: Fundamental Mechanisms, Prevalence and Challenges in Their Identification** 51
Jose M. Prieto and Andre L. D. A. Mazzari

6 **Safety and Pharmacovigilance of Herbal Medicines in Pregnancy** 77
Sally Stephens

7 **Advances in Methods and Techniques in Pharmacovigilance for Herbal and Traditional Medicines and Other Natural Health Products** 93
Joanne Barnes

8 **Botanical Nomenclature for Herbal Medicines and Natural Products: Its Significance for Pharmacovigilance** 141
Bob Allkin and Kristina Patmore

9 Coding Reports Involving Herbal Medicines in a Pharmacovigilance Database 177
Florence van Hunsel and Souad Skalli

10 Causality Assessment in Pharmacovigilance for Herbal Medicines 189
Rolf Teschke and Gaby Danan

11 The Value of Complementary Approaches to Causality Assessment for Individual Case Safety Reports: The Example of *Artemisia annua* and Hepatotoxicity 211
Ruth L. Savage

12 Development of a Natural Health Product Active Surveillance Method in Outpatient Centers in Canada 221
Morgan Bharadia, Candace Necyk, and Sunita Vohra

13 Ethnopharmacovigilance and Traditional Medicines 237
Eliana Rodrigues and Joanne Barnes

14 Pharmacovigilance for Herbal Medicines: A Perspective from the Herbal Medicines Industry 249
Phil Rasmussen

15 Pharmacovigilance and Risk Communication for the Safe Use of Herbal and Traditional Medicines: How to Promote Evidence and Keep People Safe 261
Bruce Hugman and Souad Skalli

Part II International Perspectives in Pharmacovigilance for Herbal Medicines

16 Reports for Herbal Medicines in the Global Suspected ADR Database VigiBase 279
Florence van Hunsel, Lucie M. Gattepaille, Camilla Westerberg, and Joanne Barnes

17 The Italian Phytovigilance Spontaneous Reporting Scheme 291
Francesca Menniti-Ippolito and Fabio Firenzuoli

18 Pharmacovigilance for Herbal and Traditional Medicines in Bosnia and Herzegovina 297
Martin Kondža and Biljana Tubić

19 Pharmacovigilance for Herbal Products: A Canadian Perspective 307
Kevin Bernardo and Shahid Perwaiz

20 Pharmacovigilance of Cannabis Products for Medical and Non-medical Purposes 317
Stephanie Jack

21 Pharmacovigilance for Herbal Medicines in Brazil 335
Marcelo Vogler de Moraes and Juliana de Castro Zoratto

22 Pharmacovigilance for Indian Traditional Medicines 347
Vivekanandan Kalaiselvan and Galib Ruknuddin

23 Pharmacovigilance for Traditional Chinese Medicinal Drugs in China 357
Li Zhang and Tianyi Yang

24 Pharmacovigilance for Herbal and Traditional Medicines in Japan 371
Lida Teng

25 Pharmacovigilance for Herbal Medicines in Iraq 385
Manal M. Younus and Inas R. Ibrahim

26 Pharmacovigilance for Herbal and Traditional Medicines in the Sultanate of Oman 395
Hussain Al Ramimmy, Shirly Varughess, and Nawal Al Alawi

27 Pharmacovigilance for Herbal Medicines in Sudan 401
Ammar Abbas

Epilogue 407

Contributors

Ammar Abbas Countess of Chester Hospital NHS Foundation Trust, Chester, UK

Pharmacovigilance Committee—National Medicines & Poisons Board, Khartoum, Sudan

Nawal Al Alawi Department of Pharmacovigilance & Drug Information, Ministry of Health, Muscat, Oman

Hussain Al Ramimmy Department of Pharmacovigilance & Drug Information, Ministry of Health, Muscat, Oman

Bob Allkin Medicinal Plant Names Services, Royal Botanic Gardens, Kew, Richmond, UK

Jeffrey K. Aronson Centre for Evidence Based Medicine, Nuffield Department of Primary Care Health Sciences, Radcliffe Observatory Quarter, Oxford, UK

Joanne Barnes School of Pharmacy, Faculty of Medical and Health Sciences, University of Auckland, Auckland, New Zealand

Kevin Bernardo Marketed Health Products Directorate, Health Canada, Ottawa, ON, Canada

Morgan Bharadia Faculty of Pharmacy and Pharmaceutical Sciences, University of Alberta, Edmonton, AB, Canada

Department of Medicine, Faculty of Medicine and Dentistry, University of Alberta, Edmonton, AB, Canada

Gaby Danan Pharmacovigilance Consultancy, Paris, France

Juliana de Castro Zoratto Regulation Specialist, Brazilian Health Regulatory Agency - Anvisa, Brasília, Brazil

Fabio Firenzuoli CERFIT, Research and Innovation Center in Phytotherapy and Integrated Medicine, Careggi, University Hospital, Florence, Italy

Yuan Gao China Military Institute of Chinese Medicine, Fifth Medical Center of Chinese PLA General Hospital, Beijing, People's Republic of China

Lucie M. Gattepaille Uppsala Monitoring Centre, Uppsala, Sweden

Yu-ming Guo China Military Institute of Chinese Medicine, Fifth Medical Center of Chinese PLA General Hospital, Beijing, People's Republic of China

Bruce Hugman Uppsala Monitoring Centre (at the time of writing), Uppsala, Sweden

Inas R. Ibrahim College of Pharmacy, Uruk University, Baghdad, Iraq

Stephanie Jack, BSc, MSc, MPH Office of Cannabis Science and Surveillance, Controlled Substances and Cannabis Branch, Health Canada, Ottawa, Canada

Vivekanandan Kalaiselvan Indian Pharmacopoeia Commission, Ministry of Health & Family Welfare, Government of India, Ghaziabad, India

Mina Kalantar zadeh Food Research Consultant, Wokingham, Berkshire, UK

Martin Kondža Faculty of Pharmacy, University of Mostar, Mostar, Bosnia and Herzegovina

E Lyn Lee School of Pharmacy, Faculty of Medical and Health Sciences, University of Auckland, Auckland, New Zealand

Andre L. D. A. Mazzari Centre for Pharmacognosy, UCL School of Pharmacy, London, UK

Francesca Menniti-Ippolito National Centre for Drug Research and Evaluation, National Institute of Health, Rome, Italy

Candace Necyk Faculty of Pharmacy and Pharmaceutical Sciences, University of Alberta, Edmonton, AB, Canada

Kristina Patmore Medicinal Plant Names Services, Royal Botanic Gardens, Kew, Richmond, UK

Shahid Perwaiz Marketed Health Products Directorate, Health Canada, Ottawa, ON, Canada

Jose M. Prieto Centre for Natural Products Discovery, School of Pharmacy and Biomolecular Sciences, Liverpool John Moores University, Liverpool, UK

Phil Rasmussen Phytomed Medicinal Herbs Ltd., Auckland, New Zealand

Eliana Rodrigues Center for Ethnobotanical and Ethnopharmacological Studies—Institute of Environmental, Chemical and Pharmaceutical Sciences, Universidade Federal de São Paulo, Sao Paulo, Brazil

Galib Ruknuddin All India Institute of Ayurveda, An Autonomous Organization Under the Ministry of AYUSH, Government of India, New Delhi, India

Ruth L. Savage Uppsala Monitoring Centre, Uppsala, Sweden

New Zealand Pharmacovigilance Centre, Division of Health Sciences, University of Otago, Dunedin, New Zealand

Department of General Practice, University of Otago, Christchurch, New Zealand

Souad Skalli Faculty of Science, University Mohammed V in Rabat, Rabat, Morocco

Sally Stephens UK Teratology Information Service, Regional Drug and Therapeutics Centre, Newcastle-Upon-Tyne, UK

Lida Teng Department of Health Economics and Outcomes Research, Graduate School of Pharmaceutical Sciences, The University of Tokyo, Tokyo, Japan

Rolf Teschke Division of Gastroenterology and Hepatology, Department of Internal Medicine II, Klinikum Hanau, Hanau, Teaching Hospital of the Medical Faculty, Goethe University, Frankfurt/Main, Germany

Biljana Tubić Agency for Medicines and Medical Devices of Bosnia and Herzegovina, Banja Luka, Bosnia and Herzegovina

Faculty of Medicine, University of Banja Luka, Banja Luka, Bosnia and Herzegovina

Florence van Hunsel Netherlands Pharmacovigilance Centre Lareb, 's-Hertogenbosch, The Netherlands

Shirly Varughess Department of Pharmacovigilance & Drug Information, Ministry of Health, Muscat, Oman

Marcelo Vogler de Moraes Regulation Specialist, Brazilian Health Regulatory Agency - Anvisa, Brasília, Brazil

Sunita Vohra Department of Pediatrics, Faculty of Medicine and Dentistry, University of Alberta, Edmonton, AB, Canada

Department of Psychiatry, Faculty of Medicine and Dentistry, University of Alberta, Edmonton, AB, Canada

Jia-bo Wang School of Traditional Chinese Medicine, Capital Medical University, Beijing, People's Republic of China

China Military Institute of Chinese Medicine, Fifth Medical Center of Chinese PLA General Hospital, Beijing, People's Republic of China

Camilla Westerberg Uppsala Monitoring Centre, Uppsala, Sweden

Elizabeth M. Williamson University of Reading, School of Pharmacy, Whiteknights, Berkshire, UK

Xiao-he Xiao China Military Institute of Chinese Medicine, Fifth Medical Center of Chinese PLA General Hospital, Beijing, People's Republic of China

Tianyi Yang Rollins School of Public Health Emory University, GA, Atlanta, USA

Manal M. Younus Iraqi Pharmacovigilance Center, Directorate of Technical Affairs, Ministry of Health, Baghdad, Iraq

International Society of Pharmacovigilance (ISoP), London, UK

Council For International Organizations of Medical Sciences (CIOMS), Geneva, Switzerland

Li Zhang, MD, PhD Dongfang Hospital Beijing University of Chinese Medicine, Beijing, China

Part I
Advances and Challenges in Pharmacovigilance for Herbal Medicines

Chapter 1
The Historical Development of Pharmacovigilance for Herbal Medicines

Jeffrey K. Aronson

1.1 Introduction

Herbs have been used therapeutically since it was first recognized that they might have beneficial effects. That this probably dates back at least to Neanderthal times is evidenced by the discovery of pollen grains in Iraqi burial sites, many of which are hypothesized to have been used for their medicinal properties, as they have been in that area in later times [1]. They include species of *Achillea* to treat a range of gastrointestinal complaints, including dysentery and colic, *Centaurea cyanus*, used as a diuretic, emmenagogue, tonic, astringent, and febrifuge, and *Senecio vulgaris*, used as an emetic, diuretic, and purgative. Medicinal plants were depicted on the walls of the caves at Lascaux and there is evidence of the use of opium and psychedelic drugs during the Neolithic period [2].

1.2 Early Texts

Texts describing the use of herbal medicines are also of some antiquity. However, they deal almost exclusively with beneficial effects. The Ebers papyrus, for example, named after its discoverer, the German Egyptologist Georg Ebers, a collection of therapeutic prescriptions, includes about 700 medicines but says nothing about their adverse effects [3].

The same is true of other well-known herbals. These include Περί ὑλης ἰᾶτρικής, literally "about medical stuff", written by Pedanius Dioscorides, in 50–70 AD, whose title is better known in its Latin translation, *De Materia Medica*, and the

J. K. Aronson (✉)
Centre for Evidence Based Medicine, Nuffield Department of Primary Care Health Sciences, Radcliffe Observatory Quarter, Oxford, UK
e-mail: jeffrey.aronson@phc.ox.ac.uk

J. Barnes (ed.), *Pharmacovigilance for Herbal and Traditional Medicines*,
https://doi.org/10.1007/978-3-031-07275-8_1

Herbarium of Apuleius Platonicus, or Pseudoapuleius, which was first translated into Anglo-Saxon in around the year 1000 [4].

The *Herbarium* is one of four Anglo-Saxon texts that have survived. The earliest English word for a book that described treatments was the Anglo-Saxon "laeceboc" or leechbook, actually a medical treatise, a leechdom being a medicine. *Bald's Laeceboc*, in three volumes, which Bald, presumed to be a physician, commissioned from one Cild, presumed to be a writer or compiler, was probably compiled in the early tenth century, from recipes used in the court of King Alfred and soon after his death [5, 6]. The herbal remedies were simples, preparations containing single active ingredients, soaked in water, beer, vinegar, or milk. Bald's leechbook listed remedies but without mentioning adverse effects [7]. The other two texts were *Lacnunga*, which means "remedies", a collection of medical texts and prayers, and *Peri Didaxeon*, a transliteration of the Greek περί διδαξεων, literally "instructions".

The twelfth-century Welsh Physicians of Myddfai were herbalists who lived in the Welsh village of Myddfai in Carmarthenshire, supposedly descended from the sons of the legendary Lady of the Lake. The first of their books was published in Welsh in a vellum manuscript known as *Llyfr Coch Hergest* (*The Red Book of Hergest*) in around 1382. The books contained many therapeutic recipes, including surgical and herbal remedies [8]. Many of the herbal treatments recommended were probably both ineffective and harmless; for example, the roots of comfrey, dock, and valerian are recommended for an impostume (an abscess). However, some may have been harmful. Foxglove, for example, is recommended for an abscess and for a violent headache; toxicity could easily have occurred if the dose was ill-chosen. Some of the remedies were for treating poisoning with unnamed poisons, but again there is no mention of adverse effects that might be expected.

The same is true of later herbals and formularies, at times when most medicinal interventions were based on herbs. This includes, for example, herbals kept in religious houses, such as the sixteenth-century Syon Abbey Herbal [9], and a range of continental dispensatories and pharmacopoeias, also sometimes called pharmacopinaces, compiled to serve individual European municipalities or city states. Some of these are listed in Table 1.1.

The first pharmacopoeia to be compiled to serve a whole nation, the *London Pharmacopoeia*, compiled by members of the College of Physicians of London [10], was published in Latin on 7 May 1781, preceded by a royal proclamation that "all Apothecaries of this Realme [should] follow this Pharmacopoeia … upon paine of our high displeasure". In his edition of the Pharmacopoeia, published in 1653 [11], Nicholas Culpeper included "A Key to Galen's Method of Physic", based on Hippocratic humoral theory. In it he described the uses of different types of medicines, be they possessed of hot or cold, wet or dry properties, and mentioned those to whom certain medicines may be harmful. For example, "Children, and such people whose Stomachs are weak, are easily hurt by cold Medicines". However, he did not specify the types of harms to be expected, and the individual monographs in the Pharmacopoeia did not mention adverse effects.

There is at least one notable exception, which, as it happens, includes the first recorded use of the word "pharmacopoeia" in English, citing the *Pharmacopaea* by

Table 1.1 European pharmacopoeias, dispensatories, and pharmacopinaces published between 1498 and 1666

Title	Place and date of publication
Antidotarium Florentinum	Florence, 1498
Concordia Pharmacolorum Barcelonensium	Barcelona, 1535
Dispensatorium Valerii Cordis	Nuremberg, 1546
Concordia Aromatorium Cesaraugustae Saragosa	Saragosa, 1546
Pharmacopoeia seu de medicamentorum simplicium delectu: praeparationibus, mistionis modo by Jacques Dubois	Basel, 1552
Pinax iconicus antiquorum ac variorum in sepulturis rituum by Lilius Gregorius Giraldus	Lyon, 1556
Pharmacopoeia, medicamentorum omnium, quae hodie ad publica medentium munia in officinis extant by Anutius Foesius	Basel, 1561
Augstburgensis Pharmacopoeis	Augsburg, 1564
Dispensatorium usuale pro Pharmacopoeis inclytae Reipublicae Coloniensis	Cologne, 1565
Pinax theatri botanici by Caspar Bauhin	Basel, 1623
Pharmaco-Pinax, or a Table and Taxe of the Pryces of all vsuall Medicaments, Simple and composed, contayned in D. Gordon's Apothecarie and Chymicall Shop	Aberdeen, 1625
Pinax rerum naturalium Britannicarum, by Christopher Merrett	London, 1666

Querketanus. It is a text called *Opiologia*, a translation by Thomas Bretnor of a Latin text by Angelus Sala Vincentinus Venitus (1576–1637) [12]. Chapter 3 of that work is titled "Concerning the good and euill affects which Opium may produce in mens bodies". It begins "That Opium (as aforesaid) being wisely administred [*sic*] to the diseased, produceth in them many good effects, and contrariwise abused exciteth diuers dangerous and mortall accidents …". It then discusses, in terms of humoral theory, why opium does what it does, through its hot and cold properties. Later Bretnor writes that "Wine (as all Authors affirme, and Experience her selfe declareth) being abused, doth cause in processe of time *Phrensie, Madnesse, Rage, Furie, Stupiditie, Lethargie, Palsie* and other dangerous diseases, euen [even] as Opium doth" [emphasis in the original]. However, there is no evidence of any attempt to document cases systematically, nor to investigate the incidence of adverse effects.

After the publication of the *Pharmacopoeia Londinensis*, various other pharmacopoeias appeared, including *The Edinburgh Pharmacopoeia* (1699) and *The Dublin Pharmacopoeia* (1807). The last edition of the *London Pharmacopoeia,* the 11th, appeared in 1851. By then the need for harmonization had become clear, particularly because the Poor Law Amendment Act of 1834, with the institution of infirmaries and dispensaries, had resulted in increasing demands for medicines. *The British Pharmacopoeia* (*Pharmacopoeia Britannica*), respectively recommended and announced in the Medical Acts of 1858 and 1862, appeared in 1864 and is still in use today [13]. However, by then the purposes of a pharmacopoeia had changed. We no longer expect pharmacopoeias to deal with therapeutic matters. That role has instead been adopted by formularies and textbooks.

Even so, modern formularies, which range from simple lists of approved medicines to more detailed texts, such as *The British National Formulary*, are not expected to give highly detailed information about adverse drug effects and adverse drug reactions, beyond listing those that have been reported and giving simple advice about cautions, contraindications, and adverse drug–drug interactions. Drug–herb interactions, for example, are neglected, with occasional exceptions, such as interactions with *Hypericum perforatum*, listed under "St John's wort" in the *British National Formulary*. The *British Pharmacopoeia* (BP) and the *European Pharmacopoeia* (PhEur) include monographs on herbal substances, but they deal with quality, not therapeutic uses. In any case, as time has gone by, herbal preparations have largely disappeared from official pharmacopoeias and formularies.

1.3 Awareness of Adverse Effects of Medicines and Resulting Adverse Reactions

Although the pharmacopoeias discussed above did not for the most part give information about expected adverse effects of medicines, physicians of those times were well aware of such effects.

Most famously, Paracelsus (Philippus Aureolus Theophrastus Bombastus von Hohenheim, 1493–1541), in his *Septem Defensiones* (published posthumously in 1564), wrote his famous dictum, "Was ist das nit gifft ist? alle ding sind gifft/und nichts ohn gifft/Allein die dosis macht das ein ding kein gifft ist". In English: "What is there that is not a poison? Every thing is a poison and nothing is not a poison. Only the dose determines that a thing is not a poison". A Latin translation appeared in the margin of the first edition of the text: "Nil sine veneno praesertim dosi non servari" or "nothing lacks poison[ous effects] especially if the dose is not heeded". Actually, Paracelsus's therapeutic practice was largely based on three types of medicines, none of which was herbal: mercury, sulphur, and different metallic salts, particularly potassium nitrate (sal nitri, nitre, or saltpetre). Nevertheless, there is no reason to believe that he thought that his principle would not also apply to herbal medicines.

However, it was not until the work of Guldberg and Waage, published 300 years later, in 1864, when they described what has since come to be known as the Law of Mass Action, that the principle of dose-responsiveness started to become firmly established [14]. Practical demonstrations of pharmacological dose-response curves and theoretical explanations of such curves then followed, putting this aspect of pharmacology on a firm scientific footing.

From time to time, individual physicians, when reporting the beneficial actions of medicines, also described their adverse effects. An excellent example is that of the eighteenth-century English physician William Withering who, in 1785, published his monograph *An Account of the Foxglove and Some of its Medical Uses, &c.*, in which he detailed over 150 cases of his own and several from correspondents and discussed the proper use of digitalis in the treatment of dropsies [15]. Here he is describing adverse reactions that he attributed to the medicine: "The Foxglove when given in very large and quickly-repeated doses, occasions sickness, vomiting, purging, giddiness, confused vision, objects appearing green and yellow; increased secretion of urine, with frequent motions to part with it, and sometimes inability to retain it; slow pulse, even as slow as 35 in a minute, cold sweats, convulsions, syncope, death". And in a footnote he adds, "I am doubtful whether it does not sometimes excite a copious flow of saliva". Even so, Withering made no attempt to assess the frequencies of these effects.

In contrast, while herbals might occasionally mention the adverse effects of herbs known to be poisonous, they generally restricted their descriptions of other herbal medicines to their beneficial effects. This is typical of early herbals, such as those of John Gerard (*The Herball, or Generall Historie of Plantes*, 1597), John Parkinson (*Theatrum Botanicum*, 1640) and Nicholas Culpeper (*The English Physitian*, 1652, later published as *The Complete Herbal*, 1653). An exception is *A Modern Herbal* by Mrs. M Grieve (1931). She labels the headings to some entries "(POISON)", and mentions the adverse effects of others not so labelled, and advises cautions and methods of treatment [16].

1.4 Systematic Approaches to Gathering Reports of Adverse Drug Effects and Reactions

1.4.1 Louis Lewin

Adverse effects of medicines were first catalogued systematically by the Berlin toxicologist and pharmacologist Louis Lewin (pronounced "Leveen") [17]. Lewin was born on 9 November 1850 in Tuchel (Konitz) in West Prussia. His parents, Rahel and Hirsch, originally came from Suwalki, a Polish province in Russia, but they fled westward during the Russian pogroms, and changed their name from Appelbaum to Lewin. In 1856 the Lewin family moved to Berlin, where Lewin attended a Jewish school, later graduating to the Friedrich–Werderschen Gymnasium. He then read medicine at the Friedrich–Wilhelm Universität (now the Humboldt University), where he later became a titular professor at the Pharmacological Institute.

In 1881 Lewin published *Die Nebenwirkungen der Arzneimittel. Pharmakologisch–klinisch Handbuch (The Adverse Effects of Drugs—a Clinical*

Pharmacological Handbook). Three subsequent editions appeared in 1893, 1899, and 1909. In 1883 the book appeared in a so-called "second edition" as *The Untoward Effects of Drugs*, having been translated into cumbersome English by JJ Mulheron, Professor of the Principles of Medicine, Materia Medica, and Therapeutics in the Michigan College of Medicine in Detroit [18]. Of the just over 100 entries in the book, 53% are devoted to substances derived from plants; almost all of the rest are metallic elements or their salts. Extracts of cubebs, the berries of a climbing shrub, *Piper cubeba* L.f. (synonym: *Cubeba officinalis* Raf.), a native of Java and adjacent islands, rub shoulders with limewater and sodium nitrate. However, this was a time during which the use of herbal products was gradually being replaced by the use of inorganic chemical compounds, with organic medicaments starting to emerge.

In his introduction to the 1881 edition, Lewin had written "The records of the individual facts here indicated—the appearance of abnormal effects of drugs—are scattered throughout the most diverse parts of medical literature, and are either not at all or but superficially considered in the manuals or textbooks of material medica. For this reason I have for a long time been making a collection of these facts, examining them critically, and making additions to this collection from my own personal experience".

A year later, Mulheron, in his translator's preface, wrote, "The necessity of a treatise on the subject indicated by this title must have been felt by all practitioners, for previous to the appearance of this book by Dr Lewin there was no systematic work of this nature".

At the end of his preface, Lewin wrote, "I have presented the results of this labour in this book in the hope that they will meet a practical want, and at the same time stimulate others to further observations in the same direction". Lewin has been called the father of toxicology, and we might also call him the father of pharmacovigilance.

1.4.2 Secret Remedies

Despite the excellent example set by Lewin, it continued to be difficult to discover anything about adverse effects of medicines. The example of secret remedies demonstrates this [19]. In the UK, patent medicines had been specifically excluded from the Pharmacy Act of 1868 and the Sale of Food and Drugs Act of 1875, and their contents could therefore be kept secret, as could their adverse effects when they were advertised to the general public. Such medicines might contain chemical or plant-derived ingredients, sometimes both. A preparation of Wood's cure for tobacco habit, for instance, contained phenolphthalein, quassia, aloin, and strychnine; the product was advertised with the rubric "Tobacco habit conquered in 3 days". In 1909, the British Medical Association (BMA) gathered a set of articles about patent medicines that had previously been published in the *British Medical Journal* into a single volume, *Secret Remedies*. Public interest was enormous. The book sold

62,000 copies by June 1910, and a second collection was published in 1912. The volumes included information on the contents of the products and the costs of their ingredients, compared with the much larger prices they commanded; adverse reactions were not mentioned.

1.4.3 The US Federal Food, Drug, and Cosmetic Act, 1938

On 6 January 1937 the Democrat Senator for the State of New York, Royal S Copeland, introduced a new act into the Senate. In late 1937, an elixir of sulphanilamide, which contained diethylene glycol as a solvent, caused the deaths of more than 100 people across 15 states in the USA. Hastened by this, the new act, the Federal Food, Drug, and Cosmetic Act, which replaced the Pure Food and Drug Act of 1906, was signed into law by President Franklin D Roosevelt in 1938. The Act required manufacturers to prove that the drug was safe, although not that it was effective. A drug could not be removed from the market unless it was proved to be unsafe. It was not until the 1962 Kefauver–Harris Amendment to the Act, introduced following the thalidomide affair, that adverse reactions ("side effects") were specifically dealt with. Manufacturers were then required, in the case of prescription drugs, to provide proof of both the effectiveness and the safety [i.e. lack of harm] of their drugs before approval and to disclose accurate information about adverse reactions in their advertisements.

By that time, however, most therapeutic interventions were modern pharmacological agents. Herbal products were not included in the Act, although they were later included in a section on dietary supplements.

1.4.4 Meyler's Side Effects of Drugs

Apart from Otto Seifert's 1915 book on the adverse effects of a few medicines [20], no texts appeared to replace Lewin's until 1951. The Dutch physician Leopold Meyler underwent treatment for tuberculosis during the late 1940s, and experienced adverse reactions to the antituberculosis drugs. He discovered that there was no current single text to which medical practitioners could look for information about unwanted effects of drug therapy. He therefore determined to make such information available and published a book, in Dutch, entirely devoted to descriptions of the adverse effects that drugs could cause and the adverse reactions that could result. The first edition of 192 pages (*Schadelijke Nevenwerkingen van Geneesmiddelen*) appeared in 1951 [21] and an English version (*Side Effects of Drugs*) a year later [22].

Then, in 1957, Meyler started to publish what he called surveys of unwanted effects of drugs, each covering a period of 2–4 years. After having published seven volumes, Meyler died unexpectedly, and the publishers invited Graham Dukes to take over the editing of Volume VIII.

By this time, pharmacological interventions largely involved the use of organic chemical medicaments, and Meyler's volumes did not cover herbal extracts, although purified plant-derived compounds, such as quinine, were included. However, after publishing Volume VIII, Dukes replaced the intermittent updates with a series of four-yearly encyclopaedic versions (labelled the ninth edition and so on) and a parallel series of regular annual updates, called *Side Effects of Drugs Annuals* (SEDA). In SEDA-1 (1977) he introduced a chapter titled "Treatments used in non-orthodox medicine", which included herbal extracts, and in SEDA-20 (1997) that was changed to "Treatments used in complementary medicine" [23]. The chapter is now called "Safety of complementary and alternative medicine treatments and practices" [24]. The 16th edition of the encyclopaedia [25] contained a section on herbal medicines sufficiently long for it to be published separately as a stand-alone volume [26]. A 17th edition is in preparation.

Other texts cataloguing adverse drug reactions have since appeared, but Meyler remains the most thorough and up-to-date.

1.4.5 The UK Medicines Act, 1968

The UK's Therapeutic Substances Act of 1925 had covered the manufacture of medicinal products, following problems with the antisyphilitic drug arsphenamine (Salvarsan). Then, following the thalidomide affair, the UK Government instituted the Committee on Safety of Drugs (CSD), whose remit was to scrutinize new drugs before they were marketed and to promote post-marketing surveillance of adverse drug reactions. The activities of the CSD resulted in annual reports, which led to a White Paper titled "Forthcoming Legislation on the Safety, Quality and Description of Drugs and Medicines 1967" [27]. The 1968 Medicines Act followed, and established a Licensing Authority and the Medicines Commission. The Commission in turn, under section 4 of the Act, established the Committee on Safety of Medicines. In 1964 the CSD also put in place the 'Yellow Card' scheme, whereby suspected adverse reactions could be reported. Suspected reactions to herbal medicines can be reported through this scheme, although the database contains very few such reports.

The Medicines Act introduced licences, so-called Marketing Authorizations, granting permission to manufacturers to market compounds as medicinal products; these authorizations are granted to the manufacturer, who is licensed, not the product [28]. In the UK, the Act allowed herbal products to be marketed as licensed herbal medicines, as herbal medicines exempt from licensing, or as unlicensed food supplements without medicinal claims [29]. When licensing was introduced, the 600 or so products already on the market were granted product licences of right, although they had not undergone the stringent testing required to obtain full marketing authorization today [30]. These products were included in the Yellow Card scheme, which was extended to unlicensed herbal products in 1996.

The Medicines Act also introduced restrictions on drug advertising and required manufacturers to include adequate information about such things as adverse reactions [30]. This was probably one factor that led to the large reduction in advertisements in UK general medical journals during the 1970s [31]. In the UK, licensed prescription-only medicinal products, herbal or otherwise, may not be advertised direct to the general public. In some other countries, e.g. in the USA, such restrictions do not apply.

1.5 Conclusions

In recent years, regulatory authorities have started paying more attention to herbal products. A 2004 EU Directive on Traditional Herbal Medicinal Products required member states to set up a traditional herbal registration scheme, under which marketing authorization was available for traditional herbal remedies if they were to be used in minor conditions for which medical supervision was not required. In the UK, for example, the Medicines and Healthcare products Regulatory Agency (MHRA) did so in 2005. Eligibility for registration includes a requirement that the herbal medicinal product has been traditionally used to treat the stated condition for a minimum of 30 years, 15 years of which must have been in the European Union [32]. Manufacturers of herbal products so registered are required to comply with most of the current legislation on pharmacovigilance.

However, as the history outlined above shows, pharmacovigilance of herbal products did not begin until the use of herbal products was starting to wane compared with the rise of inorganic and later organic chemical medicaments, and has been brought late to modern methods of surveillance. It needs to be further improved. Without better information we cannot be sure that the benefit to harm balance that attends the use of herbal products will always be favourable.

References

1. Solecki RS (1975) Shanidar IV, a Neanderthal flower burial in Northern Iraq. Science 190(4217):880–881. https://doi.org/10.1126/science.190.4217.880
2. Stephens MDB (2012) The Dawn of Drug Safety, 2nd edn. George Mann Publications, Hampshire
3. Bryan CP, Joachim H (1930) The Papyrus Ebers. G Bles, London
4. Blunt W, Raphael S (1979) The Illustrated Herbal. Frances Lincoln Publishers Limited, London
5. Payne JF (1903) The Fitzpatrick lectures on the history of medicine. Br Med J 1(2217):1477–1480. https://doi.org/10.1136/bmj.1.2217.1477
6. Voigts LE (1979) Anglo-Saxon plant remedies and the Anglo-Saxons. Isis 70(252):250–268. https://doi.org/10.1086/352199
7. Cockayne TO (1864–1866) Leechdoms, Wortcunning, and Starcraft of Early England. Three volumes. Longman, Green, Longman, Roberts, and Green, London
8. Beverton T (2012) The Physicians of myddfai: cures and remedies of the mediaeval world. Cambria Books, Carmarthenshire

9. Adams J, Forbes S (eds) (2015) The syon abbey herbal—the last monastic herbal in england c.AD 1517. AMCD (Publishers) Ltd, London
10. Fowler CJ (2018) Pharmacopoeia londinensis 1618 and its descendants. Royal College of Physicians, London
11. Culpeper N (1653) Pharmacopoeia Londinensis: or the London dispensatory further adorned by the studies and collections of the fellows, now living of the said Colledg: being that book by which all apothecaries are bound to make up all the medicines in their shops. Peter Cole, London
12. Sala, A (1576–1637), Bretnor T (1607–1618) (translator and editor). Opiologia: or, A treatise concerning the nature, properties, true preparation and safe vse and administration of opium For the comfort and ease of all such persons as are inwardly afflicted with any extreame griefe, or languishing paine, especially such as depriue the body of all naturall rest, and can be cured by no other meanes or medicine whatsoeuer. Dedicated to the illustrious, high and mighty lords, the estates generall of the vnited prouinces in the Netherlands. By Angelus Sala Vincentinus Venitus. And done into English, and something inlarged by Tho. Bretnor. M. M. Nicholas Okes, London
13. Cartwright AC (2015) The british pharmacopoeia, 1864 to 2014: medicines, international standards, and the state. Ashgate, UK
14. Aronson JK, Ferner RE (2018) It does all depend on the dose. Understanding beneficial and adverse drug effects since 1864: clinical and experimental attitudes to the Law of Mass Action and concentration–effect curves. In: Grell O, Cunningham A, Arrizabalaga J (eds) 'It all depends on the dose': poisons and medicines in european history. Taylor & Francis Ltd., pp 210–239
15. Aronson JK (1985) An Account of the Foxglove and its Medical Uses 1785–1985. Oxford University Press, Oxford
16. Grieve Mrs M (1931) A modern herbal. Jonathan Cape, London. https://www.botanical.com/botanical/mgmh/mgmh.html. Accessed 30 October 2020
17. Aronson JK (2004) Louis lewin—meyler's predecessor. In: Aronson JK (ed) Side effects of drugs annual 27. Elsevier, Amsterdam, pp xxv–xxix
18. Mulheron JJ (translator) (1883) The Untoward Effects of Drugs. A Pharmacological and Clinical Manual by Dr L Lewin, 2nd ed., revised and enlarged. George S Davis, Detroit, Michigan
19. Aronson JK (2009) Patent medicines and secret remedies. BMJ 339:b5415. https://doi.org/10.1136/bmj.b5415
20. Seifert O (1915) Die Nebenwirkungen der modernen Arzneimittel. Verlag von Curt Kabitzsch, Würzburg
21. Meyler L (1951) Schadelijke Nevenwerkingen van Geneesmiddelen. Elsevier Publishing Company, Amsterdam
22. Meyler L (1952) Side Effects of Drugs. Translated by PH Vuijsje & W Mulhall Corbet. Elsevier Publishing Company, Amsterdam
23. Aronson JK (ed) (1997) Side Effects of Drugs Annual 20. Elsevier, Amsterdam
24. Ray SD (ed) (2019) Side Effects of Drugs Annual 41. Elsevier, Amsterdam
25. Aronson JK (ed) (2016) Meyler's side effects of drugs—The international encyclopedia of adverse drug reactions and interactions, 16th edn. Elsevier, Amsterdam
26. Aronson JK (ed) (2008) Meyler's side effects of herbal medicines. Elsevier, Amsterdam
27. Anonymous (1967) Proposed legislation on drugs. Br Med J 3:734
28. Aronson JK, Ferner RE (2017) Unlicensed and off-label uses of medicines: definitions and clarification of terminology. Br J Clin Pharmacol 83(12):2615–2625
29. Barnes J (2011) Adverse drug reactions and pharmacovigilance of herbal medicines. In: Talbot J, Aronson JK (eds) Stephens' detection and evaluation of adverse drug reactions: principles and practice, 6th edn. Wiley-Blackwell, Oxford, pp 645–683
30. Aronson JK (2018) Licensing authorization and the Medicines Act 1968. https://blogs.bmj.com/bmj/2018/10/05/jeffreyaronson-when-i-use-a-word-licensing-authorization-and-the-medicines-act-1968.

31. Aronson JK, Green AR (2019) The british pharmacological society's WDM paton memorial lecture 2019: how doctors were informed about pharmaceutical products through advertising in the *British Medical Journal* from 1955/6 to 1985/6. Br J Clin Pharmacol 85(9):1901–1906. https://doi.org/10.1111/bcp.14016
32. GOV. UK. Guidance—Apply for a traditional herbal registration (THR). https://www.gov.uk/guidance/apply-for-a-traditional-herbal-registration-thr. Accessed 30 October 2020

Chapter 2
Prevalence of Use of Herbal and Traditional Medicines

E Lyn Lee and Joanne Barnes

2.1 Herbal and Traditional Medicines: Descriptions and Characteristics

Globally, there is no consensus on how herbal and traditional medicines are defined or described. Typically, herbal medicines (also known as phytomedicines, phytotherapeutic preparations, or botanicals) are described as "*herbs, herbal materials, herbal preparations and finished herbal products that contain, as active ingredients, parts of plants, other plant materials or combinations thereof*" [1]. The World Health Organization (WHO) recognizes that herbal medicines "*may contain, by tradition, natural organic or inorganic active ingredients that are not of plant origin (e.g., animal and mineral materials)*" [1], i.e., the term may also be used to describe traditional medicines that contain non-herbal (non-plant) ingredients. In Europe, the Traditional Herbal Medicinal Products Directive defines the term "herbal medicinal product" as "*any medicinal product, exclusively containing as active ingredients, one or more herbal substances, or one or more herbal preparations, or one or more such herbal substances in combination with one or more such herbal preparations*" [2]. Here, herbal substances refer to unprocessed plants or plant parts, whereas herbal preparations are obtained by subjecting herbal substances to treatments, such as extraction, distillation, and purification [2]. Generally, herbal medicines contain active ingredients from crude/processed plants or plant parts; an isolated chemical entity (constituent) isolated from plant material is not considered to be a herbal medicine. Herbal medicines typically encompass a range of dose forms from relatively crude preparations, such as tinctures and extracts that are supplied by herbal medicine practitioners, to manufactured or finished products, usually formulated as

E. L. Lee (✉) · J. Barnes
School of Pharmacy, Faculty of Medical and Health Sciences, University of Auckland, Auckland, New Zealand
e-mail: elyn.lee@auckland.ac.nz; j.barnes@auckland.ac.nz

J. Barnes (ed.), *Pharmacovigilance for Herbal and Traditional Medicines*,
https://doi.org/10.1007/978-3-031-07275-8_2

tablets and capsules [3]. In many countries, most herbal medicines can be purchased without a prescription [3], although there are some herbal substances that have prescription-only status, or other restrictions, in some countries.

Use of the term "traditional medicine" (TM) to describe indigenous medicines and practices has a long history. The WHO defines TM as the "*sum total of the knowledge, skill, and practices based on the theories, beliefs, and experiences indigenous to different cultures, whether explicable or not, used in the maintenance of health as well as in the prevention, diagnosis, improvement or treatment of physical and mental illness*" [1]. Traditional medicine approaches include the use of traditional medicines (TMs) that are plant, animal, and/or mineral-based, and other therapies, such as spiritual healing, manual techniques, and exercises. Almost all countries and ethnic groups have a traditional system of medicine. Examples include traditional Chinese medicine (TCM), Ayurvedic medicine (India), Rongoā Māori (New Zealand), and Kampo medicine (Japan). For some countries, traditional medicine is part of the dominant healthcare system (e.g., TCM in China), while for other countries, some "new" forms of traditional medicine became established over time through immigration and increasing interest in such health approaches (e.g., TCM in the UK) [4].

Herbal medicines are sometimes referred to collectively with other categories of products (e.g., vitamins, amino acids) under broader terminologies such as "dietary supplements" and "natural health products." The terms "complementary medicines/remedies" and "alternative medicines/remedies" are also often used to describe a variety of products, including herbal medicines, traditional medicines (many of which are herbal medicines), homeopathic remedies, dietary supplements, flower essences, and anthroposophic medicines, many of which either consist of or originate from plant material [4].

Unlike conventional medicines, which typically consist of a single chemical entity/drug, herbal medicines are chemically rich, complex mixtures comprising a range of potentially active constituents [3, 5]. The chemical constituents in an herbal product/preparation, which could number up to several hundred or more, are not known, or only partly described, for many herbal medicines. Even for herbal medicines with well-documented phytochemistry, the active constituent(s) responsible for pharmacological activity are often unknown [5, 6]. The specific profile of constituents is not uniform throughout all parts of a plant (e.g., roots, leaves), and differences in herbal raw materials exist due to a range of factors, such as inter- and intraspecies variation, environmental effects (e.g., climate), time of harvesting, and post-harvesting conditions such as storage, drying, and processing [6]. Adding to the complexity, many preparations/mixtures prescribed by herbal medicine practitioners, and manufactured or finished herbal medicine products, contain combinations of herbal ingredients, making it more difficult to determine their pharmacokinetics, pharmacodynamics, pharmacology, and toxicology.

2.2 Regulation of Herbal and Traditional Medicines

Since 1999, the number of WHO member states with herbal medicines regulations and traditional medicine policies has increased from 65 to 124 (2018) and 25 to 69 (2012), respectively [1, 7]. National policy, and regulatory approaches, however, differ between countries and continue to develop over time.

In several countries, herbal medicines are regulated under a broader umbrella term alongside other (related) products. For instance, herbal medicines are included as "natural health products" in Canada, and as "dietary supplements" in the United States of America (USA), together with other products, such as vitamins, minerals, and amino acids, that are regulated under the Natural Health Products Regulations [8] and Dietary Supplement Health and Education Act (1994), respectively [9]. In some other countries, such as those in the European Union (EU), specific regulations have been developed for specific categories of complementary medicines. For example, according to the European Union directive 2001/83/EC, herbal medicinal products, traditional herbal medicinal products, and homeopathic medicinal products adhere to different specific requirements for product registration [10].

In other countries, herbal medicines, whether as a stand-alone category, or included as part of a broader group (e.g., dietary supplements, complementary medicines, or natural health products) alongside other products, are regulated under (or captured by) existing food or drug regulations in the country [5]. There are countries where some herbal medicines are regulated as prescription medicines or non-prescription medicines, or as a variety of categories under foods such as health foods, general food products, and functional foods. From 2005 to 2012, it was noted that there was a decrease in the number of WHO member states regulating herbal medicines as "non-prescription medicines" (from 137 to 79), but an increase in those regulating such products as "herbal medicines" (from 25 to 77) [1], implying the change in regulatory approach over time and an increasing trend to regulate herbal medicines as an independent category of products.

At present, most countries implement regulations on herbal medicines that are updated periodically to reflect national priorities and needs. Despite differences in approach, the common goal for regulation is to ensure that herbal medicines are safe and of appropriate quality, with the ultimate aim of protecting consumer health. In the current market, where products are often manufactured/produced in a country other than where they are sold, ensuring the safety and quality of products becomes (more) challenging. In some regions, this has driven interest among regulatory agencies in harmonizing regulations on herbal medicines regionally and internationally [7]. For example, the European Union implemented a legal framework underpinning uniform regulations on herbal medicinal products and traditional herbal medicinal products across the region [2]. In South-East Asia, efforts to harmonize regulations are ongoing through an Association of Southeast Asian Nations (ASEAN) Product Working Group for Traditional Medicines and Health Supplements [11, 12].

2.3 Global Use of Herbal and Traditional Medicines

Due to the differences in regulations across countries and regions, it is difficult to accurately assess the prevalence and patterns of use of herbal and traditional medicines worldwide. Also, most countries report the use of these products collectively with other products/therapies (e.g., dietary supplements), making it difficult to obtain accurate data on prevalence of use of herbal medicines and making comparisons across countries almost impossible. Available data, however, suggest that the use of these herbal and traditional medicines is substantial internationally.

Market growth for herbal and traditional medicines is evident internationally and is projected to rise further in the coming years [13]. Currently, there are over 85,000 supplement products (which includes herbal medicines and others) in the US market, a substantial increase from an estimated 4000 products when the Dietary Supplement Health and Education Act became law in 1994 [5]. New products enter the market every year, and the herbal medicines market is large, with a global market share valued at USD 5.26 billion in 2017 [13]. From retail sales data, consumers in the USA spent USD 9.602 billion on herbal supplements in 2019, an 8.6% increase from 2018 [14]. In Canada, consumers spent approximately CAD 700 million (~USD 528 million) on herbs and vitamins in 2015–2016 [15]. For the Australian population, annual expenditure on western or Chinese herbal medicines was estimated at over AUD 270 million (~USD 209 million) in 2017 [16]. In the Asia-Pacific region, the dietary supplements market is rapidly growing due to increased demand for herbal and traditional medicines in countries like China, India, and Japan [17]. The traditional Chinese medicine pharmaceutical industry in China had a total output value of over RMB 786 billion (~USD 125 billion), accounting for over a quarter of the total generated by the country's pharmaceutical industry in 2015 [1]. In Japan, revenue from herbal medicine (Kampo medicine) increased from USD 1.42 billion to USD 1.47 billion in a year from 2007 to 2008 [18].

Market research and other data indicate there has been a shift in the way herbal medicines are accessed by consumers in some countries. According to market sales data in 2017, herbal medicines sold through direct sales channels (e.g., multilevel marketing companies, mail-order sales) outperformed mainstream and natural/health food channels for the first time since 2012 [19]. This trend continued in 2018 [20]. Specific herbal medicine product sales trends have also changed over time. According to a market analysis report, echinacea had the largest revenue share, taking up over one-third of the global market, in 2017 [13]. In the USA, sales of herbs for immune support have increased substantially; sales for elderberry and echinacea grew by more than 50% in 2020 [14]. Cannabis-derived products are also becoming popular. In 2018, sales of cannabidiol, although not considered a dietary supplement by the US Food and Drug Administration, increased 332% from the previous year, overtaking turmeric which had been the top-selling herbal ingredient in natural retail stores since 2013 [20].

More than 80% of WHO member states reported the use of herbal and traditional medicines in their respective populations [1]; however, the prevalence of use across countries differs (Table 2.1) due to several factors, such as ease of access, regulations, cultural aspects, and historical influence [7]. The difference in prevalence can be observed in three general patterns: (1) use in developed countries where the

Table 2.1 Prevalence of herbal and traditional medicine use in selected countries; data are extracted from the WHO global report on traditional and complementary medicine [1]

Country	Year	Variable	Estimated prevalence[a] (%)
WHO African Region			
Congo	NR	Indigenous TM	80–99
	2006	Ayurvedic medicine, chiropractic, and herbal medicines	80–99
	NR	Traditional Chinese medicine products	1–19
Ethiopia	NR	Indigenous TM	60–79
South Africa	2010	T&CM practices, including acupuncture, ayurvedic medicine, chiropractic, herbal medicines, homeopathy, naturopathy, osteopathy, traditional Chinese medicine, Unani medicine, and other practices such as therapeutic aromatherapy, therapeutic massage therapy, and therapeutic reflexology.	1–19
United Republic of Tanzania	NR	Indigenous TM	60–79
	NR	Herbal medicines	60–79
WHO Regions of the Americas			
Brazil	2007	Indigenous TM	1–19
Canada	2005	Acupuncture, chiropractic, herbal medicines, homeopathy, and naturopathy	1–19
Cuba	2010	Indigenous TM	80–99
	2010	Herbal medicines	80–99
Mexico	2009-2010	Indigenous TM and herbal medicines	20–39
United States of America		*Data not available*	
WHO Eastern Mediterranean Region			
Bahrain	NR	Indigenous TM	60–79
	NR	Herbal medicines	80–99
Oman	NR	Indigenous TM	80–99
Pakistan	NR	Indigenous TM	40–59
	NR	Herbal medicines	40–59
Saudi Arabia	2010	Indigenous TM	40–59
	NR	Herbal medicines	40–59
United Arab Emirates	2012	Indigenous TM practices	20–39

(continued)

Table 2.1 (continued)

Country	Year	Variable	Estimated prevalence[a] (%)
WHO European Region			
Czech Republic	NR	Indigenous TM	1–19
Germany	2000	Indigenous TM practices	60–79
	2004	Herbal medicines	20–39
Switzerland	2007	Ayurvedic medicine, traditional Chinese medicine, neural therapy, and anthroposophic medicine	<1%
	2007	Acupuncture, chiropractic, herbal medicines, homeopathy, and osteopathy	1–19
United Kingdom of Great Britain and Northern Ireland	NR	Herbal medicines	20–39
WHO South-East Asia Region			
Bangladesh	2007	Indigenous TM	20–39
	NR	Herbal medicines, homeopathy, and Unani medicine	1–19
Bhutan	NR	Herbal medicines	20–39
Indonesia	2010	Indigenous TM and herbal medicines	40–59
Myanmar	2009	Indigenous TM	80–99
		Herbal medicines	80–99
Thailand	2010	Thai traditional medicine	1–19
	2010	Traditional Chinese medicine	1–19
WHO Western Pacific Region			
Australia		*Data not available*	
China		*Data not available*	
Japan		*Data not available*	
Malaysia	2015	T&CM with consultation within the past 12 months	21.51
Mongolia	NR	Indigenous TM	40–59
New Zealand	NR	Consults T&CM practitioners	Up to 20
Papua New Guinea	NR	Indigenous TM	80–99
Singapore	2013	Consulted a traditional Chinese medicine practitioner at least once in their lives	26.5

NR = not reported, TM = traditional medicine
[a]Prevalence timeframe not reported

conventional healthcare system is well developed; (2) use in some developed and developing countries with fairly developed conventional healthcare systems, and where traditional medicine has significant importance due to cultural and historical influences; (3) use in countries, typically developing countries, where conventional healthcare is limited and traditional medicine is one of the primary sources, or sometimes the only accessible source, of healthcare [7].

Countries in the first group (above) include developed countries, such as the USA, Canada, Australia, and many European countries, where herbal and traditional medicines are mostly considered to be complementary to the mainstream conventional healthcare system. Data from the US National Health Interview Surveys from 2002, 2007, and 2012 indicate that "non-vitamin, non-mineral" supplements, which includes herbal medicines, were the most commonly used "complementary health approach" throughout the 10 years. The prevalence of use in the 12 months preceding the surveys was unchanged at 18.9% (2002), 17.7% (2007), and 17.7% (2012) [21]. In Canada, the prevalence of use of herbal therapies over the previous 12 months also remained consistent at 10% from 2006 to 2016 [15]. A similar prevalence of use (9.5%) of western and Chinese herbal medicines over the previous 12 months was observed in Australia [22].

In the second group of countries, traditional medicine plays a prominent role in healthcare owing to its strong historical and cultural roots. Despite established conventional healthcare systems in countries such as China, South Korea, and Singapore, the prevalence of traditional medicine (including use of herbal medicines and other therapies/practices, such as acupuncture) use in these countries remains relatively high at more than 90%, 86%, and 53%, respectively, in 2008 [18]. Although national data on herbal medicines alone were not reported, the extent of use of these products is likely to be considerable based on reports from smaller localized studies. For instance, a cohort study of 3420 (response rate = 85%) Chinese older adults (aged 65 years and above) in a residential town in Singapore reported that 25.3% of the participants had used Chinese herbal medicines over the past year [23]. A 2015 consumer survey (n = 1134) using an online research panel in South Korea found that 61.1% of respondents had taken herbal medicines within the past year [24].

In Africa, the use of traditional medicine is widespread due to its availability and (relative) affordability, and people's limited access to conventional healthcare [1]. The ratio of medical doctors to the African population is 1:40,000, whereas the ratio for traditional healers is 1:500; hence, for millions of African people, particularly those living in rural areas, traditional medicine practitioners are, in fact, sometimes their only source of healthcare [25]. A systematic review of studies exploring traditional, complementary, and alternative medicine product use among the general population in sub-Saharan Africa reported prevalence rates ranging from 4.6% in an urban settlement in Ethiopia to 94% in semi-urban settlements in Nigeria and Ethiopia [26]. In the review, the prevalence of herbal medicines use was not reported in the overall population, however, utilization rates were high in subpopulations: 76.2% among women seeking infertility care; 40–60% among surgical patients during their preoperative period; 42.4% among patients with mycetoma [26].

2.4 Issues Relating to the Use of Herbal and Traditional Medicines

Herbal and traditional medicines are used by a wide range of individuals for a variety of health reasons, including maintenance of health and well-being, and prevention, and treatment of minor ailments and chronic conditions. These individuals include older adults, pregnant or breastfeeding women, children, as well as people with serious chronic diseases such as cancer, AIDS, and multiple sclerosis [4]. Herbal and traditional medicines can be used in preference to, or (more usually) concurrently with, conventional medicines. Many reasons contribute to the popularity of herbal and traditional medicines, but one of the key factors is consumers' inclination towards natural products [13], which many consumers believe are safe and free from risk of adverse effects.

Users of herbal and traditional medicines usually select these products without seeking professional advice, mainly relying on friends' and relatives' recommendations, and information from the media [27, 28]. These products are easily accessed through multiple avenues, including purchase over the internet and from retail outlets where no trained healthcare professional is present. In pharmacies, herbal and traditional medicines are commonly available for purchase without the need to interact with a pharmacist or pharmacy assistant. Even if a consultation occurs, healthcare professionals may not have the knowledge and understanding about herbal and traditional medicines to provide evidence-based advice [29, 30]. A small proportion of users obtains herbal medicines as (part of) treatment from an herbal or traditional medicine practitioner; however, in many countries, these practitioners are not regulated and do not require a license or certificate to practise. There is often no legal requirement for these practitioners to undertake training in herbal and/or traditional medicine, and while many practitioners will have undergone some informal or formal training, some will not [1]. In other countries like China, many health professionals are formally trained in both "western" and traditional medicine, and use these approaches alongside one another in hospitals and primary care facilities [31].

Another concern with the use of herbal and traditional medicines is (lack of) disclosure of use to healthcare professionals. It is estimated that only one-third of complementary medicine users disclose their use of these products/therapies to their (conventional) healthcare providers [32, 33], partly due to lack of enquiry by those healthcare providers [32]. Healthcare professionals receiving reports of suspected adverse drug reactions (ADRs) associated with conventional medicines rarely note information on herbal and traditional medicine use on patient records [34]. Hence, the undisclosed herbal and traditional medicine use is not considered as a possible contributor to, or cause of, the ADRs.

Disclosure of herbal and traditional medicines use to healthcare professionals is essential, particularly where such products/preparations are started, stopped, or used concurrently with conventional medicines. To the same extent, it is important to disclose the use of conventional medicines to herbal or traditional medicine practitioners as there may be potential drug-herb interactions [4]. Existing data indicate that a substantial proportion of patients co-uses herbal medicines with conventional medicines. In a national survey on complementary medicine use among US adults, 18.4% of prescription-medicine users took herbal remedies and/or high-dose vitamins concurrently [35]. A systematic review of 22 studies reported that the prevalence of concurrent use of herbal medicine products and prescription medicines among older adults varied between 5.3 and 88.3% [36].

There is a paucity of information regarding the safety of many herbal and traditional medicines. Contrary to consumers' belief that these products/preparations are safe, serious ADRs, including drug interactions, have been reported in association with some herbal and traditional medicines [4, 6]. Some well-documented examples include hepatotoxicity associated with the use of black cohosh (*Actaea racemosa* L.), kava kava (*Piper methysticum* G.Forst.), Chinese knotweed (he shou wu, *Reynoutria multiflora* (Thunb.) Moldenke, synonym: *Polygonum multiflorum* Thunb.), and nephrotoxicity associated with aristolochic acids found in *Aristolochia manshuriensis* Kom. [4, 6]. Drug-herb interactions described for herbal and traditional medicines include increased bleeding risk with the use of antiplatelet or anticoagulant drugs in combination with ginkgo (*Ginkgo biloba* L.) and reduction in the effects of cyclosporin, tacrolimus, warfarin, digoxin, and certain other medicines when used concurrently with St. John's wort (*Hypericum perforatum* L.) [37].

Although some ADR reports are well documented, there is inadequate safety information for most herbal and traditional medicines. This is partly due to the current regulatory frameworks where there is little incentive for manufacturers to conduct preclinical tests, clinical trials, and post-marketing surveillance. Hence, limited information is available on the types and frequencies of adverse effects, including interactions with drugs, foods, alcohol, and disease states, as well as other aspects relevant to safety, such as active constituents of herbal medicines, pharmacokinetics, pharmacodynamics, and use in specific population groups (e.g., older adults, children, pregnant women) [3]. A "light-touch" regulatory approach also contributes—in some countries—to reports of low-quality products in the market where these medicines were adulterated with various prescription-only medicines[38, 39], or sub-standard raw materials are used [6, 40].

In conclusion, the ways in which herbal and traditional medicines are marketed and easily accessed, together with consumers' perceptions of these products/preparations as free from the potential to cause harm, as well as issues related to healthcare professionals' and herbal and traditional practitioners' practice present opportunities for inappropriate and unsafe use of herbal and traditional medicines and the potential for ADRs to go undetected. The global acceptance and extensive use of herbal and traditional medicines in the context of the current regulatory landscape, and ubiquitous reports of quality issues with HTM raw materials and finished products, raise questions around whether there is adequate protection of the public health from poor-quality and unsafe products.

References

1. World Health Organisation (2019) WHO Global report on traditional and complementary medicine. https://www.who.int/traditional-complementary-integrative-medicine/WhoGlobalReportOnTraditionalAndComplementaryMedicine2019.pdf?ua=1. Accessed 24 May 2020
2. European Commision (2004) Directive 2004/24/EC of the European parliament and of the council of 31 March 2004 amending, as regards traditional herbal medicinal products,

Directive 2001/83/EC on the Community code relating to medicinal products for human use. Official Journal of the European Union, Brussels
3. Barnes J (2003) Pharmacovigilance of herbal medicines: a UK perspective. Drug Saf 26(12):829–851. https://doi.org/10.2165/00002018-200326120-00001
4. Barnes J (2012) Adverse drug reactions and pharmacovigilance of herbal medicines. In: Talbot J, Aronson J (eds) Stephens' detection and evaluation of adverse drug reactions, 6th edn. Wiley, Chichester, pp 645–683
5. Dwyer JT, Coates PM, Smith MJ (2018) Dietary supplements: regulatory challenges and research resources. Nutrients 10(1). https://doi.org/10.3390/nu10010041
6. Barnes J, Anderson LA, Phillipson JD (2007) Herbal medicines. A guide for health-care professionals, 3rd edn. Pharmaceutical Press, London
7. World Health Organisation (2013) WHO traditional medicine strategy: 2014–2023. https://www.who.int/medicines/publications/traditional/trm_strategy14_23/en/. Accessed 16 Mar 2020
8. Health Canada Natural health products—the evolution of regulations. https://www.canada.ca/en/health-canada/services/drugs-health-products/natural-non-prescription/regulation/43-000-products-authorized-sale-growing.html. Accessed 16 Mar 2020
9. US Food and Drug Administration Dietary Supplements. https://www.fda.gov/food/dietary-supplements. Accessed 16 Mar 2020
10. European Commision (2001) Directive 2001/83/EC of the European parliament and of the council of amending, as 6 November 2001 on the Community code relating to medicinal products for human use
11. Association of Southeast Asian Nations (2019) ASEAN holds training on traditional medicines and health supplements. https://asean.org/asean-holds-training-traditional-medicines-health-supplements/?highlight=Traditional%20Medicines%20and%20Health%20Supplement%20Product%20Working%20Group. Accessed 16 Mar 2020
12. Policy and guidelines-traditional medicines and health supplement product working group association of southeast Asian nations. https://asean.org/asean-economic-community/sectoral-bodies-under-the-purview-of-aem/standards-and-conformance/policy-and-guidelines/. Accessed 16 Mar 2020
13. Grand View Research (2018) Herbal supplements market size, share & trends analysis report by product (turmeric, echinacea), by formulation (capsules, powder and granules), by consumer (pregnant women, adults), and segment forecasts, 2018–2025. https://www.grandviewresearch.com/industry-analysis/herbal-supplements-market. Accessed 16 Mar 2020
14. Smith T, May G, Eckl V, Morton C (2020) US sales of herbal supplements increase by 8.6% in 2019. HerbalGram J Am Bot Council 127:54–69
15. Esmail N (2017) Complementary and alternative medicine: use and public attitudes 1997, 2006 and 2016. Fraser Institute, Vancouver
16. Harnett JE, McIntyre E, Steel A, Foley H, Sibbritt D, Adams J (2019) Use of complementary medicine products: a nationally representative cross-sectional survey of 2019 Australian adults. BMJ Open 9(7):e024198. https://doi.org/10.1136/bmjopen-2018-024198
17. Mordor Intelligence. Dietary supplements market - growth, trends, and forecast (2019–2024). https://www.mordorintelligence.com/industry-reports/dietary-supplement-market. Accessed 16 Mar 2020
18. World Health Organisation (2012) The Regional Strategy for Traditional Medicine in the Western Pacific (2011–2020). https://iris.wpro.who.int/bitstream/handle/10665.1/5538/9789290615590_eng.pdf. Accessed 16 Mar 2020
19. Smith T, Kawa K, Eckl V, Morton C, Stredney R (2018) Herbal supplement sales in US increased 8.5% in 2017, topping $8 Billion. HerbalGram J Am Bot Council 119:62–71
20. Smith T, Gillespie M, Eckl V, Knepper J, Reynolds CM (2019) Herbal supplement sales in US increase by 9.4% in 2018. HerbalGram J Am Bot Council 123:62–73
21. Clarke TC, Black LI, Stussman BJ, Barnes PM, Nahin RL (2015) Trends in the use of complementary health approaches among adults: United States, 2002–2012. Natl Health Stat Rep 79:1–16

22. Steel A, McIntyre E, Harnett J et al (2018) Complementary medicine use in the Australian population: results of a nationally-representative cross-sectional survey. Sci Rep 8(1):17325. https://doi.org/10.1038/s41598-018-35508-y
23. Ng TP, Tan CH, Kua EH, Singapore Chinese Longitudinal Aging S (2004) The use of Chinese herbal medicines and their correlates in Chinese older adults: the Singapore Chinese Longitudinal Aging Study. Age Ageing 33(2):135–142. https://doi.org/10.1093/ageing/afh016
24. Jang S, Kim KH, Sun SH et al (2017) Characteristics of herbal medicine users and adverse events experienced in South Korea: a survey study. Evid Based Complement Alternat Med 2017:4089019. https://doi.org/10.1155/2017/4089019
25. Abdullahi AA (2011) Trends and challenges of traditional medicine in Africa. Afr J Tradit Complement Altern Med 8(Suppl. 5):115–123. https://doi.org/10.4314/ajtcam.v8i5S.5
26. James PB, Wardle J, Steel A, Adams J (2018) Traditional, complementary and alternative medicine use in Sub-Saharan Africa: a systematic review. BMJ Glob Health 3(5):e000895. https://doi.org/10.1136/bmjgh-2018-000895
27. Barnes J, Mills SY, Abbot NC, Willoughby M, Ernst E (1998) Different standards for reporting ADRs to herbal remedies and conventional OTC medicines: face-to-face interviews with 515 users of herbal remedies. Br J Clin Pharmacol 45(5):496–500. https://doi.org/10.1046/j.1365-2125.1998.00715.x
28. Brevoort P (1998) The booming U.S. botanical market: anew overview. HerbalGram J Am Bot Council 44:33–46
29. Chang ZG, Kennedy DT, Holdford DA, Small RE (2000) Pharmacists' knowledge and attitudes toward herbal medicine. Ann Pharmacother 34(6):710–715. https://doi.org/10.1345/aph.19263
30. Gall A, Anderson K, Adams J, Matthews V, Garvey G (2019) An exploration of healthcare providers' experiences and perspectives of traditional and complementary medicine usage and disclosure by indigenous cancer patients. BMC Complement Altern Med 19(1):259. https://doi.org/10.1186/s12906-019-2665-7
31. Xu J, Yang Y (2009) Traditional Chinese medicine in the Chinese health care system. Health Policy 90(2-3):133–139. https://doi.org/10.1016/j.healthpol.2008.09.003
32. Foley H, Steel A, Cramer H, Wardle J, Adams J (2019) Disclosure of complementary medicine use to medical providers: a systematic review and meta-analysis. Sci Rep 9(1):1573. https://doi.org/10.1038/s41598-018-38279-8
33. Kennedy J, Wang CC, Wu CH (2008) Patient disclosure about herb and supplement use among adults in the US. Evid Based Complement Alternat Med 5(4):451–456. https://doi.org/10.1093/ecam/nem045
34. Barnes J (2001) An examination of the role of the pharmacist in the safe, effective and appropriate use of complementary medicines. University of London
35. Eisenberg DM, Davis RB, Ettner SL et al (1998) Trends in alternative medicine use in the United States, 1990–1997: results of a follow-up national survey. JAMA 280(18):1569–1575. https://doi.org/10.1001/jama.280.18.1569
36. Agbabiaka TB, Wider B, Watson LK, Goodman C (2017) Concurrent use of prescription drugs and herbal medicinal products in older adults: a systematic review. Drugs Aging 34(12):891–905. https://doi.org/10.1007/s40266-017-0501-7
37. Asher GN, Corbett AH, Hawke RL (2017) Common herbal dietary supplement-drug interactions. Am Fam Physician 96(2):101–107
38. Ko RJ (1998) Adulterants in Asian patent medicines. N Engl J Med 339(12):847. https://doi.org/10.1056/nejm199809173391214
39. Koh HL, Woo SO (2000) Chinese proprietary medicine in Singapore: regulatory control of toxic heavy metals and undeclared drugs. Drug Saf 23(5):351–362. https://doi.org/10.2165/00002018-200023050-00001
40. Xu M, Huang B, Gao F et al (2019) Assessment of adulterated traditional Chinese medicines in China: 2003–2017. Front Pharmacol 10:1446. https://doi.org/10.3389/fphar.2019.01446

Chapter 3
Pyrrolizidine Alkaloids in Herbal Medicines and Food: A Public Health Issue

Mina Kalantar zadeh and Elizabeth M. Williamson

3.1 Pyrrolizidine Alkaloids and Their Occurrence in Herbal Medicines

Pyrrolizidine alkaloids (PAs) are toxic compounds that occur naturally in several plant families and may be present in some herbal medicines. In China, hepatic sinusoidal obstruction syndrome presents as abdominal distension, pain, ascites, jaundice, and hepatomegaly and is associated with the oral intake of plants containing pyrrolizidine alkaloids [1].

PAs do not have any known medicinal value and are responsible for many cases of poisoning, and therefore most regulatory agencies take a strict approach to controlling their availability in herbal products that are taken internally. Not all pyrrolizidine alkaloids are toxic, only those that are unsaturated at the 1,2-position (e.g. senecionine; Fig. 3.1), and these are the PAs referred to in this chapter. They cause veno-occlusive disease and are hepato-carcinogenic, and their effects are cumulative [2–4]. A limit of exposure of 1.0 μg per day for PAs has been set by the Federal Institute for Drugs and Medical Devices (Bundesinstitut für Arzneimittel und Medizinprodukte, (BfArM)) and adopted elsewhere in the European Union [5], but risk assessment strategies are continually being updated (e.g. Chen et al. [6]). Comfrey, *Symphytum officinale* L., has a long history of herbal use and may contain toxic PAs, although the more commonly used medicinal variety *Symphytum* x *uplandicum* Nyman does not, at a level of detection of 8 μg/kg [7]. External preparations containing comfrey are used to treat skin and joint inflammation and may have traditional herbal registration (THR) status. A recent study suggests that these

M. Kalantar zadeh (✉)
Food Research Consultant, Wokingham, Berkshire, UK

E. M. Williamson
University of Reading, School of Pharmacy, Whiteknights, Berkshire, UK
e-mail: e.m.williamson@reading.ac.uk

J. Barnes (ed.), *Pharmacovigilance for Herbal and Traditional Medicines*,
https://doi.org/10.1007/978-3-031-07275-8_3

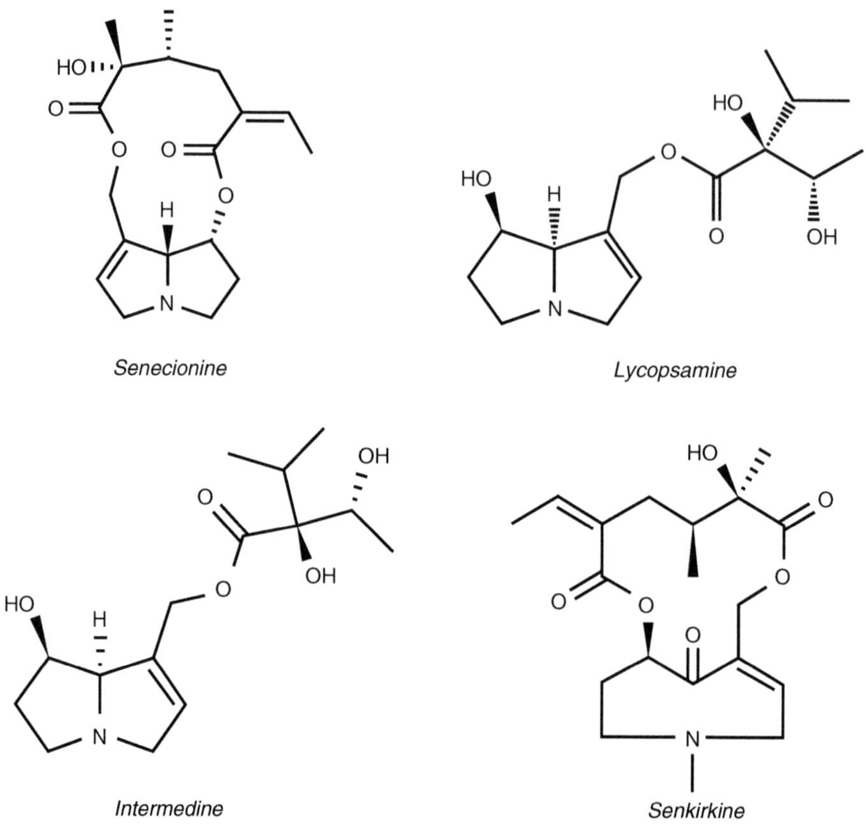

Fig. 3.1 Structures of commonly occurring pyrrolizidine alkaloids in medicinal plants

topical applications are safe, as the main PA, lycopsamine, is poorly absorbed through human skin, and that regulatory limits for systemic exposure do not apply [7].

There are reports of PAs being discovered in herbal products made from plants that do not naturally contain them, such as some St John's wort (*Hypericum perforatum* L.) THR products [5, 8]. A separate study found that not only were *H. perforatum*-containing products frequently contaminated with PAs from *Echium* spp., but that *Cynara cardunculus* L. and fixed-combination products of *Gentiana lutea* L., *Rumex acetosa* L., *Verbena officinalis* L., *Sambucus nigra* L., and *Primula veris* L. were commonly contaminated with PAs from *Senecio* spp. Several products contained PA concentrations above the recommended values of both the German and European Medicines Agencies [9]. This contamination and inadequate control monitoring is highly damaging to registered herbal products since the THR mark/logo is intended to indicate safety and quality.

PAs have been found widely in food, in culinary herbs and spices, including oregano, parsley, and cumin, in herbal teas [10, 11] and in milk, eggs, salads, and meat [12]. Honey made from *Echium* and other PA-containing flowers contains significant concentrations of PAs [11], and particular concerns have been linked to regional honey collected in areas where *Senecio jacobaea* infestation is high [13].

PA-containing plants may be used to make compost as a mulch and fertiliser for cultivating other crops, but it has now been shown that interspecific transfer of PAs can occur in the soil, for example, the leaching of PAs from *Senecio jacobaea* compost and their uptake by non-PA containing herbs such as chamomile (*Matricaria chamomilla* L.) [14].

These hidden sources of contamination add to the overall exposure to PAs, but there is currently no way of knowing to what extent. As the toxicity of PAs is not in any doubt, it is not justifiable to use PA-containing herbs taken orally. Their absence in herbal products must be monitored if there is any chance of contamination, for example, in wild-collected plant species, collection from fields that may also host PA-containing weeds, and where the provenance of the herbal material is not known. The problem is exacerbated by poor regulatory control in many countries where raw material is sourced, and lack of traceability of supply lines. If considered as food products, herbal materials are even less likely to be scrutinised for the presence of PAs.

3.2 Plant Species Containing Pyrrolizidine Alkaloids

PAs are found more frequently in Asteraceae (37 species, particularly in the genera *Senecio* L., *Petasites* Mill., *Tussilago* L. and *Eupatorium* L.); they also occur in Boraginaceae (9 species, including comfrey, *Symphytum* L.,), Fabaceae (5 species, including *Crotalaria* L.), and less frequently in the Poaceace and Convolvulaceae. The occurrence of PAs in a genus may not be consistent: for example, *Eupatorium cannabinum* L. contains toxic PAs that have not been found in *E. perfoliatum* L. or *E. purpureum* L., and where they are considered absent [15].

3.3 Chemistry of Pyrrolizidine Alkaloids

A great deal of information is available on the toxicity and chemistry of the PAs, and research continues intensively (e.g. Xu et al. [4], Robertson and Stevens [16], Schrenk [17]). Approximately 500 potentially toxic PAs have been described and classified into 4 types: monoesters, open-chain diesters, macrocyclic diesters and seco-alkaloids, which occur in the plant as free bases and as *N*-oxides. The free bases are pro-toxins which, after absorption, are activated in the liver by CYP3A and CYP2B enzymes to dihydropyrrolizine ester metabolites (DHP esters). These pyrrole metabolites are alkylating agents capable of causing tissue damage and inducing genetic mutations. Ingested pyrrolizidine alkaloid *N*-oxides are reduced to their free bases during passage through the gut and in the liver, and via hepatic activation, to the toxic DHP esters. Some of the most commonly occurring PAs in the medicinal herbs shown in Table 3.1 are senecionine, lycopsamine, intermedine and senkirkine, and their structures are shown in Fig. 3.1. In the case of food, the PA alkaloid content will depend upon the species of the contaminating weed. There are many others [4, 16–18].

The significant features of the toxic PAs are that all commonly have a double bond in the ring nucleus, an esterified hydroxyl group, and a branched carbon in at least one of the ester side chains. There are other PAs that do not have these structural components and they are not considered to be toxic.

Table 3.1 Medicinal plant species containing unsaturated pyrrolizidine alkaloids

Family	Species. Common name(s). Part(s) used	Region or tradition. Medicinal use(s).	PA constituents described (examples)
BORAGINACEAE	*Alkanna tinctoria* L. Tausch. Alkanet, alkanna, dyer's bugloss. Root.	Europe, Ayurveda. Diarrhoea, gastric ulcers; externally for skin conditions.	O-angeloylretronecine, triangularine, dihydroxytriangularine.
	Borago officinalis L. Borage, burrage, starflower. Leaf.	Europe. Demulcent, emollient, in fevers and colds.	Intermedine, lycopsamine.
	Cordia dichotoma Forst. Indian cherry. All parts.	Ayurveda, Unani. Cold, cough, fever, skin diseases; fruits used for colic, weakness.	Macrophylline
	Heliotropium indicum L., others. Heliotrope. Herb.	Africa, Asia. Inflammation, tumours; externally for skin conditions.	Indicine, acetyl-indicine, indicinine, heleurine, heliotrine, supinine, supinidine.
	Symphytum officinale L., others and hybrids. Comfrey, bruisewort, consolida, knitbone. Herb and root.	Europe, Asia. Bruising, bone and wound healing, internally and externally.	Intermedine, lycopsamine, symphytine, echimidine, symglandine.

Table 3.1 (continued)

Family	Species. Common name(s). Part(s) used	Region or tradition. Medicinal use(s).	PA constituents described (examples)
ASTERACEAE	*Chromolaena odorata* (L.) King & Robinson. Siam weed, devil weed. Leaf.	Pan-tropical. Ayurveda. Wounds, pain, fever.	Intermedine, rinderine, diacetylrinderine, supinine.
	Cynoglossum officinale L. Hound's tongue. Leaf.	Europe. Coughs, piles; externally for skin conditions.	Cynoglossine, consolidine, echinatine, heliosupine.
	Echium vulgare L. Viper's bugloss. Herb.	Europe, Asia. Inflammation, as an expectorant in cough.	Asperumine, echimidine, echimiine, heliosupine.
	Emilia sonchifolia (L.) DC. Ex DC. Lilac tasselflower, Leaf, juice.	Asia, Ayurveda, TCM. Dysentery; externally for cuts and wounds.	Doronine, senkirkine.
	Eupatorium cannabinum L. Hemp agrimony. Herb.	N. America. Inflammation, fever.	Amabiline, intermedine, lycopsamine, rinderine, echinatine, supinine.
	Petasites hybridus L.Butterbur. Herb, root.	Europe, Asia. TCM. Asthma, colds, fevers, urinary complaints.	Senecionine, integerrimine, retrorsine, seneciphylline, jacobine, senkirkine.
	Senecio aureus L. (now known as *Packera aurea* (L.) Á. Löve & D. Löve[a]. Life root, squaw root, golden ragwort. Herb, root.	N. America, Europe, Asia. Coughs, colds, amenorrhoea, menopause.	Senecionine, riddelline, retrorsine, floridanine, monocrotaline, otosenine.
	Senecio jacobaea L. (now known as *Jacobaea vulgaris* Gaertn)[a]. Ragwort. Herb.	Europe, Ayurveda. Coughs, colds.	Integerrimine, usamarine, senecionine, retrorsine, seneciphylline, riddelliine.
	Senecio scandens Buch.-Ham. ex D. Don. Climbing Senecio. Herb.	Asia, TCM. Boils, diarrhoea, eczema, influenza; external use in eye conditions.	Jacobine, senecionine, seneciphylline, senkirkine, usaramine.
	Tussilago farfara Coltsfoot, coughwort. Leaf, flower.	Europe, TCM. Demulcent. in cough or digestive disorders.	Senkirkine, senecionine, tussilagine.

(continued)

Table 3.1 (continued)

Family	Species. Common name(s). Part(s) used	Region or tradition. Medicinal use(s).	PA constituents described (examples)
FABACEAE	*Crotalaria retusa* and many others. Rattlebox, rattleweed.	Ayurveda, others. Cough, fever, diarrhoea; externally for skin conditions.	Monocrotaline, spectabiline.

[a]*Senecio aureus* L. has recently been botanically revised and is now known as *Packera aurea* (L.) Á. Löve & D. Löve, and *S. jacobaea* L. has been revised to *Jacobaea vulgaris* Gaertn). However, a search of the scientific literature using these new names gives no citations to date (March 2017) and illustrates the challenges for retrieving complete and accurate information even if botanical names are provided

3.4 Pyrrolizidine Alkaloid Toxicity

PAs cause acute liver injury, including necrosis, steatosis, and hepatic veno-occlusive disease. Chronic exposure leads to the development of cirrhosis and hepatocellular carcinoma, due to injury of the hepatic parenchyma and vasculature by pyrrole derivatives which react with DNA [2, 4, 12, 17]. PAs commonly induce lung injury, which is dependent on metabolic activation mediated by functional hepatic cytochrome enzymes [19]. The effects of PAs are cumulative, not immediate, and may be via metabolic activation [20].

Research into the toxicity of PAs started initially because of their importance as contaminants in cattle forage. Many cases of livestock poisoning from eating plants containing PAs have been recorded, some of these during periods of drought where other food was not available (see, e.g. Neuman et al. [2]). However, it is now recognised that many localised human epidemics of veno-occlusive disease and liver injury may be due to ingestion of PA-contaminated food crops, and significant, low-concentration PAs have also been found in honey, milk, eggs, salads, and meat [11–13].

Exposure to PAs causes other cancers, and tumours in the adrenal glands, bladder, intestines, kidneys, lungs, muscle, nervous system, pancreas and skin, as well as leukaemia, have been seen in experimental animals. One of the known genetic targets of PAs is TP53, which encodes the tumour suppressor protein p53. Extended, intermittent low-dose exposure to PAs leads to DNA injury and inhibition of mitoses, but the antimitotic activity of PAs has not been well investigated [18].

PAs are also linked to teratogenicity and have been shown to cause fatal veno-occlusive disease (also known as hepatic sinusoidal obstruction syndrome) and cirrhosis in neonates born to pregnant women who had been exposed to PAs from herbal teas or medicines. PAs and their DHP metabolites are teratogenic in rats and cross the placenta to form DHP adducts in embryos. These and other toxic effects of PAs, such as pulmonary arterial hypertension, are discussed in detail by Edgar et al. [12] and Schrenk [17].

3.5 Quality Control Methods Used to Detect PAs in Herbal Materials

Analytical methods for measuring PA concentrations are available and are now being developed into a monograph by the European Pharmacopoeia. Since PA-containing herbs are not allowed in THR products, their presence is not tested for routinely, but in the light of new evidence as to a lack of a safe limit (e.g. Committee on Toxicity of Chemicals in Food, Consumer Products and the Environment [21]), these tests are now deemed necessary.

3.5.1 Botanical Identification

Many cases of poisoning by herbal medicines and teas are due to misidentification of the plant species. In the case of plants of the Asteraceae and Fabaceae families especially, identification by a non-expert can be difficult. In folk-lore reports and traditional herbal medicine texts, delayed symptoms, such as liver toxicity, are rarely recognised or mentioned. Colloquial plant names differ between regions and ethnic groups regarding their scientific names. Sometimes, case histories do not provide the botanical identity of the suspect plant, or it cannot be confirmed, it may not be known at all, or it may have been misidentified. The co-administration of other drugs or herbal products may not be mentioned, so a judgement of causality cannot easily be made for the involvement of the plant species. The issues of reporting accurately on herbal medicines and their identity have been discussed by Chan et al. [22].

Botanical identification begins with authentication of the raw material using its morphological and microscopical characteristics. Pharmacopoeial monographs for medicinal plants include descriptions, which aid identification, but may not provide conclusive evidence alone, especially in closely related species. These are backed up by further diagnostic tests and, increasingly, DNA profiles are used to identify species of importance. These may provide a rapid way of detecting PA-containing plant species, but they are not suitable for plant extracts which contain little or no DNA.

3.5.2 DNA Methods of Detecting PA-Containing herbs

Molecular methods, based on genomic variation between plant species, are effective diagnostic tools. DNA-based markers used for identity confirmation have the advantage that they are not affected by physiological or environmental factors and can be extracted from dried or fresh materials. They are not age- or tissue-specific and only a small quantity of starting materials is required for laboratory analysis.

DNA barcoding is a molecular technique that uses a standard short sequence in the genome to identify a species. DNA barcoding was initially proposed to be applied for identification of plant species as part of a global collaborative initiative, known as the Barcode of Life, which aims to use a unique sequence gene to identify all living species on earth. Sequences are amplified using universal primers in polymerase chain reaction techniques and compared with a reference database (e.g. Kress et al. [23]). The method has a high accuracy in detecting contamination with toxic plant species in herbal materials and food (e.g. Barcaccia et al. [24], Barthelson et al. [25]).

However, the chloroplast regions maturase K (matK) and large subunit of ribulose biphosphate carboxylase (rbcL), recommended by the Consortium Barcode of Life plant working groups as universal barcodes for identification of plant species, were found to not always bear sufficient genetic diversity in many plant species. Therefore, the search for potential DNA barcodes with discriminatory power for identification of medicinal plant species has not been limited to the two chloroplast gene candidates [26]. The internal transcribed spacer ITS2 was found to be a potential DNA barcode for species identification of the family Fabaceae [27]. In some cases, using multiple barcodes was required for differentiating between species. Thongkhao et al. [28] used 4 DNA barcodes of rbcL, *mat*K, ITS2 and trnH-psba, the intergenic spacer region to differentiate *Cyanthillium cinereum* from its PA-containing adulterant *Emilia sonchifolia* [28].

In recent years, DNA barcoding methods have been increasingly used for species identification of medicinal plants and are found to be most powerful when used for authentication of single plant species using reliable reference sequences [29] and differentiating medicinal plants from their known adulterants [30]. DNA barcoding coupled with High Resolution Melting analysis is demonstrated to detect and differentiate the PA-containing herb *Crotalaria spectabilis* from *Thunbergia laurifolia* in Thailand, where it is commonly used, and shares the name "Rang Chute" [30].

As with any other analytical method, DNA barcoding has drawbacks, mainly because it is prone to error when DNA is degraded through processing of medicinal plants and may fail to identify unknown or non-targeted adulterant species. Multiple component natures of most medicinal drugs introduce extra challenges in successful isolation and amplification of genomic DNA that is needed for genome sequencing and their accurate interpretations. In the last decade, technological advancements achieved in genome sequencing, such as pyrosequencing and next generation sequencing, have enabled researchers to overcome the limitations of the Sanger sequencing traditionally used for barcoding [31]. High-throughput sequencing methods have allowed for simultaneous DNA amplifications and identification of multicomponent herbal drugs [32, 33]. As DNA sequencing methods become cost-efficient, more research studies can test larger genomes as super barcodes for differentiating low divergent species. A recent study successfully identified 6 species of *Ligularia,* a PA-containing herb based on phylogenetic evaluation of their whole chloroplast genome sequences [34].

The current challenge of using DNA-based methods is the absence of a global plant DNA reference database to rely on for accurate identification of plant species [35]. The National Institute of Health database GenBank is an open access repository used for data comparison, but the quality of some of its data is in question due to inconsistencies created by assignment of misidentified species [36]. The Barcode of Life Database (BOLD) and the Medicinal Materials DNA Barcodes Database (MMDBD) provide more reliable resources, including voucher samples for some species.

PA-containing plants are contaminants of some Chinese herbs: for example, the seed of *Crotalaria* spp., which contain retrorsine, is a known adulterant of *Astragalus complanatus* Bunge; *Emilia sonchifolia* DC, which contains senkirkine, is an adulterant of *Taraxacum* L. species; *Gynura segetum* Merr., which contains seneciphylline, has been found as an adulterant of *Atractylodes macrocephala* Koidz [37]. Mistaking *Arenebia euchroma* (Ruan Zicao), which is low in toxic pyrrolizidine alkaloids, with high PA-containing species of *Lithospermum erythrorhizon* or *Onosma paniculata* (Dian Zicao), introduces toxicological risks. The toxic herbs listed above are botanically related to the medicinal herb and have many morphological characters in common, and in such conditions specific tests should be performed to check for the absence of the toxic adulterant. In these cases, DNA methods would be especially useful. It is important to note that DNA barcoding does not provide information on chemical composition of herbal drugs, but it can be used in combination with appropriate chemical analysis to ensure their quality [38].

3.5.3 Chemical Tests and Assays

Chemical tests in a pharmacopcial monograph are used to detect the presence or absence of individual, or classes of, chemical constituents. Assays are designed to measure the concentration of known active or marker substances as a measure of quality. PAs are used as chemical markers for safety monitoring of herbal drugs if contamination is suspected, mainly using HPLC methods. Currently, there is no official test method for determination of PAs in herbal products.

PAs are extracted from plant material using hot or cold alcohol. *N*-oxide forms are usually converted to their corresponding basic forms at this stage. Alternatively, PAs can be extracted in dilute aqueous acid. A clean-up step, using solid phase extraction cartridges, increases the recovery (e.g. Bundesinstitut für Risikobewertung [39]).

Early detection methods were based on distinguishing different PAs by their differences in polaritics and oxidation properties. A simple qualitative method that was specific for unsaturated pyrrolizidine alkaloids was developed by Mattocks [40] and is the most useful colorimetric method for detection of unsaturated pyrrolizidines, which can measure as little as 5 μg of most alkaloids in this group.

Thin-layer chromatography (TLC): The most sensitive TLC methods use Ehrlich reagent for detecting PAs. The plates are first sprayed with *ortho*-chloranil and then with Ehrlich reagent for visualising unsaturated alkaloids. The corresponding N-oxides are treated with acetic anhydride before spraying. For saturated pyrrolizidine alkaloids, the plates are sprayed with an iodobismuth reagent [40].

Gas Chromatography (GC): Coupled with mass spectrometry (MS), GC allows detection of PAs either in a mixture or as separate components [41]. Derivatization may be required due to thermal instability and the *N*-oxide alkaloids should be converted to the corresponding basic PAs or derivatized before GC analysis. Stelljes et al. [42] have identified PAs from *Senecio serra*, *S. dimophophyllus* and S. *hydrophyllus* using GC-MS analysis, which allowed comparison of the *Senecio* species in terms of PA content [42].

High-Performance Liquid Chromatography (HPLC) and Liquid Chromatography Mass Spectrometry (LC): coupled with mass spectrometry (MS) are the most widely used analytical methods for determination of PAs [3, 41, 43–46]. Both basic alkaloids and their corresponding *N*-oxides can be detected simultaneously without any prior derivatization. At present, the SPE-LC-MS/MS method published by the German Federal Institute of Risk assessment [39] is considered the best available. The plant material is extracted and purified as described above and an RP-HPLC column with a binary gradient system is used for analysis. A methanol and water mobile phase, containing ammonium formate and formic acid, is given as a suitable example. PAs are detected by mass spectrometry. The method allows sufficient quantification (below 1 mg/kg) and high specificity by multiple reaction monitoring analysis with 28 internal standards.

In response to the demands of European regulators concerning reports of trace contaminations of herbal medicinal products (HMPs) and foods with pyrrolizidine alkaloids in some Ph. Eur. member states, the European Pharmacopoeia Commission adopted a new general chapter, Contaminant pyrrolizidine alkaloids (2.8.26), at its 168th session in November 2020. This chapter sets out the validation requirements and describes a successful analytical method for the determination of 28 target PAs, as an example. Owing to the substantial variations in the content of herbal drugs and HMPs and the applicable limits of measurement, the new chapter recognises any analytical method that consists of chromatography coupled with MS/MS or high-resolution MS if they meet the validation requirements set out in the chapter. Contaminant pyrrolizidine alkaloids (2.8.26) was due to be published in Supplement 10.6 of the European Pharmacopoeia on 1 July 2021 [47].

3.6 Pharmacovigilance for Pyrrolizidine Alkaloid Toxicity

The pharmacovigilance of PA-containing herbs is carried out by assessing reports of suspected adverse drug reactions (ADRs) associated with species of plants that contain them, rather than reports of PA toxicity that have come from agricultural and food studies and reports. An important issue from a public health point of view is the lack of information regarding the extent of human exposure to PAs in the diet. Pharmacovigilance in the case of PA-containing herbs is not needed to provide evidence of toxicity, but to try to measure the extent to which these herbs are still being used and the impact of that use.

The pharmacovigilance of herbal medicines for reports associated with possible ingestion of PA-containing herbs provides further challenges due to the lack of relevant information in many such case reports regarding the herbal product taken, such as the chemical composition, species identification and possible adulteration. In Europe and most other countries it is illegal to market PA-containing herbs for medicinal use and, unlike foods, medicines have legal standards of quality that must be adhered to before sale. The focus of the herbal and regulatory industries is, therefore, on improving quality control methods to prevent these toxins from reaching patients in the first place, and pharmacovigilance reports are crucial in assessing the impact of the use of species containing PAs where known.

References

1. Zhuge Y, Liu Y, Xie W et al (2019) Expert consensus on the clinical management of pyrrolizidine alkaloid-induced hepatic sinusoidal obstruction syndrome. J Gastroenterol Hepatol 34(4):634–642. https://doi.org/10.1111/jgh.14612
2. Neuman MG, Cohen L, Opris M, Nanau RM, Hyunjin J (2015) Hepatotoxicity of pyrrolizidine plkaloids. J Pharm Pharm Sci 18(4):825–843. https://doi.org/10.18433/j3bg7j
3. Chen J, Lü H, Fang L et al (2017) Detection and toxicity evaluation of pyrrolizidine alkaloids in medicinal plants *Gynura bicolor* and *Gynura divaricata* collected from different Chinese locations. Chem Biodivers 14(2):e1600221. https://doi.org/10.1002/cbdv.201600221
4. Xu J, Wang W, Yang X, Xiong A, Yang L, Wang Z (2019) Pyrrolizidine alkaloids: an update on their metabolism and hepatotoxicity mechanism. Liver Res 3(3-4):176–184. https://doi.org/10.1016/j.livres.2019.11.004
5. European Medicines Agency (2016) Public statement on contamination of herbal medicinal products/traditional herbal medicinal products with pyrrolizidine alkaloids. EMA/HMPC/328782/2016. https://www.ema.europa.eu/en/documents/public-statement/public-statement-contamination-herbal-medicinal-products/traditional-herbal-medicinal-products-pyrrolizidine-alkaloids_en.pdf. Accessed 3 Feb 2021
6. Chen L, Mulder PPJ, Louisse J, Peijnenburg A, Wesseling S, Rietjens IMCM (2017) Risk assessment for pyrrolizidine alkaloids detected in (herbal) teas and plant food supplements. Regul Toxicol Pharmacol 86:292–302. https://doi.org/10.1016/j.yrtph.2017.03.019
7. Kuchta K, Schmidt M (2020) Safety of medicinal comfrey cream preparations (*Symphytum officinale*): the pyrrolizidine alkaloid lycopsamine is poorly absorbed through human skin. Regul Toxicol Pharmacol 118:104784. https://doi.org/10.1016/j.yrtph.2020.104784
8. Medicines and Healthcare Products Regulatory Agency (2016) Precautionary recall six batches of St John's Wort tablets. https://www.gov.uk/government/news/precautionary-recall-six-batches-of-st-johns-wort-tablets. Accessed 3 Feb 2021
9. Letsyo E, Jerz G, Winterhalter P, Lindigkeit R, Beuerle T (2017) Incidence of pyrrolizidine alkaloids in herbal medicines from German retail markets: risk assessments and implications to consumers. Phytother Res 31(12):1906–1909. https://doi.org/10.1002/ptr.5935
10. Kaltner F, Rychlik M, Gareis M, Gottschalk C (2020) Occurrence and risk assessment of pyrrolizidine alkaloids in spices and culinary herbs from various geographical origins. Toxins 12(3):155. https://doi.org/10.3390/toxins12030155

11. Bodi D, Ronczka S, Gottschalk C et al (2014) Determination of pyrrolizidine alkaloids in tea, herbal drugs and honey. Food Addit Contam 31(11):1886–1895. https://doi.org/10.1080/19440049.2014.964337
12. Edgar JA, Molyneux RJ, Colegate SM (2015) Pyrrolizidine alkaloids: potential role in the etiology of cancers, pulmonary hypertension, congenital anomalies, and liver disease. Chem Res Toxicol 28(1):4–20. https://doi.org/10.1021/tx500403t
13. Gottschalk C, Kaltner F, Zimmermann M et al (2020) Spread of *Jacobaea vulgaris* and occurrence of pyrrolizidine alkaloids in regionally produced honeys from northern Germany: inter- and intra-site variations and risk assessment for special consumer groups. Toxins 12(7):441. https://doi.org/10.3390/toxins12070441
14. Nowak M, Wittke C, Lederer I, Klier B, Kleinwächter M, Selmar D (2016) Interspecific transfer of pyrrolizidine alkaloids: an unconsidered source of contaminations of phytopharmaceuticals and plant derived commodities. Food Chem 213:163–168. https://doi.org/10.1016/j.foodchem.2016.06.069
15. Röder E, Wiedenfeld H (2013) Plants containing pyrrolizidine alkaloids used in the Traditional Indian medicine—including Ayurveda. Pharmazie 68(2):83–92
16. Robertson J, Stevens K (2017) Pyrrolizidine alkaloids: occurrence, biology, and chemical synthesis. Nat Prod Rep 34:62–89
17. Schrenk D (2020) Toxicology of pyrrolizidine alkaloids. Food Chem Toxicol 135:1190938. https://doi.org/10.1016/j.fct.2019.110938
18. Kristanc L, Kreft S (2016) European medicinal and edible plants associated with subacute and chronic toxicity part II: plants with hepato-, neuro-, nephro- and immunotoxic effects. Food Chem Toxicol 92:38–49. https://doi.org/10.1016/j.fct.2016.03.014
19. He Y, Song Z, Fan X et al (2020) Investigation on pneumotoxicity induced by pyrrolizidine alkaloids. FASEB J 34(S1):1–1. https://doi.org/10.1096/fasebj.2020.34.s1.08849
20. Yang M, Ma J, Ruan J, Ye Y, Fu PP, Lin G (2019) Intestinal and hepatic biotransformation of pyrrolizidine alkaloid N-oxides to toxic pyrrolizidine alkaloids. Arch Toxicol 93(8):2197–2209. https://doi.org/10.1007/s00204-019-02499-2
21. Committee on Toxicity of Chemicals in Food, Consumer Products and the Environment (2017) COT statement on pyrrolizidine alkaloids in food. http://cot.food.gov.uk/sites/default/files/cot/cotstatementpa200806.pdf. Accessed 3 Feb 2021
22. Chan K, Shaw D, Simmonds MSJ et al (2012) Good practice in reviewing and publishing studies on herbal medicine, with special emphasis on traditional Chinese medicine and Chinese materia medica. J Ethnopharmacol 140(3):469–475. https://doi.org/10.1016/j.jep.2012.01.038
23. Kress JW, Wurdack KJ, Zimmer EA, La W, Janzen DH (2005) Use of DNA barcodes to identify flowering plants. Proc Natl Acad Sci U S A 102(23):8369–8374. https://doi.org/10.1073/pnas.0503123102
24. Barcaccia G, Lucchin M, Cassandro M (2016) DNA barcoding as a molecular tool to track down mislabelling and food piracy. Diversity 8(1):2. https://doi.org/10.3390/d8010002
25. Barthelson RA, Sundareshan P, Galbraith D, Woosley RL (2006) Development of a comprehensive detection method for medicinal and toxic plant species. Am Bot 93(4):566–574. https://doi.org/10.3732/ajb.93.4.566
26. Zhang D, Jiang B (2019) Species identification in complex groups of medicinal plants based on DNA barcoding: a case study on Astragalus spp. (Fabaceae) from southwest China. Conserv Genet Resources 12:469–478. https://doi.org/10.1007/s12686-019-01119-6
27. Gao T, Yao H, Song J et al (2010) Identification of medicinal plants in the family Fabaceae using a potential DNA barcode ITS2. J Ethnopharmacol 130(1):116–121. https://doi.org/10.1016/j.jep.2010.04.026
28. Thongkhao K, Pongkittiphan V, Phadungcharoen T et al (2020) Differentiation of *Cyanthillium cinereum*, a smoking cessation herb, from its adulterant *Emilia sonchifolia* using macroscopic and microscopic examination, HPTLC profiles and DNA barcodes. Sci Rep 10:14753. https://doi.org/10.1038/s41598-020-71702-7

29. Han J, Pang X, Liao B, Yao H, Song J, Chen S (2016) An authenticity survey of herbal medicines from markets in China using DNA barcoding. Sci Rep 6:18723. https://doi.org/10.1038/srep18723
30. Singtonat S, Osathanunkul M (2015) Fast and reliable detection of toxic *Crotalaria spectabilis* Roth. in *Thunbergia laurifolia* Lindl. herbal products using DNA barcoding coupled with HRM analysis. BMC Complement Altern Med 15:162. https://doi.org/10.1186/s12906-015-0692-6
31. Sarwat M, Yamdagni MM (2016) DNA barcoding, microarrays and next generation sequencing: recent tools for genetic diversity estimation and authentication of medicinal plants. Crit Rev Biotechnol 36(2):191–203. https://doi.org/10.3109/07388551.2014.947563
32. Coghlan L, Haile J, Houston J et al (2012) Deep sequencing of plant and animal DNA contained within traditional Chinese medicines reveals legality issues and health safety concerns. PLoS Genet 8(4):e1002657. https://doi.org/10.1371/journal.pgen.1002657
33. Gao Z, Liu Y, Wang X, Wei X, Han J (2019) DNA mini-barcoding: a derived barcoding method for herbal molecular identification. Front Plant Sci 10:987. https://doi.org/10.3389/fpls.2019.00987
34. Chen X, Zhou J, Cui Y, Wang Y, Duan B, Yao H (2018) Identification of *Ligularia* herbs using the complete chloroplast genome as a super-barcode. Front Pharmacol 9:695. https://doi.org/10.3389/fphar.2018.00695
35. Kress WJ (2017) Plant DNA barcodes: applications today and in the future. J Syst Evol 55(4):291–307. https://doi.org/10.1111/jse.12254
36. Hinchliff CE, Smith SA (2014) Some limitations of public sequence data for phylogenetic inference (in plants). PLoS One 9(7):e98986. https://doi.org/10.1371/journal.pone.0098986
37. Zhou J, Ouedraogo M, Qu F, Duez P (2013) Potential genotoxicity of traditional Chinese medicinal plants and phytochemicals: an overview. Phytother Res 27(12):1745–1755. https://doi.org/10.1002/ptr.4942
38. Palhares RM, Drummond MG, Brasil BD, Cosenza GP, Brandão MGL, Oliveira G (2015) Medicinal plants recommended by the world health organization: DNA barcode identification associated with chemical analyses guarantees their quality. PLoS One 10(5):e0127866. https://doi.org/10.1371/journal.pone.0127866
39. Bundesinstitut für Risikobewertung (2014) Method protocol BFR-PA-Tea-2.0/2014: Determination of pyrrolizidine alkaloids (PA) in plant material by SPE-LCMS/MS. http://www.bfr.bund.de/cm/349/determinationof-pyrrolizidine-alkaloids-pa-in-plant-material.pdf. Accessed 3 Feb 2021
40. Mattocks AR (1968) Toxicity of pyrrolizidine alkaloids. Nature 217(5130):723–728. https://doi.org/10.1038/217723a0
41. Kowalczyk E, Kwiatek K (2020) Application of the sum parameter method for the determination of pyrrolizidine alkaloids in teas. Food Addit Contam Part A Chem Anal Control Expo Risk Assess 37(4):622–633. https://doi.org/10.1080/19440049.2020.1719284
42. Stelljes ME, Kelley RB, Molyneux RJ, Seiber JN (1991) GC-MS determination of pyrrolizidine alkaloids in four *Senecio* species. J Nat Prod 54(3):759–773. https://doi.org/10.1021/np50075a004
43. Li SL, Lin G, Fu PP et al (2008) Identification of five hepatotoxic pyrrolizidine alkaloids in a commonly used traditional Chinese medicinal herb, Herba Senecionis scandentis (Qianliguang). Rapid Commun Mass Spectrometry 22(4):591–602. https://doi.org/10.1002/rcm.3398
44. Griffin CT, Danaher M, Elliott CT, Kennedy DG, Furey A (2013) Detection of pyrrolizidine alkaloids in commercial honey using liquid chromatography–ion trap mass spectrometry. Food Chem 136(3-4):1577–1583. https://doi.org/10.1016/j.foodchem.2012.02.112
45. Griffin CT (2014) Investigation of pyrrolizidine alkaloids in foods using liquid chromatography mass spectrometry. Dissertation. Cork Institute of Technology

46. Picron JF, Herman M, Van Hoeck E, Goscinny S (2018) Analytical strategies for the determination of pyrrolizidine alkaloids in plant-based food and examination of the transfer rate during the infusion process. Food Chem 266:514–523. https://doi.org/10.1016/j.foodchem.2018.06.055
47. European Directorate for the Quality of Medicines (2021) European Pharmacopoeia Commission adopts new general chapter Contaminant pyrrolizidine alkaloids (2.8.26). https://www.edqm.eu/en/news/european-pharmacopoeia-commission-adopts-new-general-chapter-contaminant-pyrrolizidine. Accessed 3 Feb 2021

Chapter 4
Pharmacovigilance for Herbal and Traditional Medicine-Induced Liver Injury

Jia-bo Wang, Yuan Gao, Yu-ming Guo, and Xiao-he Xiao

4.1 Introduction

Herbal and traditional medicines (HTMs) are widely used throughout the world; however, HTM-related adverse reactions, including HTM-induced liver injury, have not been fully investigated. This chapter reviewed the current status and epidemiology of HTM-induced liver injury around the world, which revealed divergent data from different countries. Not surprisingly, the most implicated species of HTM-induced liver injury are remarkably distinct between Western countries and Eastern countries, namely China; this warrants specific considerations in pharmacovigilance for HTM-induced liver injury in different areas. The main risk factors and mechanisms for HTM-induced liver injury are also summarized and discussed using *Polygonum multiflorum* (heshouwu; accepted name: *Reynoutria multiflora* (Thunb.) Moldenke) as an example.

Throughout human history, traditional medicines (TMs) have significantly contributed to the prevention and treatment of diseases in different countries and regions. The efficacy and safety of several TMs have been shown by tradition and by history, although not as definitively as for conventional drugs. According to figures from the World Health Organization (WHO), 40 to 80% of the population in developing countries has experienced some form of traditional medicine therapy [1,

J.-b. Wang (✉)
School of Traditional Chinese Medicine, Capital Medical University, Beijing, People's Republic of China

China Military Institute of Chinese Medicine, Fifth Medical Center of Chinese PLA General Hospital, Beijing, People's Republic of China
e-mail: jiabo_wang@ccum.edu.cn

Y. Gao · Y.-m. Guo · X.-h. Xiao
China Military Institute of Chinese Medicine, Fifth Medical Center of Chinese PLA General Hospital, Beijing, People's Republic of China

J. Barnes (ed.), *Pharmacovigilance for Herbal and Traditional Medicines*,
https://doi.org/10.1007/978-3-031-07275-8_4

2]. In China, TM accounts for around 40% of all healthcare delivered and is used to treat roughly 200 million patients annually [3]. Meanwhile, in many developed countries, complementary and alternative therapies (CAM) are becoming more and more popular [1, 4, 5]. The WHO has introduced TM into the eleventh edition of the global influential medical compendium in 2018 for the first time [6], as part of achieving global healthcare. However, compared with growing demands for herbal and traditional medicines (HTMs), attention and research on safety of these preparations have lagged far behind those for conventional drugs.

With the trend in widespread application of HTMs all over the world, as well as ongoing improvements in drug administration systems, there are new challenges arising relating to safety problems associated with HTMs. In recent years, HTM-related adverse events, particularly those represented by HTM-induced liver injury, have occurred frequently [7–9], and this has become a crucial issue for public health.

4.2 An Epidemiological Overview of Herbal and Traditional Medicines (HTMs) and Drug-Induced Liver Injury (DILI)

4.2.1 Description and Epidemiology of HTM DILI

Drug-induced liver injury (DILI) is a common adverse drug reaction (ADR) and can lead to liver failure and even death [9]. It is increasingly appreciated to be one of the most challenging diseases for physicians and gastroenterologists. HTM DILI is a type of adverse drug reaction related to HTM, and now represents a growing segment of DILI worldwide.

According to epidemiological data, the incidence of DILI in the general population is estimated to be between 1/100,000 and 20/100,000 [10, 11]. At present, the annual incidence in the general population is estimated to be 23.80 per 100,000 persons in China; the leading single classes of implicated drugs were traditional Chinese medicines (TCMs) or herbal and dietary supplements (HDS) [12, 13]. Research data from the USA Drug-Induced Liver Injury Network (DILIN) showed that liver injury associated with HDSs has increased rapidly, with the percentage of liver injuries involving these substances rising dramatically from 7% in 2005 to 19% in 2012 [14]. Data from the Asia-Pacific region showed that Chinese herbal medicine (CHM) was the main cause of DILI in Korea and Singapore [15]. Single- and multicenter retrospective clinical studies in China with large samples showed that the composition ratio of HTM DILI to the total DILI was approximately 20% [16]. The current data mainly consider trends in HTM DILI by analyzing the composition ratio of TCM DILI versus the total DILI; however, statistical data vary substantially across different countries and regions [17–21].

4.2.2 *Proportions of HTM DILI in All-Cause DILI*

Although racial differences exist in various populations, the current understanding of genetic factors associated with liver injury risks for diverse herbal substances is still ambiguous [22]. Nongenetic factors, including culture and dietary preference, medication profiles, underlying disease spectrum, etc., can be important in different liver injury risks for diverse herbal substances.

A retrospective study to determine the incidence and causes of DILI in mainland China showed CHM was implicated in 26.81% of all cases of patients with DILI [12], which is higher than that in Western countries (Fig. 4.1). However, several other Eastern countries, namely Japan, South Korea, and Singapore, had reported the extremely divergent proportions, 1.0%, 30.8%, and 73%, respectively, of HTM DILI in all-cause DILI [23–25]. Notably, data from Singapore also reported 29% of products involved in causal HDS cases contained adulterants, including hepatotoxic synthetic drugs, such as acetaminophen [24]. In summary, these data varied in the studies from different regions. The possible reasons might be that the statistical methods and dimensions varied in these studies. Besides, the shortage of accurate and consistent diagnostic method for HTM DILI might also contribute to this inconsistency between countries. Thus, there is a fundamental need to carry out multicountry surveys of HTM DILI according to uniform diagnostic criteria.

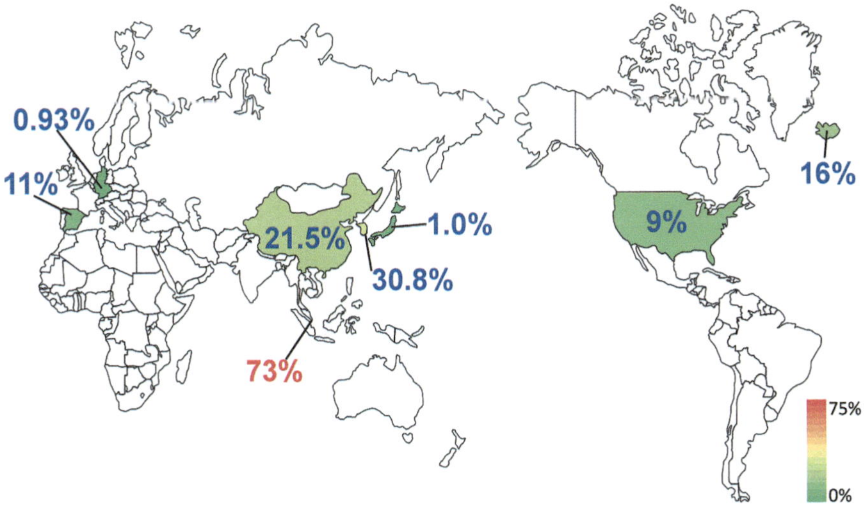

Fig. 4.1 Proportions of HTM DILI in all-cause DILI from studies in different regions

4.3 Species Implicated in HTM DILI

The use of herbs for medical or healthcare purposes has a long history in Southeast Asia, Africa, South America, etc., and is also widespread in Western countries. There have been some publications summarizing the species of herbs associated with HTM DILI [26]. However, it should be noted that distinctions exist within China, Western countries, and the other regions using folk medicines (Fig. 4.2), due to the different cultures of medication use and diverse geographical distribution of plants. In Chinese literature, the most reported herbs associated with HTM DILI include *Polygonum multiflorum* Thunb. (roots or stems and leaves), *Dictamnus dasycarpus* Turcz. (cortex of roots), *Tripterygium wilfordii* Hook. f. (roots), *Dioscorea bulbifera* L. (roots), *Psoralea corylifolia* L. (fruits), *Gynura japonica* (Thunb.) Juel (roots), *Rheum palmatum* L. (roots), and *Senna alexandrina* Mill. (leaves) [7–9, 27]. In English biomedical literature, the herbs most reported to be associated with HTM DILI include extract of green tea (leaves of *Camellia sinensis* (L.) Kuntze), *Atractylis gummifera* Salzm. ex L. (roots), black cohosh (roots and rhizomes of *Actaea racemosa* L.), cascara (usually the bark of *Rhamnus purshiana* DC.), and ephedra (aerial parts of *Ephedra sinica* Stapf) [28]. In Mexico, herbs associated with HTM DILI include *Scoparia dulcis* L. (leaves), *Citrus aurantium* L. (immature fruits), *Rosmarinus officinalis* L. (leaves), and *Equisetum hyemale* L. (stems) [29].

According to the latest nationwide survey on TCM resources, there are over 11,000 species of plants being used as herbal medicines in China. The plant families

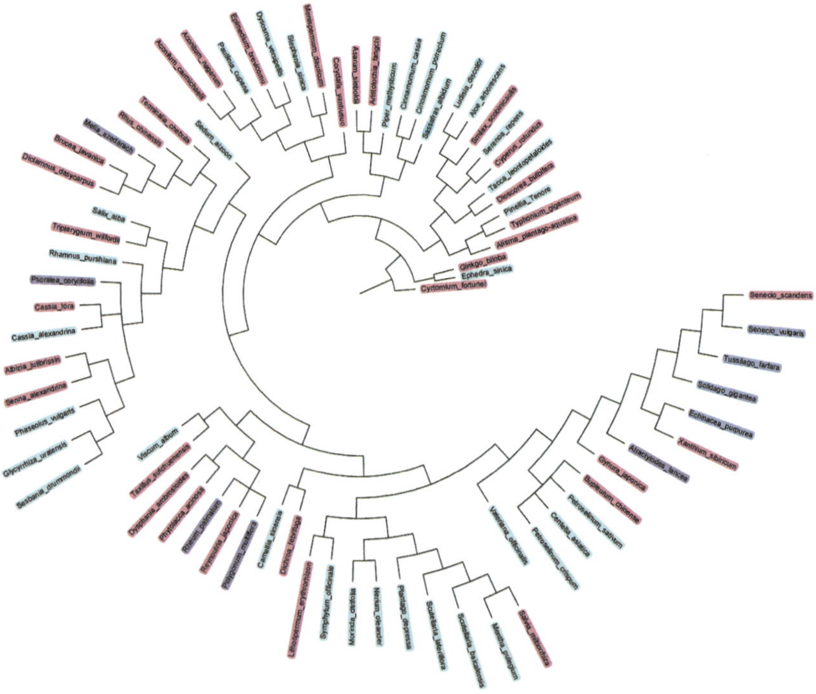

Fig. 4.2 Differences between herbs associated with HTM DILI reported in China and Western countries

Asteraceae (Compositae), Leguminosae, and Polygonaceae are those usually reported in association with HTM DILI in Chinese literature. By contrast, some plant families, such as *Ephedraceae*, have been reported in association with liver injury in western literature, but they are not considered hepatotoxic in Chinese TCM knowledge. The potential reason for this divergence might rely on the different uses between China and Western countries, such as the USA: in China, Ephedra is used as an anti-influenza therapy for short-term use, while in the USA, it is used for weight loss with long-term use. Another example is green tea (*Camellia sinensis*), which is also used in Western countries for weight loss and is the subject of emerging reports associated with liver injury; this herb is used as a daily drink in China and Southeastern Asia countries and used without concerns about hepatotoxicity. Besides, there are some plant species peculiar to the original production area, which have special cultural uses and hepatotoxic concerns. For example, black cohosh (*Actaea racemosa* L.), a folk medicinal herb native to North America and originally used by Native American Indians, is currently popular in European countries and the USA, but is barely used in China. Kava-kava (*Piper methysticum* G.Forst.) is a native plant in islands of the western Pacific Ocean and has now been used as a sedative in Western countries. These species of herbs are seldom reported in association with hepatotoxicity in China because they are seldom used there. Thus, there is an obligatory need to establish an open-source database of HTM DILI-associated herb lists and translated literature in English regarding TCM and other traditional and folk medicines by cooperation across the world.

The herbs associated with HTM DILI were clustered according to their plant family and genus. There was a distinct difference between the reports from China (labelled as) and the species from Western countries (labelled as). A small number of species were the same ones reported in either China or Western countries (labelled as).

4.4 Main Risk Factors for HTM DILI

The risk factors for HTM DILI are relatively complex and should be analyzed from the perspectives of drugs, the human body, and their interactions. In particular, for idiosyncratic liver injuries, the influence of immunity, metabolism, heredity, and other factors should be considered to more purposefully obtain information regarding liver injury risk factors. When evaluating HTM DILI, interference factors, such as poor drug quality and medication errors, should first be excluded.

4.4.1 Risk Factors Related to Herbs and Drugs Combination

Beyond the recognition of herb species associated with liver injury, there are still many compound formulae consisting of different herbs where it is not possible to clarify the culprit herb or herbs. Synergistic effects in hepatotoxicity might also be a confounding factor and should be considered. In TCM theory, incompatibilities and compatibilities between specific pairs of herbs are usually described to avoid unwanted synergistic effects on toxicity and vice versa. For example, *Poria cocos* (Schw.) Wolf, also known as fuling, a kind of fungal medicine used in TCM, is traditionally considered as a "good" combination when used with *Polygonum multiflorum* (PM), and this has been demonstrated with respect to the hepatoprotective effect of *Poria cocos* against *Polygonum multiflorum*/lipopolysaccharide (LPS)-induced liver injury in rats for the first time [30]. For those compound formulae, when they consist of both a suspected hepatotoxic herb and a hepatoprotective one, limited knowledge has been obtained to evaluate their safety risks.

4.4.2 Risk Factors Related to Quality Problems

Different plant origins, harvest sources, medicinal plant parts as well as harvesting time and processing are often important interfering factors that affect the assessment of HTM DILI. When evaluating risk factors for HTM DILI, comprehensive inspections should be conducted as follows.

(1) The plant origin may result in the use of homonyms due to mistakes in translating names or aliases of different countries or regions. For example, Tusanqi, a common name for *Gynura segetum*, is a considerable herb associated with HTM DILI. This herb contains unsaturated pyrrolizidine alkaloids which have been well documented in association with hepatic sinusoidal obstruction syndrome/hepatic veno-occlusive disease (HSOS/HVOD) [31]. However, some literature has confused this herb with another herb—*Panax notoginseng*—which is called Sanqi in Chinese. Tusanqi is a misused substitute for Sanqi in some folk regions in China; these herbs should not be confused in clinical reports.
(2) The place of origin, medicinal plant parts used, harvesting time, processing and formulation method, and exogenous contaminants, such as impurities, agricultural and farm chemicals and heavy metal residues, could increase related risks.
(3) Adulteration with intentionally added chemicals or synthetic drugs increases the risks of harms associated with HTMs. For example, a Chinese herbal ointment, known as shen-fu-cao (with the ingredients *Typhonium giganteum* Engl., *Phellodendron chinense* C.K. Schneid., *Stemona japonica* (Blume) Miq., et al.), was recently reported in Denmark to contain an illegally added potent corticosteroid (clobetasol propionate) and antifungals (ketoconazole and miconazole), and caused a rash that appeared to be caused by hormone drugs [32].

4.4.3 Risk Factors Related to HTM Product Use

Changes in dose form and the route of drug administration may increase the risk of harms associated with HTMs; this is especially so with changes from external use to internal administration, from topical to systemic use and from oral administration to injection. Changes in the usual dosage and course or duration of treatment may also significantly increase the risk of harms associated with HTMs.

Changes in indications, often accompanied by changes in dosage, duration of treatment, extracting procedure, or drug administration route, can also lead to liver injury risks of HTM. For instance, *Artemisia annua* and its water—or ethanol—extraction products containing artemisinin derivatives are mainly used for treatment of malaria in China while, in New Zealand, it was recently reported that use of a supercritical carbon dioxide extract from *A. annua*, promoted to maintain and support joint health and movement, was associated with a series of liver injury cases [33]. In most countries, it is common to find concurrent use of HTMs with conventional medicines. As HTMs are accessible without a physician's prescription in most countries, there may be instances where conventional drugs and HTMs are used concurrently and for which physicians are unaware, and lack of involvement of traditional medicine practitioners, which all increase the risk of inappropriate use. Some previous studies have shown that the combination of HTMs with conventional medicines can be an important risk factor for adverse drug reactions, including those resulting from drug interactions [34]. However, the contribution that concurrent use of HTMs and conventional drugs makes to the development of DILI is still lacking in research.

4.5 Mechanisms of HTM DILI

DILI is generally classified as intrinsic DILI and idiosyncratic DILI [35]. Most toxic HTMs, such as *Tripterygium wilfordii* Hook.f., are associated with intrinsic DILI, which generally has a significant dose-time-toxicity relationship. HTMs not known to be toxic usually have no significant dose-time-toxicity relationship with DILI and reveal substantial differences between individuals in occurrence of DILI. This suggests that liver injury related to HTMs not known to be hepatotoxic is more likely to be idiosyncratic DILI. However, there is little attention and research on idiosyncratic DILI mediated by individual characteristics at present.

It has been demonstrated that the immunological, stress-mediated, tri-element injury hypothesis supports the idiosyncratic toxicity of *Polygonum multiflorum* (heshouwu; accepted name: *Reynoutria multiflora* (Thunb.) Moldenke) [36]. When the body is in an immune stress state, the hepatotoxicity thresholds for *cis*-stilbene glycoside (*cis*-SG) are altered to induce liver injury; meanwhile, *trans*-stilbene glycoside (*trans*-SG) enhances the immune response to further aggravate *cis*-SG-induced liver injury (Fig. 4.3).

It is further confirmed that *Polygonum multiflorum*-induced liver injury was related to organism factors, especially immune-related genetic differences, and *HLA-B*35:01* as the susceptibility gene was found [37]. In addition, TNF-α, MCP-1, VEGF, etc., and endogenous metabolites of phenyllactic acid, crotonoyl-CoA, indole-5,6-quinone, and other biomarkers found to be expressed abnormally may be risk factors for liver injury [38–42]. In addition, the mechanism of liver injury induced by fructus psoraleae (buguzhi; accepted name: *Cullen corylifolium* (L.) Medik.) and epimedium (yinyanghuo; accepted name: *Epimedium brevicornu* Maxim.) was also revealed. It is confirmed that the injury was caused by a combination of direct toxic ingredients and by NLRP3-activating ingredients. NLRP3 activation may be a risk factor for fructus psoraleae or epimedium related liver injury, and compatibility use of fructus psoraleae and epimedium may increase this risk [43, 44]. Furthermore, cortex dictamni (baixianpi; accepted name: *Dictamnus albus*

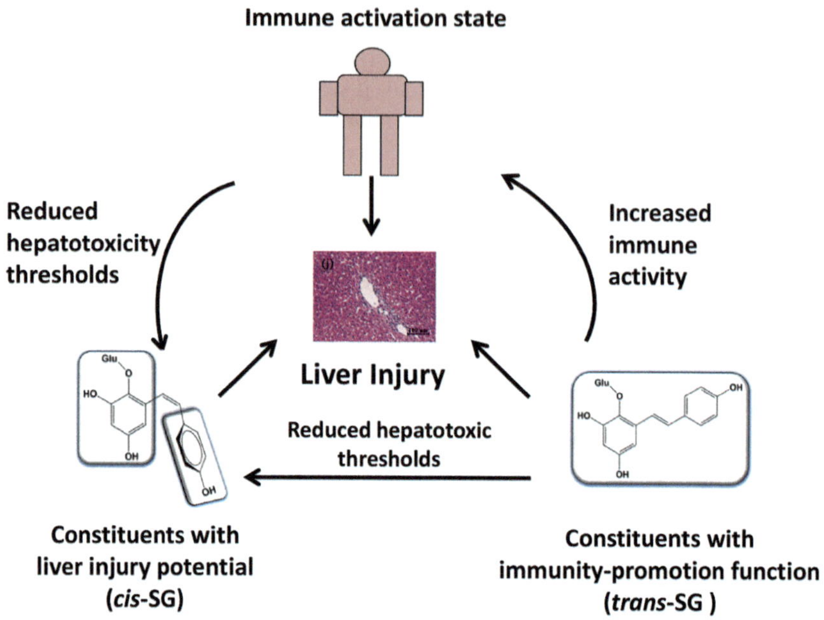

Fig. 4.3 The hypothesis of immunological stress-mediated tri-element idiosyncratic DILI [36]

L.) induced liver injury has also been confirmed to be immunological idiosyncratic DILI [45].

In conclusion, HTM-induced liver injury is an important, yet unsolved, problem and needs further research to emphasize this public challenge. Owing to the different cultures of using HTMs, there are great distinctions between the most implicated species of HTM-induced liver injury. International collaborations are needed to overcome the language and cultural barriers within countries throughout the world to improve the ability of pharmacovigilance for HTM-induced liver injury.

References

1. World Health Organization (2002) WHO Traditional medicine strategy 2002–2005. https://apps.who.int/iris/bitstream/handle/10665/67163/WHO_EDM_TRM_2002.1.pdf;jsessionid=2E85DFAFA0599960A11316226A83EE6D?sequence=1. Accessed 12 Apr 2021
2. Regional Committee for Africa (2000) Promoting the role of traditional medicine in health systems: a strategy for the African Region. World Health Organization. https://apps.who.int/iris/handle/10665/95467. Accessed 12 Apr 2021
3. World Health Organization Regional Office for the Western Pacific (1999) Traditional and modern medicine: harmonizing the two approaches: a report of the consultation meeting on traditional and modern medicine: harmonizing the two approaches, 22–26 November 1999, Beijing, China. https://apps.who.int/iris/handle/10665/207162. Accessed 12 Apr 2021
4. Moschik EC, Mercado C, Yoshino T, Matsuura K, Watanabe K (2012) Usage and attitudes of physicians in Japan concerning traditional Japanese medicine (Kampo medicine): A descriptive evaluation of a representative questionnaire-based survey. Evid Based Complement Alternat Med 2012:139818. https://doi.org/10.1155/2012/139818
5. Frass M, Strassl RP, Friehs H, Müllner M, Kundi M, Kaye AD (2012) Use and acceptance of complementary and alternative medicine among the general population and medical personnel: a systematic review. Ochsner J 12(1):45–56

6. Cyranoski D (2018) Why Chinese medicine is heading for clinics around the world. Nature 561(7724):448–450. https://doi.org/10.1038/d41586-018-06782-7
7. Wang JB, Zhu Y, Bai ZF, Wang FS, Li XH, Xiao XH (2018) Guidelines for the diagnosis and management of herb-induced liver injury. Chin J Integr Med 24(9):696–706. https://doi.org/10.1007/s11655-018-3000-8
8. Xiao XH, Tang J, Mao Y et al (2019) Guidance for the clinical evaluation of traditional Chinese medicine-induced liver injury issued by China Food and Drug Administration. Acta Pharm Sin B 9(3):648–658. https://doi.org/10.1016/j.apsb.2018.12.003
9. Council for International Organizations of Medical Sciences (2020) Drug-induced liver injury (DILI): current status and future directions for drug development and the post-market setting. https://cioms.ch/wp-content/uploads/2020/06/CIOMS_DILI_Web_16Jun2020.pdf. Accessed 12 Apr 2021
10. Larrey D (2002) Epidemiology and individual susceptibility to adverse drug reactions affecting the liver. Semin Liver Dis 22(2):145–155. https://doi.org/10.1055/s-2002-30105
11. Bjornsson ES (2014) Epidemiology and risk factors for idiosyncratic drug-induced liver injury. Semin Liver Dis 34(2):115–122. https://doi.org/10.1055/s-0034-1375953
12. Shen T, Liu YX, Shang J et al (2019) Incidence and etiology of drug-induced liver injury in mainland China. Gastroenterology 156(8):2230–2241.e11. https://doi.org/10.1053/j.gastro.2019.02.002
13. Zhu Y, Niu M, Chen J et al (2016) Comparison between Chinese herbal medicine and Western medicine-induced liver injury of 1985 patients. J Gastroenterol Hepatol 31(8):1476–1482. https://doi.org/10.1111/jgh.13323
14. Navarro VJ, Khan I, Björnsson E, Seeff LB, Serrano J, Hoofnagle JH (2017) Liver injury from herbal and dietary supplements. Hepatology 65(1):363–373. https://doi.org/10.1002/hep.28813
15. Cho JH, Oh DS, Hong SH et al (2017) A nationwide study of the incidence rate of herb-induced liver injury in Korea. Arch Toxicol 91(12):4009–4015. https://doi.org/10.1007/s00204-017-2007-9
16. Xu JM (2007) A multicenter survey on hospital inpatients with drug-induced acute liver injury in China. Chin J Digestion 27(7):439–442
17. López-Gil S, Nuño-Lámbarri N, Chávez-Tapia N, Uribe M, Barbero-Becerra VJ (2017) Liver toxicity mechanisms of herbs commonly used in Latin America. Drug Metab Rev 49(3):338–356. https://doi.org/10.1080/03602532.2017.1335750
18. Valdivia-Correa B, Gómez-Gutiérrez C, Uribe M, Méndez-Sánchez N (2016) Herbal medicine in Mexico: a cause of hepatotoxicity. A critical review. Int J Mol Sci 17(2):235. https://doi.org/10.3390/ijms17020235
19. Medina-Caliz I, Garcia-Cortes M, Gonzalez-Jimenez A et al (2018) Herbal and dietary supplement-induced liver injuries in the Spanish DILI registry. Clin Gastroenterol Hepatol 16(9):1495–1502. https://doi.org/10.1016/j.cgh.2017.12.051
20. Vega M, Verma M, Beswick D et al (2017) The incidence of drug- and herbal and dietary supplement-induced liver injury: preliminary findings from gastroenterologist-based surveillance in the population of the state of Delaware. Drug Saf 40(9):783–787. https://doi.org/10.1007/s40264-017-0547-9
21. Jing J, Teschke R (2018) Traditional Chinese medicine and herb-induced liver injury: comparison with drug-induced liver injury. J Clin Transl Hepatol 6(1):57–68. https://doi.org/10.14218/JCTH.2017.00033
22. Chen M, Borlak J, Tong W (2014) Predicting idiosyncratic drug-induced liver injury - some recent advances. Expert Rev Gastroenterol Hepatol 8(7):721–723. https://doi.org/10.1586/17474124.2014.922871
23. Mantani N, Kogure T, Sakai S et al (2002) Incidence and clinical features of liver injury related to Kampo (Japanese herbal) medicine in 2496 cases between 1979 and 1999: problems of the lymphocyte transformation test as a diagnostic method. Phytomedicine 9(4):280–287. https://doi.org/10.1078/0944-7113-00141
24. Suk KT, Kim DJ, Kim CH et al (2012) A prospective nationwide study of drug-induced liver injury in Korea. Am J Gastroenterol 107(9):1380–1387. https://doi.org/10.1038/ajg.2012.138

25. Wai CT, Tan BH, Chan CL et al (2007) Drug-induced liver injury at an Asian center: a prospective study. Liver Int 27(4):465–474. https://doi.org/10.1111/j.1478-3231.2007.01461.x
26. Navarro VJ, Lucena MI (2014) Hepatotoxicity induced by herbal and dietary supplements. Semin Liver Dis 34(2):172–193. https://doi.org/10.1055/s-0034-1375958
27. Xu CY, Wang FW, Sun ZG, Chen LP (2014) Meta analysis on rates of drug-induced liver injury by Chinese herbal medicines in China. J Emerg Tradit Chin Med 11(1988–1989):2043
28. Maddukuri VC, Bonkovsky HL (2014) Herbal and dietary supplement hepatotoxicity. Clin Liver Dis 4(1):1–3. https://doi.org/10.1002/cld.364
29. Valdivia-Correa B, Gómez-Gutiérrez C, Uribe M, Méndez-Sánchez N (2016) Herbal medicine in Mexico: a cause of hepatotoxicity. a critical review. Int J Mol Sci 17(2):235–244. https://doi.org/10.3390/ijms17020235
30. Gao D, Pang JY, Zhang CE et al (2016) Poria attenuates idiosyncratic liver injury induced by polygoni multiflori radix praeparata. Front Pharmacol 7:386. https://doi.org/10.3389/fphar.2016.00386
31. Lin G, Wang JY, Li N et al (2011) Hepatic sinusoidal obstruction syndrome associated with consumption of Gynura segetum. J Hepatol 54(4):666–673. https://doi.org/10.1016/j.jhep.2010.07.031
32. Mose KF, Bygum A (2019) Chinese herbal remedy found to contain steroids and antifungals. Lancet 393(10170):446. https://doi.org/10.1016/S0140-6736(19)30116-3
33. Stebbings S, Beattie E, McNamara D, Hunt S (2016) A pilot randomized, placebo-controlled clinical trial to investigate the efficacy and safety of an extract of Artemisia annua administered over 12 weeks, for managing pain, stiffness, and functional limitation associated with osteoarthritis of the hip and knee. Clin Rheumatol 35(7):1829–1836. https://doi.org/10.1007/s10067-015-3110-z
34. Henderson L, Yue QY, Bergquist C, Gerden B, Arlett P (2002) St John's wort (*Hypericum perforatum*): drug interactions and clinical outcomes. Br J Clin Pharmacol 54(4):349–356. https://doi.org/10.1046/j.1365-2125.2002.01683.x
35. Chalasani NP, Hayashi PH, Bonkovsky HL et al (2014) ACG clinical guideline: the diagnosis and management of idiosyncratic drug-induced liver injury. Am J Gastroenterol 109(7):950–966. https://doi.org/10.1038/ajg.2014.131
36. He LZ, Yin P, Meng YK et al (2017) Immunological synergistic mechanisms of trans−/cis-stilbene glycosides in Heshouwu related idiosyncratic liver injury. Sci Bull 62(11):748–751
37. Li CP, Rao T, Chen X et al (2019) HLA-B*35:01 Allele is a potential biomarker for predicting polygonum multiflorum-induced liver injury in humans. Hepatology 70(1):346–357. https://doi.org/10.1002/hep.30660
38. Tu C, He Q, Li CY et al (2019) Susceptibility-related factor and biomarkers of dietary supplements Polygonum multiflorum-induced liver injury in rats. Front Pharmacol 10:335. https://doi.org/10.3389/fphar.2019.00335
39. Li CY, Tu C, Gao D et al (2016) Metabolomic study on idiosyncratic liver injury induced by different extracts of polygonum multiflorum in rats integrated with pattern recognition and enriched pathways analysis. Front Pharmacol 7:483. https://doi.org/10.3389/fphar.2016.00483
40. Zhang CE, Niu M, Li Q et al (2016) Urine metabolomics study on the liver injury in rats induced by raw and processed Polygonum multiflorum integrated with pattern recognition and pathways analysis. J Ethnopharmacol 194:299–306. https://doi.org/10.1016/j.jep.2016.09.011
41. Zhang L, Niu M, Wei AW et al (2019) Risk profiling using metabolomic characteristics for susceptible individuals of drug-induced liver injury caused by Polygonum multiflorum. Arch Toxicol 94(1):245–256. https://doi.org/10.1007/s00204-019-02595-3
42. Tu C, Niu M, Wei AW et al (2021) Susceptibility-related cytokine panel for prediction of Polygonum multiflorum-induced hepatotoxicity in humans. J Inflamm Res 14:645–655. https://doi.org/10.2147/JIR.S299892
43. Gao Y, Wang Z, Tang J et al (2020) New incompatible pair of TCM: Epimedii Folium combined with Psoraleae Fructus induces idiosyncratic hepatotoxicity under immunological stress conditions. Front Med 14(1):68–80. https://doi.org/10.1007/s11684-019-0690-z
44. Tang JF, Wang XY, Wen Q et al (2017) Idiosyncratic hepatotoxicity evaluation of Zhuangguguanjie wan mediated by immune stress. Acta Pharm Sin 52(7):1033–1040
45. Shi W, Gao Y, Guo Y et al (2019) Idiosyncratic hepatotoxicity evaluation of Cortex Dictamni based on immune stress. Acta Pharm Sin 54(4):678–686

Chapter 5
Herb-Drug Interactions: Fundamental Mechanisms, Prevalence and Challenges in Their Identification

Jose M. Prieto and Andre L. D. A. Mazzari

5.1 General Introduction

The impact of 'drug interactions' in patient safety is a key concern for public health systems globally. Minimising this impact calls for concerted efforts in preclinical, clinical and post-marketing research towards informing practice and regulations. Under the 'umbrella term' drug interactions, there are several distinctive aspects, such as drug-drug interactions, drug-disease interactions, drug-laboratory test interactions, herb-drug interactions and food-drug (or drug-diet or drug-nutrient) interactions.

Drug interactions usually translate to an abnormal bioavailability, unbalanced distribution, higher or lower clearance rates of the drug, and even changes in concentrations of disease biomarkers within the patient. The overall effect is in some cases an 'adverse effect'—or adverse drug reaction(s) (ADR)—where the patient experiences symptoms of toxicity, and/or the therapeutic value of the drug is voided [1]. There are cases when this interaction may be positive for the patient. Indeed, clinicians use positive interactions everyday as illustrated with the classic example of whether a drug must be taken with a full stomach to minimise its adverse effects by virtue of a drug-food interaction.

This chapter concerns herb-drug interaction(s) (HDI). These are easier to distinguish in countries where herbal medicines are regulated as such. However, the distinction from 'food-drug interactions' may be unclear in countries where 'herbal

J. M. Prieto (✉)
Centre for Natural Products Discovery, School of Pharmacy and Biomolecular Sciences, Liverpool John Moores University, Liverpool, UK
e-mail: j.m.prietogarcia@ljmu.ac.uk

Andre L. D. A. Mazzari
Centre for Pharmacognosy, UCL School of Pharmacy, London, UK
e-mail: andre.mazzari.12@ucl.ac.uk

J. Barnes (ed.), *Pharmacovigilance for Herbal and Traditional Medicines*,
https://doi.org/10.1007/978-3-031-07275-8_5

medicines' are legally considered as 'food/dietary supplements', such as in the United States.

Like drug-drug interactions, herb-drug interactions may be pharmacodynamic, pharmacokinetic and/or pharmaceutical:

- A pharmacodynamic herb-drug interaction may occur when taking one or more herbal substances of equal or opposing pharmacological effects to the drug.
- A pharmacokinetic interaction occurs when taking one or more herbal substances that alter the normal Absorption-Distribution-Metabolism-Elimination/Excretion (ADME) processes of a drug. The most recognisable of these are cytochrome P450-mediated drug interactions, although many other types of enzymes, receptors and transporter proteins have been discovered to contribute to pharmacokinetic interactions.
- Pharmaceutical interactions derive from changes in the release, solubility and/or stability of a drug when one or more herbal substances alter the behaviour of the pharmaceutical form in which it is formulated.

A well-known example of a 'negative' herb-drug interaction is the concomitant use of St John's wort (*Hypericum perforatum* L.) flowering aerial parts with selective serotonin reuptake inhibitors, such as citalopram, as both have the same pharmacological effect, thus potentially resulting in the patient experiencing a 'serotonin syndrome' (excess serotonin causes signs and symptoms that can range from mild (shivering and diarrhoea) to severe (muscle rigidity, fever and seizures). Severe serotonin syndrome can cause death if not treated.

In contrast, a case of 'positive' herb-drug interaction is the use of an ethanol extract of *Schisandra sphenanthera* Rehder & E.H.Wilson (fruits) co-administered with tacrolimus for the treatment of drug-induced hepatitis in organ transplant recipients in China. This herbal extract significantly increases the blood concentration of tacrolimus, minimises its ADRs and improves liver function [2]. However, this is an intervention that requires medical supervision. Gerber and co-workers have reviewed many other potentially beneficial herb-drug interactions [3].

5.2 Origin and Evolution of Awareness and Research on HDI

A search by the MESH term 'Herb-Drug Interactions' in PubMed evidences a low but steady interest of the scientific and clinical communities in the subject during the last quarter of the last century and a sudden increase in attention by the turn of the millennium (Fig. 5.1).

The first record of a publication on HDI was in 1967. This single work assessed potential interactions between herbal medicines and anaesthetic drugs [4]. This was prompted by an accumulation of clinical observations by anaesthesiologists of cases where patients were responding in unexpected ways to anaesthesia and other emergency drugs. Retrospective studies linked these unexpected responses to the

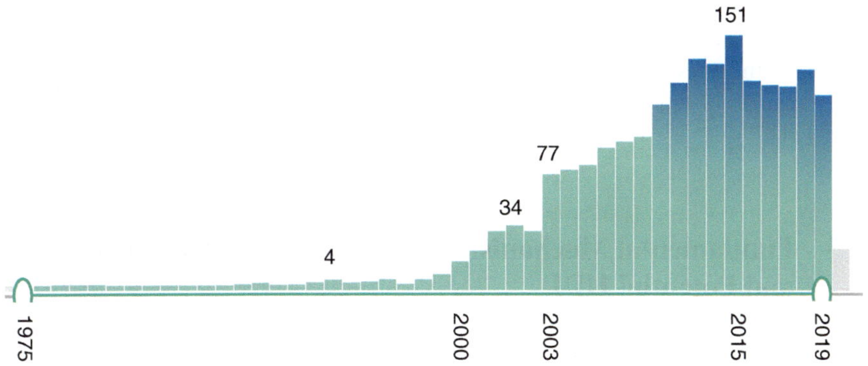

Fig. 5.1 Number of publications per year using the keyword 'herb-drug interactions' in PubMed (Source: U.S. National Library of Medicine)

consumption of herbal supplements. In 1974, a publication on the interaction between cannabinoids and phenytoin [5] triggered more interest in the subject [6–8]. By the 1990s, herbal medicines start appearing consistently in review articles on drug interactions [9–11] as well as in phytotherapy reference books [12]. Clinical findings on HDI involving grapefruit (*Citrus paradisi* Macfad.) juice and ginkgo (*Ginkgo biloba* L.) leaves led to experimental work on HDI [13–16]. This decade also witnessed the interest of Chinese researchers on HDI between Western drugs and Chinese herbal drugs [17, 18].

The turn of the century sees both a marked increase in publications and a shift of focus to herbals such as St John's wort aerial parts (the topic of about 35% of papers) and kava (*Piper methysticum* G.Forst.) roots. This followed clinical reports of potential pharmacodynamic interactions. On the one hand, a case report showed that consumption of St John's wort together with other prescription antidepressants was able to cause severe ADRs due to HDI, particularly in older people [19]. On the other hand, the hospital admission of a 54-year-old man in a lethargic and disoriented state after being co-administered kava and alprazolam suggested that kava might have additive effects with benzodiazepines [20].

Subsequent cases of HDI with St John's wort made it the main subject of HDI studies during the first decade of the twenty-first century [21–27]. The first attempt to elucidate the possible interaction mechanisms of St John's wort was published in 2000, suggesting that the herbal medicine was able to induce CYP3A4 and consequently reduce the efficacy of certain drugs, including oral contraceptives [28]. This stimulated further research and, eventually, St John's wort was found to increase the expression of intestinal P-glycoprotein (P-gp) and CYP3A4 in the liver, therefore potentially interfering with the action of drugs that are also metabolised by CYP3A4 [29]. The effects of St John's wort on phase II metabolic enzymes, such as glutathione-*S*-transferase (GST), were also investigated [30].

Besides CYP3A4, other phase I metabolising enzymes such as the CYP isoforms CYP1A2, CYP2C9, CYP2C19, CYP2D6, and CYP2E1 are nowadays the subject of metabolic studies involving herbal medicines. Herbal medicines have increasingly become the subject of metabolic and transporter studies using methods and targets similar to the ones used for synthetic drugs, but with one difference: such studies are not a regulatory requirement to market traditional herbal medicines in many countries [31].

5.3 Fundamental Mechanisms of Herb-Drug Interaction and Potential HDI

5.3.1 Overview

HDI may occur at any point of the ADME processes for any drug. Metabolism is perhaps the most scrutinised aspect in literature, followed by absorption processes. Excretion and distribution studies are, comparatively, lacking. Preclinical research models for the activity of both cytochromes and P-glycoprotein are relatively easy and cheap to implement contributing to most of the data available.

Metabolism (or biotransformation) processes convert xenobiotics into metabolites with lower bioactivity and increased polarity (Fig. 5.2) and involve numerous mechanisms and enzymes that are ubiquitous in the body [32]. Still, some less lipid-soluble drugs, such as atenolol, can be cleared from the body unchanged, i.e. without being significantly affected by phase I and/or phase II metabolism. However, other drugs remain in the body and can exert effects for substantial periods of time [33].

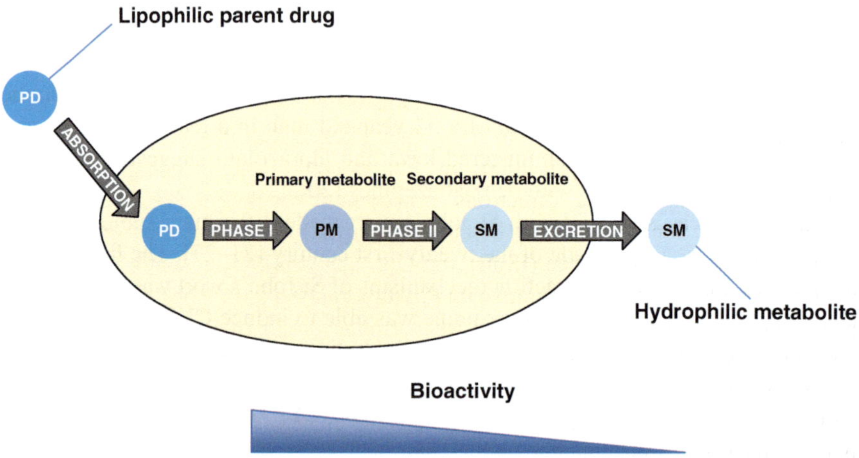

Fig. 5.2 Scheme of sequential drug metabolism leading to more hydrophilic and less active metabolites

5.3.2 *Absorption*

Drug-transporter proteins are known to allow xenobiotics to cross biological membranes, the most well-known one being P-gp. P-gp was first discovered in 1986 as a product of the multidrug resistance gene (MDRI) in cancer cells, therefore reducing the intracellular accumulation of drugs [34].

This protein plays a role as an efflux pump: it pushes metabolites and drugs out of cells, which can result in absorption alterations [35]. P-gp contains two adenosine triphosphate (ATP) binding sites, where ATP will bind in the presence of a P-gp substrate. As a consequence, ATPase is activated in order to hydrolyse ATP. The energy produced by the ATP hydrolysis allows the P-gp to transport numerous substrates across cellular membranes [34]. In the intestine, P-gp can be found on the apical surface of epithelial cells. When a drug is taken up by an enterocyte, the substance can be either metabolised by CYP3A4 or pumped back into the lumen [36]. Due to its high expression and activity in several tissues (kidney, liver, intestine and blood-brain barrier), which are pharmacologically important epithelial barriers, the bioavailability of a drug could be compromised if P-gp activity is affected [37]. Inhibition or induction of P-gp activity results in altered absorption and bioavailability of P-gp substrates [38]. Hence, oral delivery can be compromised [39].

Modulation of drug transporters like P-gp by (constituents of) herbal medicines has been explored in many studies. St John's wort displays inductive effects in P-gp efflux activity. Chronic treatment with St John's wort increases the P-gp efflux in healthy subjects, which can also be observed *in vitro* by a reduced intracellular accumulation of the P-gp substrate Rh-123 [40]. This inductive effect on P-gp has been demonstrated to attenuate the efficacy of a potent P-gp inhibitor drug, ritonavir, which potentially leads to treatment failure [41]. In contrast, commercial garlic (*Allium sativum* L.) bulb extracts, standardised on alliin, increased duodenal P-glycoprotein by 131% when administered to healthy volunteers for 21 days, but this induction was not correlated with changes in AUCs of reference drugs [42].

5.3.3 *Drug Metabolism*

Drug metabolism is normally divided into two phases: phase I metabolism (functionalisation reactions), which results in metabolites with higher polarity (usually inactive), and phase II metabolism (conjugative reactions), which results in the phase I metabolites becoming even more polar. Phase I reactions prepare the drug for phase II metabolism by adding polar functional groups to the xenobiotic [32]. Several comprehensive reviews of effects of herbal substances and herbal medicines on cytochromes are available [43–46].

Metabolism may be affected by herbal-medicine constituents able to inhibit the activity or induce the expression of drug-metabolising enzymes. In addition to the induction and inhibition of drug-metabolising enzymes, several factors can contribute to the variability in biotransformation. They include genetic polymorphism, disease, age, sex and environmental factors. Age is probably the single most important of such factors. In this regard, the long-term supplementation (28 days) of standardised commercial supplements containing either St John's wort, garlic (*Allium sativum* L.) oil, ginseng (*Panax ginseng* C.A.Mey.) root or ginkgo leaves affected CYP1A2, CYP2D6, CYP2E1 or CYP3A4 activity was studied in healthy older volunteers (aged 60 to 76 years). Probe drug cocktails of midazolam, caffeine, chlorzoxazone and debrisoquine were administered before and at the end of administration of the herbal substances to calculate phenotypic ratios for CYP3A4, CYP1A2, CYP2E1 and CYP2D6. The results revealed significant induction of CYP3A4 (approximately 140%) and CYP2E1 activity (approximately 28%). Garlic oil inhibited CYP2E1 activity by approximately 22%. Inhibition of CYP2D6 by ginseng was statistically significant, but the magnitude of the effect (*c.a.* 7%) did not appear to be clinically relevant. None of the herbal substances tested in this study appeared to affect CYP1A2 activity [47]. An example of a sex-related potential HDI is that St John's wort appears to induce CYP1A2 in females only [48].

5.3.3.1 Phase I Metabolism (Cytochrome P450)

Human drug-metabolising enzymes are ubiquitous in the body, mostly in the smooth endoplasmic reticulum of the liver and other extrahepatic tissues, such as the kidneys, skin, gastrointestinal tract and lungs. The CYP monooxygenase enzymes CYP450 is a family of enzymes responsible for the metabolism of xenobiotics through oxidation reactions such as aromatic and aliphatic hydroxylation, *N*-dealkylation, *O*-dealkylation, deamination, oxidation and sulphoxidation. The products of these functional chemical reactions are compounds with chemically reactive functional groups that will be further targeted by phase II enzymes [32]. Over 50 human CYPs have already been isolated; the major ones found in the liver include CYP1A2, CYP2C9/19, CYP2D6 CYP2E1 and CYP3A4/5/7 [49]. The whole CYP family is responsible for metabolising about 90% of commonly used drugs. The CYP 1, 2 and 3 families are the most abundant families of CYP metabolising enzymes, and the CYP1A2, CYP2C and CYP3A4 isoforms account for the metabolism of most prescription drugs [50]. Among the main isoforms, most of the currently marketed medicines share drug metabolism with CYP3A4, which can result in severe drug interactions [51]. Table 5.1 shows selected preclinical and clinical data on potential HDI due to modulation of the activity or expression of cytochromes by some of the most popular herbal medicines.

Table 5.1 Preclinical and clinical data on potential HDI of selected herbal medicines [45]

CYP isoform:	CYP2C9	CYP1A2	CYP2D6	CYP2E1	CYP3A4	CYP2C19
% Drugs metabolised (approx.): Examples: Herbal medicine	**15%** Ibuprofen Losartan Fluoxetine Phenytoin Fluvastatin	**20%** Acetaminophen Propranolol Clomipramine Warfarin	**25%** Propafenone Timolol amitriptyline Haloperidol Risperidone chlorphenamine	**10%** Acetaminophen Halothane	**50%** Macrolide antibiotics Antiarrhythmics Benzodiazepines HIV antivirals Antihistamines Calcium channel blockers	**10%** Omeprazole Amitriptyline Fluoxetine Diazepam Phenobarbital
Garlic (bulbs) (*Allium sativum* L.)	–, +	+	NE	–	NE, –	–
Chamomile (inflorescences) (*Matricaria chamomilla* L.)					–	
Turmeric (tuber) (*Curcuma longa* L.)		+		NE	NE	
Eucalyptus (leaves, oil) (*Eucalyptus globulus* Labill.)	–		–		–	–
Fennel (seeds) (*Foeniculum vulgare* Mill.)					–	
Soya (beans) (*Glycine max* (L.) Merr.)		–		NE		
Devil's claw (rhizome) (*Harpagophytum procumbens* (Burch.) DC. ex Meisn.)	NE, –	–	NE, –		NE, –	NE
Mint (aerial parts) (*Mentha* x *piperita* L.)	–	–	–		–	–
St John's wort (flowering aerial parts) (*Hypericum perforatum* L.)					+	

(continued)

Table 5.1 (continued)

CYP isoform:	CYP2C9	CYP1A2	CYP2D6	CYP2E1	CYP3A4	CYP2C19
Pomegranate (juice) (*Punica granatum* L.)	–	–			–	
Clover (aerial part) (*Melilotus officinalis* (L.) Pall.)	–	–			–	–
Ginger (tubers) (*Zingiber officinale* Roscoe)	–				(*)	

(+) enzyme induction; (–) enzyme inhibition; (NE) no effect

Induction of Drug-Metabolising Enzymes

The main types of drug induction are substrate-dependant induction and receptor-mediated and inhibitor-mediated interaction. Substrate-dependant induction is where the herbal drug influences the metabolism and duration of action of numerous other drugs. Receptor-mediated is characterised by interactions with important regulator pathways, which is the case of the human nuclear pregnane X (hPXR) and the constitutive androstane (CAR) receptors. These are well known also as xenobiotic sensors, which are activated by numerous compounds leading to the activation of their downstream target genes [52]. The binding of xenobiotics to the receptor directly affects the clearance of those compounds and, consequently, protects the body from foreign chemicals. Many drug-metabolising enzymes involved in the metabolism of endogenous cellular regulators (steroids, eicosanoids) can be induced by hormones. For example, the growth hormone has been also proven to alter CYP expression [53, 54].

Due to the increased popularity of St John's wort as an antidepressant, numerous interactions with CYP3A4 have been detected *in vitro* and *in vivo*, the interaction with oral contraceptives being one of the most remarkable cases of HDI [29]. Effects of St John's wort in other cytochromes (1A2, 2C9, 2C19, 2D6, 2E1) have been also evidenced [45]. The numerous publications alerting to the potential interactions of St John's wort involving this and other CYPs due to enzyme induction and/or inhibition raised awareness that consumption of herbal metabolic enzyme inducers increases the chances of inefficacy of the co-administered drug in usual therapeutic doses due to HDI [55].

Other popular herbal medicines such as ginkgo (*Ginkgo biloba* L.) leaves and milk thistle (*Silybum marianum* (L.) Gaertn.) fruits have been extensively studied to understand their potential to cause HDI due to their effects on CYP activity and/or expression. Published studies showed that both are able to modulate the effect of the main CYP isoforms due to the inducible and/or inhibitory effects of their constituents [56, 57]. A drug interaction study demonstrated that 140 mg of silymarin, the main active principle of milk thistle, given three times a day inhibited hepatic clearance of losartan in Chinese subjects. The active metabolite of losartan (E-3174) is formed by CYP2C9. Due to the inhibition of the enzyme caused by the milk thistle constituent silymarin, the amount of E-3174 found in the subjects treated also with the herbal medicine decreased compared to the control group [58]. Another example of a metabolic HDI involving CYP was indicated that the intake of noni juice, an herbal remedy made from the fruit of noni (*Morinda citrifolia* L.), induces CYP2C9 and interacts with the anti-epileptic drug phenytoin by decreasing its bioavailability [59].

Inhibition of Drug-Metabolising Enzymes

An inhibitor is a substance that interferes with the action of the enzyme, which will slow down the speed of the reaction. Inhibitor-mediated interaction involves a stabilisation mechanism, i.e. the drug decreases the degradation of CYP and therefore the concentration of the enzyme is increased [60]. In contrast to the effects observed in the induction of drug metabolic enzymes, inhibition delays xenobiotic biotransformation resulting in a higher concentration of the compound in the bloodstream. The consequence of such inhibition is the increased adverse reactions due to exacerbated pharmacological and toxicological effects [61]. A reversible inhibitor will bind to an enzyme and will subsequently be released, whereas an irreversible inhibitor reacts with the enzyme producing a protein that will not be enzymatically active anymore and the original enzyme cannot be regenerated [62].

5.3.3.2 Phase II Metabolism (Conjugation Reactions)

Phase II metabolism reactions (or conjugation reactions) occur when metabolic enzymes react with functional groups of a drug that were formed during the phase I process. Endogenous species, such as a sugar or an amino acid, are added to the drug in order to increase the polarity to allow its elimination. The two main phase II biotransformation reactions are glutathione (GSH) conjugation by glutathione-*S*-transferase (GST) and glucuronidation by UDP-glucuronosyltransferase (UGT). However, other conjugative reactions, such as sulphation, methylation and acetylation, are also relevant [50]. Phase II metabolising enzymes such as UGTs and GSTs can also be inducible [63–66]. Like phase I enzymes, phase II enzymes can also be induced and/or inhibited by xenobiotics. One study by Liu and co-workers [67]demonstrated that UGT isoforms can be inhibited by vitamin A [67], whereas Chang and co-workers [68] showed that oestrogens are able to inhibit GSTs [68]. Table 5.2 shows some examples of potential Phase 2 metabolic HDI.

5.3.4 Distribution and Plasma Concentrations

Drug transporters, which play significant roles in the absorption and elimination of drugs, also affect their uptake by other tissues (distribution). Herbs that alter absorption processes potentially may alter their internal distribution in the body. A case in point is the blood-brain barrier (BBB). A typical endothelial BBB cell depends on multidrug resistance protein transporter (MRP), P-glycoprotein (P-gp), organic anion transporting polypeptide (OATP), and organic anion transporter (among others) to keep away from or selectively allow drugs into the central nervous system (CNS). St John's wort was suspected to cause increased transport of hydrocortisone and corticosterone across the BBB of rats by inducing P-gp. Repeated administration of 300 mg St John's wort three times daily maintains clinically significant plasma concentrations of hypericin which may explain P-gp induction in vivo [69, 70].

Plasma concentrations of drugs may be also altered by the interaction of herbal components with serum albumin, leading to toxicity by displacing other drugs. This is the case with many narrow 'therapeutic window' medicines. For example,

Table 5.2 Examples of drugs and herbal medicines involved in/modulating main phase II enzymes

Phase II mechanism enzyme/ substrate	Effect on phase II metabolic mechanisms Herbal drugs/phytochemicals
Glucuronidation UDP-glucuronosyltransferase (UGT) Glucuronic acid	*Inhibition of UGT activity*: Milk thistle (*Silybum marianum* L.) flavonolignans [74] Valerian (*Valeriana officinalis* L.) root [75] *Increase of UGT expression*: Garlic (*Allium sativum* L.) bulb [45] Turmeric (*Curcuma longa* L.) tuber
Glutathione conjugation Glutathione-*S*-transferase (GST) Glutathione (GSH)	*Increase of GSH expression* [45]: Aloe vera (*Aloe vera* (L.) Burm.f.) juice, Artichoke (*Cynara cardunculus* L.) leaf Calendula (*Calendula officinalis* L.) flower extract Chamomile (*Matricaria chamomilla* L.) flower Clover (*Melilotus officinalis* (L.) Pall.) blossom Devil's claw (*Harpagophytum procumbens* (Burch.) DC. ex Meisn.) rhizome Eucalyptus (*Eucalyptus globulus* Labill.) leaf Fennel (*Foeniculum vulgare* Mill.) fruit Garlic (*Allium sativum* L.) bulb Ginger (*Zingiber officinale* Roscoe) tuber Mint (*Mentha* x *piperita* L.) leaves Pomegranate (*Punica granatum* L.) juice Soya (*Glycine max* (L.) Merr.) beans St John's wort (*Hypericum perforatum* L.) flowering aerial parts Turmeric (*Curcuma longa* L.) tuber
Sulphation Sulphotransferases (SULT)	*Inhibition of SULT activity*: Curcumin [76], Quercetin [77]
Methylation Methyltransferases (MT) *S*–adenosylmethionine (SAM)	*Inhibition of catechol-O-MT activity*: St John's wort (*Hypericum perforatum* L.) flowering aerial parts [78] Myricetin, dihydromyricetin, and myricitrin from Chinese bayberry (*Myrica rubra* (Lour.) Siebold & Zucc.) [79]

competitive displacement of warfarin from the binding site by phenylbutazone or bucolome occurred resulting in an increased free fraction for warfarin resulting in increased bleeding times [71]. Many ubiquitous phytochemicals are able to bind to albumin, thus explaining their interactions with such drug [72].

5.3.5 *Excretion*

Excretion/elimination processes are the ultimate mechanism of disposal of xenobiotics. These may happen in several ways, the most important being renal clearance and biliary excretion. Other pathways of excretion include the lungs, breast milk, sweat, saliva and tears.

It is, therefore, not surprise that most of the HDI at this level have been classically assigned to plants with diuretic properties that may increase the renal elimination of other drugs. The most popular ones include bearberry leaf (*Arctostaphylos uva-ursi* (L.) Spreng.), goldenrod herb (*Solidago virgaurea* L.), dandelion leaf and root (*Taraxacum campylodes* G.E.Haglund), juniper berry (*Juniperus communis* L.), horsetail herb (*Equisetum arvense* L.), lovage root (*Levisticum officinale* W.D.J.Koch), parsley (*Petroselinum crispum* (Mill.) Fuss), asparagus root (*Asparagus officinalis* L.), stinging nettle leaf (*Urtica dioica* L.) and alfalfa leaf (*Medicago sativa* L.).

Modern molecular targets include the organic anion transporter 1 (OAT1) and 3 (OAT3). These are highly expressed in the kidney and play a key role in the renal elimination of substrate drugs. Studies on HDI at the level of excretion processes are still scarce. So far, components in the herbal medicine red sage (*Salvia miltiorrhiza* Bunge, roots), including lithospermic acid, rosmarinic acid, salvianolic acid A, salvianolic acid B and tanshinol, have been shown to inhibit human OAT1 and OAT3 (Wang & Sweet (2012)). The anthraquinones in rhubarb have been identified as strong inhibitors of human OAT1 and OAT3 (Ma et al. 2014). A recent study has systematically investigated the interactions of over 170 herbal extracts with human OATs identifying licorice roots (*Glycyrrhiza glabra* L.) as a strong in vitro inhibitor of OAT3 [73].

5.4 Clinically Relevant Interactions

Despite the already large and growing body of literature about HDI, most of the research available is either experimental/preclinical (using *in vitro* and *in vivo* models), or largely theoretical (trying to extrapolate these data to clinical situations). There is much less clinical information. An extensive analysis of clinical literature made publicly available by the European Scientific Cooperative on Phytotherapy (ESCOP) showed that out of 125 relevant herbal medicinal products only 26 (21%) have been possibly linked to causing potential or real HDI [80].

The major groups of herbal drugs with well-recognised pharmacodynamic clinical interactions profile are described below.

1. **Herbal drugs inducing hypokalaemia:** This includes classic 'poo-softener laxatives' such as aloes (*Aloe vera* (L.) Burm.f.) juice, cascara (*Frangula purshiana* Cooper) bark, senna (*Senna alexandrina* Mill.) leaf and/or fruit and rhubarb (*Rheum palmatum* L. or *Rheum officinale* Baill.) roots, as well as liquorice (*Glycyrrhiza glabra* L.) root. Their chronic, long-term use potentiates cardiac glycosides, interacts with antiarrhythmic drugs and drugs which induce reversion to sinus rhythm (e.g. quinidine), and induces serious electrolyte imbalance if concomitantly used with other drugs inducing hypokalaemia (e.g. thiazide diuretics, adrenocorticosteroids).

2. **Herbal drugs altering coagulation:** A heterogenous group of herbal medicines, including willow (*Salix alba* L.) bark, garlic, ginkgo and ginseng (*Panax ginseng* C.A.Mey.), have been related to an increase of bleeding time mostly by means of single-patient clinical case reports. The anticoagulant principles include salicin, allicin, ginkgolides and ginsenosides, respectively. However, garlic products with less than 0.6% allicin content do not appear to have any such effects (Scharbert et al. 2007) and ginkgo did not show such interactions in controlled studies.
3. **Herbal drugs with high caffeine content:** Stimulants and thermogenic HMs, such as guarana (*Paullinia cupana* Kunth) seeds and kola (*Cola acuminata* (P.Beauv.) Schott & Endl.) nuts, may void the effect of sedative medications and interact with psychoanaleptic medicines.
4. **Herbal drugs with effects on serotonin concentrations:** Several cases of serotonergic effects after concomitant use of St. John's wort preparations with certain antidepressants have been reported. It seems that a daily dose of more than 1000 mg drug equivalent or more than 1 mg hyperforin is necessary to induce such events. The attribution of some of these cases to St. John's wort remains unclear.

The major groups of herbal drugs with well-recognised pharmacokinetic clinical interactions profile are described below.

1. **Herbal drugs with cytochrome induction effects:** such as goldenseal (CYP3A4/5 and CYP2D6) and St. John's Wort (CYP3A4).
2. **Mucilage-containing herbal drugs**: This includes a group of 'bulk-forming' laxatives such as ispaghula/psyllium (*Plantago indica* L. or *Plantago ovata* Forssk.), linseed (*Linum usitatissimum* L.) seeds and marshmallow (*Althaea officinalis* L.) roots. They impair and/or delay the enteral absorption of concomitantly administered minerals (e.g. calcium, iron, lithium, zinc), vitamins (B12) and drugs. Therefore, it is recommended that medications should be taken at least 30–60 min before the administration of the herbal drug. They also may alter absorption of glucose and therefore insulin-dependent diabetics may need to adjust the insulin dose.

5.5 Prevalence of Herb-Drug Interactions

According to the World Health Organization (WHO), it is estimated that up to four billion people depend on medicinal plants for their primary healthcare due to poverty or lack of access to modern medicine; this constitutes between 65 and 80% of the world's population in developing countries [81].

In contrast, in the developed world, herbal medicines are mostly used due to the belief that they promote healthy living [82]. Taking the USA as an example we can see that during the 1990s the prevalence of use of herbal medicines saw a marked decline from 12% down to 2.5% [83] despite alternative medicine use and expenditure increasing substantially due to rise in the proportion of the population seeking alternative therapies. However, non-vitamin/non-mineral dietary supplements in the USA came back to be the most popular complementary health approach during the first decade of this century among adults [84] with herbs being the most common ingredient of this category. Indeed, over 38 million (18.6%) USA adults chose phytotherapy [85] with up to 8% of those aged 65 and older consuming herbal remedies [86]. Nowadays, these products are currently the most commonly recommended complementary health intervention by physicians in the USA [87], thus resulting in *c.a.* $13 billion a year in 'out-of-pocket' purchases of natural product supplements [88]. Overall, a large majority of patients (80%) disclose the use of such health interventions to their physicians [89], but there are certain ethnic groups more prone to non-disclosure, and this correlates with lower use of such complementary practices [90]. Although supplements are also the most popular complementary health intervention in USA children [91], only 2% of the products they consume are 'non-vitamin/non-mineral' [92].

Although anecdotal evidence suggests that use of alternative treatments may be common in certain chronic patients, a study found that the major driver of the USA population is not a response to illness, but a desire to stay healthy [93]. Moreover, it was found that the use of specific types of CAM therapies is associated with specific personality styles [94] and ethnicities [90]. Still, there are specific trends in the use of complementary health approaches within different groups of patients depending on their condition/s. Patients with chronic fatigue syndrome tend to consume more herbal products than their healthy family members [95] while those with diabetes mellitus use herbal medicines in a comparable manner to the general population [96].

Other studies in Anglophone countries, such as Australia, show slightly different results, but still evidence substantial use: complementary medicines were used by over 40% of Australians aged 50 years and older, 87.4% of whom used both conventional and complementary medicines [97]. The most popular choice is natural remedies (naturopathy, homoeopathy, Chinese medicine and herbalism) often chosen for chronic health problems, such as headaches and migraines, women's health problems of fertility and hormone imbalance, or mental health problems of stress, anxiety and depression. The users were much more likely to have higher levels of education and interestingly the authors did not find any significant differences between rural and metropolitan respondents.

With such an enormous 'consumer base', it is expected that HDI are a significant source of adverse effects in patients globally, that national healthcare systems care for patients experiencing HDI on a daily basis, and that, after scrutiny to discriminate false instances, a substantial number of reports of HDI is submitted to national pharmacovigilance centres around the world. However, in 2012, the number of reports recorded in VigiBase, the WHO global database of individual case safety reports (ICSRs) of suspected adverse drug reactions (ADRs) associated with medicines, was small, raising concerns about underreporting [98].

Although most herbal medicines traditionally used are known for their effectiveness on numerous conditions, safety concerns are rarely effectively disclosed to the population that uses them. Cases of acute toxicity caused by intake of herbal medicines, for example, have led to acute hepatitis [99] and nephrotoxicity in patients [100]. Also, chronic toxicity has been associated with herbal preparations, such as the Nigerian DAS-77 [101]. Cases of toxicity involving herbal medicines rarely impact the public and global consumption of natural products remains stable: despite the risks of consuming many herbal medicines (particularly with excessive chronic use), the general public continues to perceive such products as 'naturally safe' without being properly informed about the associated risks [82].

Cases of HDI may be underreported for several reasons. First, regulation of herbal medicines varies from country to country. In certain places, they can be categorised as food supplements and, therefore, they are subject to different regulations. In some cases, herbal medicines are not regulated. Second, there might be cases where HDI are not properly reported to the pharmacovigilance system. We reported in 2014 that Brazil did not have officially reported any HDI during the first 10 years of herbal medicines safety monitoring. With 90% of the Brazilian population using herbal medicines at least once a year, cases of HDI might be occurring without being noticed or formally reported [102]. This is mostly caused by lack of knowledge on herbal medicines by physicians and other health professionals. Third, in many countries (and especially in developing ones), pharmacovigilance systems are non-existent or in the early stages of development. Thus, HDI cases in these places may be ignored.

Another approach to ascertain the prevalence/incidence of HDI is epidemiological studies. However, these are scarce and mostly local in nature. Table 5.3 summarises some of them. The overall assessment is that the real incidence of HDI is much lower than expected. It can be hypothesised that the impact of herbal medicines on ADME processes is not as strong as expected, and that ADRs may only clearly appear in extreme cases after long-term intake of herbal medicines, when a patient's ADME processes are immature or seriously compromised (for example, in newborns and elderly), when very narrow therapeutic window drugs are co-administered, and in emergency situations (Accident and Emergency settings or during surgery).

Table 5.3 Examples of epidemiological studies exploring prevalence of HDI

	US [103]	US [104]	UK [105]	Nigeria [106]	Brazil [107]	Turkey [108]
Design	Secondary analysis 657 patients 52 ± 17 yrs. Primary Care Practices	804 patients 24–103 yrs. Secondary Care Practices	Cross-sectional survey 155 patients >65 yrs. Primary Care Practices	Cross-sectional survey 112 diabetic patients >30 yrs. Secondary Care Practices	Cross-sectional survey 280 patients on anticoagulation therapy >40 yrs. Secondary Care Practices	Cross-sectional survey 343 patients with cardiovascular diseases 54 + 16 yrs. Secondary Care Practices
Total % use of HMs	15.4%	15%	33.6%	56%	16.4%	82.5%
% Potential HDI	14%	40%	32.6%	28%	33.3%	>70%(1)
% ADR	0%(2)	7% (all mild)	Unknown	Unknown	Unknown	6.7%(3)
Main HMs[a] found by Survey/ Interviews in order of decreasing frequency	Echinacea, Garlic, Ginkgo, Ginseng, St. John's wort	Prickly pear, Chamomile, Ginkgo.	Primrose oil, Valerian, Hops, Gentian and Passionflower	Undefined botanicals,(4) Mango leaf, Bitter leaf, Scent leaf, Akuamma plant, Okro, Avocado pear, Tropical almond	Lemon, Lemon balm, Plantain.	Garlic, Onion, Walnut, Mint, Pomegranate, Thyme, Pomegranate sour syrup, Sesame, Pumpkin seed

[a]We here use English common names as for many of these herbal medicines/food/herbal supplements the correct botanical species is unclear due lack of documentation as to what the patients were exposed; (1) based on garlic potential HDI; (2) author's calculated increase in ADE incidence vs. non-HM consumers; (3) patient's self-reported data. (4) In this study a 60% of products taken by patients were of unknown composition

5.6 Sources for information on Herb-Drug Interactions

Authoritative reference textbooks, such as Martindale [109] and Stockley's Drug interactions [33], are long-standing references in this field. The significant increase in information on HDI prompted the latter to create a dedicated companion volume, Stockley's Herbal Medicine Interactions [110], which is continuously updated in its modern online version.

With the advent of the internet, several clinical databases now offer and cover HDI reports. Some are public [111–114] and others commercial [115–118]. A study on the performance of subscription versus free drug interaction checkers concluded that all of them had 'poor sensitivity' for detecting HDI [119]. The authors evidenced that Lexicomp had the highest positive predictive value and best overall performance score, while Medscape was the best-performing free tool. They also showed that the worst subscription tools were as good as, or better than, the best free tools, and, as a group, subscription tools outperformed free tools on all metrics: the typical subscription tool would detect one additional herb-drug interaction for every 10 herb-drug interactions screened by a free tool.

5.7 Challenges and Solutions

Preclinical and clinical safety studies with synthetic drugs follow very well-established protocols and are a regulatory requirement. However, this is not the case for traditional herbal medicines. In order to guide member states and to provide traditional herbal medicines with satisfactory quality and safety data, in 2000, the WHO published 'General Guidelines for Methodologies on Research and Evaluation of Traditional Medicine' [120]. Although the document states that no regulatory action should be taken for herbal medicines traditionally used without demonstrated harm, preclinical investigation is needed as their real safety has been contested by many scientific reports.

Even with national pharmacovigilance systems in place, and the WHO's Programme for International Drug Monitoring, difficulties remain in obtaining and disseminating data related to HDI; this raises concerns about the risks associated with the use of medicinal plants/herbal medicines in conjunction with conventional medicines.

The paucity of information about HDI in many countries may be related to some factors, such as the absence of promotion of the rational use of medicinal plants and herbal medicines and deficiency in the training of health professionals able to detect these cases. The lack of a proper integration of a minimal knowledge on medicinal plants and herbal medicines into Schools of Medicine and Pharmacy is often justified 'to not overload an already busy curriculum'. In this way, many health professionals, including doctors, dentists, nurses, pharmacists and nutritionists, lack the basis to prescribe medicinal plants and plant drugs and the awareness to detect HDI

[121]. This is further compounded by patients self-medicating with herbal medicines and other 'natural health' products. Due to the belief that 'natural is safe', many users of these products do not describe their use to doctors and other health professionals, thus contributing to the scarcity of data on HDI [122].

The pharmacological and pharmacokinetic knowledge on herbal drugs accumulated over the last decades prompted WHO to make an alert as to the potential risks of uncontrolled use of herbal medicines in conjunction with other synthetic drugs. In 2004, WHO issued the 'Guidelines on safety monitoring of herbal medicines in pharmacovigilance systems'. These guidelines indicate how member countries should include herbal medicines within existent pharmacovigilance systems in order to facilitate the exchange of information on adverse reactions and HDI between WHO member countries [123].

The WHO traditional medicine strategy, published in 2013, showed that the regulatory status of herbal medicines is quite diverse among WHO member countries. A medicinal plant can be registered as a food, a functional food, a dietary supplement, a traditional medicine or a full licensed herbal medicine depending on the country's own needs, cultural background, history of use of medicines and regulations. This can potentially cause difficulties for pharmacovigilance systems in collecting HDI data as, for example, a food interaction would be practically ignored and hardly detected by health professionals [124, 125].

In addition to the differences in the regulatory status of herbal medicines between many countries, other challenges can be highlighted in terms of pharmacovigilance of herbal medicines. The presence of numerous active compounds in herbal extracts, the lack of standardisation of many herbal products, use of incorrect nomenclature for many herbal medicinal products, and the many sources for herbal medicines used by patients are also examples that make the detection of HDI by health professionals more difficult. As a consequence, HDI are likely to be underreported to the pharmacovigilance systems [126, 127].

5.8 Conclusions

The change of millennium saw HDI become a recognisable and distinctive field of study. In the last two decades, a plethora of data on HDI, mainly involving CYP drug-metabolising enzymes, has been published, demonstrating that herbal medicines are able to affect drug metabolism and therefore their concurrent consumption should be carefully assessed and sometimes totally avoided.

Nonetheless, evaluation of this subject is very controversial and there is a lack of established protocols to follow. The fact that most of the published research on HDI is either in vitro, performed in animal models and/or undertaken with fractions (not whole herbs) at non-physiological concentrations, may explain why most of the 'theoretical' HDI have never been clinically observed. Clinically relevant HDI remain a rare instance according to many studies, and it appears they are 'visible' only under rather extreme circumstances.

Preclinical data are nevertheless very important because they serve as a guide for future clinical studies. Similarly, more efforts in pharmacovigilance are eventually the best tool to unveil and monitor HDI. Healthcare professionals need appropriate training on this topic as part of both undergraduate and postgraduate curricula and continuous professional development. These healthcare professionals should also have access to professional subscription tools providing accurate, up-to-date information on herbal-drug interactions.

References

1. Botting J (2002) The history of thalidomide. Drug News Perspect 15(9):604–611. https://doi.org/10.1358/dnp.2002.15.9.840066
2. Wei H, Tao X, Di P et al (2013) Effects of traditional Chinese medicine Wuzhi capsule on pharmacokinetics of tacrolimus in rats. Drug Metab Dispos 41(7):1398–1403. https://doi.org/10.1124/dmd.112.050302
3. Gerber W, Steyn JD, Kotze AF, Hamman JH (2018) Beneficial pharmacokinetic drug interactions: a tool to improve the bioavailability of poorly permeable drugs. Pharmaceutics 10(3):106. https://doi.org/10.3390/pharmaceutics10030106
4. Pellerano P (1967) Pharmacologic interactions of anesthesiologic importance. I. Interference between preoperative pharmacotherapy and anesthetic drugs. Acta Anaesthesiol 18(6):905–911
5. Chesher GB, Jackson DM (1974) Anticonvulsant effects of cannabinoids in mice: drug interactions within cannabinoids and cannabinoid interactions with phenytoin. Psychopharmacologia 37(3):255–264. https://doi.org/10.1007/BF00421539
6. Hine B, Torrelio M, Gershon S (1975) Interactions between cannabidiol and delta9-THC during abstinence in morphine-dependent rats. Life Sci 17(6):851–857. https://doi.org/10.1016/0024-3205(75)90435-x
7. Hollister LE, Gillespie H (1975) Interactions in man of delta-9-tetrahydrocannabinol. II. Cannabinol and cannabidiol. Clin Pharmacol Ther 18(1):80–83. https://doi.org/10.1002/cpt197518180
8. Singh PP, Das PK (1978) Studies on the interactions of copper and cannabis. Psychopharmacology 56(3):309–316. https://doi.org/10.1007/BF00432855
9. D'Arcy PF (1991) Adverse reactions and interactions with herbal medicines. Part 1. Adverse reactions. Adverse Drug React Toxicol Rev 10(4):189–208
10. D'Arcy PF (1992) Drug reactions and interactions. J Pharm Belg 47(2):154–160
11. D'Arcy PF (1993) Adverse reactions and interactions with herbal medicines. Part 2: Drug interactions. Adverse Drug React Toxicol Rev 12(3):147–162
12. Newall CA, Anderson LA, Phillipson JD (1996) Herbal medicines: a guide for healthcare professionals. Pharmaceutical Press, London
13. Ameer B, Weintraub RA (1997) Drug interactions with grapefruit juice. Clin Pharmacokinet 33(2):103–121. https://doi.org/10.2165/00003088-199733020-00003
14. Chermat R, Brochet D, DeFeudis FV, Drieu K (1997) Interactions of Ginkgo biloba extract (EGb 761), diazepam and ethyl beta-carboline-3-carboxylate on social behavior of the rat. Pharmacol Biochem Behav 56(2):333–339. https://doi.org/10.1016/s0091-3057(96)00298-5
15. Miller LG (1998) Herbal medicinals: selected clinical considerations focusing on known or potential drug-herb interactions. Arch Intern Med 158(20):2200–2211. https://doi.org/10.1001/archinte.158.20.2200
16. Fuhr U (1998) Drug interactions with grapefruit juice. Extent, probable mechanism and clinical relevance. Drug Saf 18(4):251–272. https://doi.org/10.2165/00002018-199818040-00002

17. Tam LS, Chan TY, Leung WK, Critchley JA (1995) Warfarin interactions with Chinese traditional medicines: danshen and methyl salicylate medicated oil. Aust NZ J Med 25(3):258. https://doi.org/10.1111/j.1445-5994.1995.tb01540.x
18. Wang W (1990) Interactions of Western antihypertensive drugs and Chinese drugs. Zhong Xi Yi Jie He Za Zhi 10(5):303–304
19. Lantz MS, Buchalter E, Giambanco V (1999) St. John's wort and antidepressant drug interactions in the elderly. J Geriatr Psychiatry Neurol 12(1):7–10. https://doi.org/10.1177/089198879901200103
20. Almeida JC, Grimsley EW (1996) Coma from the health food store: interaction between kava and alprazolam. Ann Intern Med 125(11):940–941. https://doi.org/10.7326/0003-4819-125-11-199612010-00023
21. Ingram KD, Dragosavac GB, Benner KG, Flora KD (2000) Risks of drug interactions with St. John's Wort. Am J Gastroenterol 95(11):3323–3324. https://doi.org/10.1111/j.1572-0241.2000.03322.x
22. Mehta U (2000) Potentially serious drug interactions between St John's Wort and other medicines. S Afr Med J 90(7):698
23. Baede-van Dijk PA, van Galen E, Lekkerkerker JF (2000) Drug interactions of *Hypericum perforatum* (St. John's wort) are potentially hazardous. Ned Tijdschr Geneeskd 144(17):811–812
24. McIntyre M (2000) A review of the benefits, adverse events, drug interactions, and safety of St. John's Wort (*Hypericum perforatum*): the implications with regard to the regulation of herbal medicines. J Altern Complement Med 6(2):115–124. https://doi.org/10.1089/acm.2000.6.115
25. Roots I (2000) Drug interactions with Hypericum. Is nature not so mild after all? Interview by Petra Eiden. MMW Fortschr Med 142(5):14
26. Jordan MA (2013) Interactions with drugs and dietary supplements used for weight loss. In: El-Shemy HA (ed) Drug discovery. INTECH Open Access Publisher
27. Piscitelli SC, Burstein AH, Chaitt D, Alfaro RM, Falloon J (2000) Indinavir concentrations and St John's wort. Lancet 355(9203):547–548. https://doi.org/10.1016/s0140-6736(99)05712-8
28. Roby CA, Anderson GD, Kantor E, Dryer DA, Burstein AH (2000) St John's Wort: effect on CYP3A4 activity. Clin Pharmacol Ther 67(5):451–457. https://doi.org/10.1067/mcp.2000.106793
29. Murphy PA, Kern SE, Stanczyk FZ, Westhoff CL (2005) Interaction of St. John's Wort with oral contraceptives: effects on the pharmacokinetics of norethindrone and ethinyl estradiol, ovarian activity and breakthrough bleeding. Contraception 71(6):402–408. https://doi.org/10.1016/j.contraception.2004.11.004
30. Mohamed ME, Frye RF (2011) Effects of herbal supplements on drug glucuronidation. Review of clinical, animal, and in vitro studies. Planta Med 77(4):311–321. https://doi.org/10.1055/s-0030-1250457
31. Tsay HS, Shyur LF, Agrawal DC, Wu YC, Wang SY (eds) (2016) Medicinal plants—recent advances in research and development. Springer, Singapore
32. Ionescu C, Caira MR (eds) (2005) Drug metabolism: current concepts. Springer, Dordrecht
33. Baxter K, Stockley IH (2008) Stockley's drug interactions: a source book of interactions, their mechanisms, clinical importance, and management, 8th edn. Pharmaceutical Press, London
34. Watanabe Y, Ikarashi N, Satoh T, Ito K, Ochiai W, Sugiyama K (2012) Inhibitory effects of daiokanzoto (da-huang-gan-cao-tang) on p-glycoprotein. Evid Based Complement Alternat Med 2012:361516. https://doi.org/10.1155/2012/361516
35. Williamson EM, Driver S, Baxter K (2009) Stockley's herbal medicines interactions: a guide to the interactions of herbal medicines, dietary supplements and nutraceuticals with conventional medicines. Pharmaceutical Press, London
36. Butterweck V, Derendorf H (2008) Potential of pharmacokinetic profiling for detecting herbal interactions with drugs. Clin Pharmacokinet 47(6):383–397. https://doi.org/10.2165/00003088-200847060-00003

37. Li X, Hu J, Wang B, Sheng L, Liu Z, Yang S, Li Y (2014) Inhibitory effects of herbal constituents on P-glycoprotein in vitro and in vivo: herb-drug interactions mediated via P-gp. Toxicol Appl Pharmacol 275(2):163–175. https://doi.org/10.1016/j.taap.2013.12.015
38. Zhou S, Lim LY, Chowbay B (2004) Herbal modulation of P-glycoprotein. Drug Metab Rev 36(1):57–104. https://doi.org/10.1081/dmr-120028427
39. Butterweck V, Derendorf H, Gaus W, Nahrstedt A, Schulz V, Unger M (2004) Pharmacokinetic herb-drug interactions: are preventive screenings necessary and appropriate? Planta Med 70(9):784–791. https://doi.org/10.1055/s-2004-827223
40. Hennessy M, Kelleher D, Spiers JP et al (2002) St Johns wort increases expression of P-glycoprotein: implications for drug interactions. Br J Clin Pharmacol 53(1):75–82. https://doi.org/10.1046/j.0306-5251.2001.01516.x
41. Drewe J, Gutmann H, Fricker G, Torok M, Beglinger C, Huwyler J (1999) HIV protease inhibitor ritonavir: a more potent inhibitor of P-glycoprotein than the cyclosporine analog SDZ PSC 833. Biochem Pharmacol 57(10):1147–1152. https://doi.org/10.1016/s0006-2952(99)00026-x
42. Hajda J, Rentsch KM, Gubler C, Steinert H, Stieger B, Fattinger K (2010) Garlic extract induces intestinal P-glycoprotein, but exhibits no effect on intestinal and hepatic CYP3A4 in humans. Eur J Pharm Sci 41(5):729–735. https://doi.org/10.1016/j.ejps.2010.09.016
43. Wanwimolruk S, Phopin K, Prachayasittikul V (2014) Cytochrome P450 enzyme mediated herbal drug interactions (Part 2). EXCLI J 13:869–896
44. Wanwimolruk S, Prachayasittikul V (2014) Cytochrome P450 enzyme mediated herbal drug interactions (Part 1). EXCLI J 13:347–391
45. Mazzari ALDA, Prieto JM (2014) Herbal Medicines in Brazil: pharmacokinetic profile and potential herb-drug interactions. Front Pharmacol 5(162):1–12. https://doi.org/10.3389/fphar.2014.00162
46. Fasinu PS, Bouic PJ, Rosenkranz B (2012) An overview of the evidence and mechanisms of herb-drug interactions. Front Pharmacol 3:69–69. https://doi.org/10.3389/fphar.2012.00069
47. Gurley BJ, Gardner SF, Hubbard MA et al (2005) Clinical assessment of effects of botanical supplementation on cytochrome P450 phenotypes in the elderly: St John's wort, garlic oil, Panax ginseng and Ginkgo biloba. Drugs Aging 22(6):525–539. https://doi.org/10.2165/00002512-200522060-00006
48. Wenk M, Todesco L, Krähenbühl S (2004) Effect of St John's wort on the activities of CYP1A2, CYP3A4, CYP2D6, N-acetyltransferase 2, and xanthine oxidase in healthy males and females. Br J Clin Pharmacol 57(4):495–499. https://doi.org/10.1111/j.1365-2125.2003.02049.x
49. Gibson GG, Skett P (2001) Introduction to drug metabolism, 3rd edn. Nelson Thornes, London
50. Atkinson AJ (2012) Principles of clinical pharmacology, 3rd edn. Academic Press, Elsevier, San Diego
51. Ogu CC, Maxa JL (2000) Drug interactions due to cytochrome P450. Proc (Bayl Univ Med Cent) 13(4):421–423. https://doi.org/10.1080/08998280.2000.11927719
52. Xie W (2009) Nuclear receptors in drug metabolism. Wiley, Oxford
53. Mode A, Tollet P, Strom A, Legraverend C, Liddle C, Gustafsson JA (1992) Growth hormone regulation of hepatic cytochrome P450 expression in the rat. Adv Enzym Regul 32:255–263. https://doi.org/10.1016/0065-2571(92)90021-q
54. Cheung C, Yu AM, Chen CS et al (2006) Growth hormone determines sexual dimorphism of hepatic cytochrome P450 3A4 expression in transgenic mice. J Pharmacol Exp Ther 316(3):1328–1334. https://doi.org/10.1124/jpet.105.094367
55. Tirirai C, Viljoen AM, Hamman JH (2010) Herb-drug pharmacokinetic interactions reviewed. Expert Opin Drug Metab Toxicol 6(12):1515–1538. https://doi.org/10.1517/17425255.2010.529129
56. von Moltke LL, Weemhoff JL, Bedir E et al (2004) Inhibition of human cytochromes P450 by components of Ginkgo biloba. J Pharm Pharmacol 56(8):1039–1044. https://doi.org/10.1211/0022357044021

57. Venkataramanan R, Ramachandran V, Komoroski BJ, Zhang S, Schiff PL, Strom SC (2000) Milk thistle, a herbal supplement, decreases the activity of CYP3A4 and uridine diphosphoglucuronosyl transferase in human hepatocyte cultures. Drug Metab Dispos 28(11):1270–1273
58. Brantley SJ, Argikar AA, Lin YS, Nagar S, Paine MF (2014) Herb-drug interactions: challenges and opportunities for improved predictions. Drug Metab Dispos 42(3):301–317. https://doi.org/10.1124/dmd.113.055236
59. Kang YC, Chen MH, Lai SL (2015) Potentially unsafe herb-drug interactions between a commercial product of noni juice and phenytoin- a case report. Acta Neurol Taiwanica 24(2):43–46
60. Watkins PB, Wrighton SA, Schuetz EG, Maurel P, Guzelian PS (1986) Macrolide antibiotics inhibit the degradation of the glucocorticoid-responsive cytochrome P-450p in rat hepatocytes in vivo and in primary monolayer culture. J Biol Chem 261(14):6264–6271
61. Kobayashi Y, Suzuki M, Ohshiro N et al (2002) Induction and inhibition of cytochrome P450 and drug-metabolizing enzymes by climbazole. Biol Pharm Bull 25(1):53–57. https://doi.org/10.1248/bpb.25.53
62. Ring B, Wrighton SA, Mohutsky M (2014) Reversible mechanisms of enzyme inhibition and resulting clinical significance. Methods Mol Biol 1113:37–56. https://doi.org/10.1007/978-1-62703-758-7_4
63. Soars MG, Petullo DM, Eckstein JA, Kasper SC, Wrighton SA (2004) An assessment of UDP-glucuronosyltransferase induction using primary human hepatocytes. Drug Metab Dispos 32(1):140–148. https://doi.org/10.1124/dmd.32.1.140
64. Hiroshi T, Shigeko S, Takaharu M (2008) Induction of human UGT1A1 by bilirubin through AhR dependent pathway. Drug Metab Lett 2(4):231–237. https://doi.org/10.2174/187231208786734120
65. Zhang Y, Munday R (2008) Dithiolethiones for cancer chemoprevention: where do we stand? Mol Cancer Ther 7(11):3470–3479. https://doi.org/10.1158/1535-7163.mct-08-0625
66. Sasaki H, Miyaura S, Horie K, Isono H (1989) Induction of glutathione s-transferase by phenobarbital in rat hepatocyte culture. J Pharmacobiodyn 12(12):775–780. https://doi.org/10.1248/bpb1978.12.775
67. Liu X, Cao YF, Dong PP et al (2016) The inhibition of UDP-glucuronosyltransferases (UGTs) by vitamin A. Xenobiotica 47(5):1–6. https://doi.org/10.1080/00498254.2016.1198841
68. Chang M, Zhang F, Shen L et al (1998) Inhibition of glutathione S-transferase activity by the quinoid metabolites of equine estrogens. Chem Res in Toxicol 11(7):758–765. https://doi.org/10.1021/tx9702190
69. Fricker G (2008) Drug interactions with natural products at the blood brain barrier. Curr Drug Metab 9(10):1019–1026. https://doi.org/10.2174/138920008786927758
70. Kibathi LW, Bae S, Penzak SR, Kumar P (2018) Potential influence of centrally acting herbal drugs on transporters at the blood–cerebrospinal fluid barrier and blood–brain barrier. Eur J Drug Metab Pharmacokinet 43(6):619–635. https://doi.org/10.1007/s13318-018-0486-6
71. Yamasaki K, Chuang VTG, Maruyama T (1830) Otagiri M (2013) Albumin–drug interaction and its clinical implication. Biochim Biophys Acta 12:5435–5443. https://doi.org/10.1016/j.bbagen.2013.05.005
72. Jiao Q, Wang R, Jiang Y, Liu B (2018) Study on the interaction between active components from traditional Chinese medicine and plasma proteins. Chem Cent J 12(1):48. https://doi.org/10.1186/s13065-018-0417-2
73. Lu H, Lu Z, Li X et al (2017) Interactions of 172 plant extracts with human organic anion transporter 1 (SLC22A6) and 3 (SLC22A8): a study on herb-drug interactions. PeerJ 5:e3333. https://doi.org/10.7717/peerj.3333
74. Gufford BT, Chen G, Vergara AG, Lazarus P, Oberlies NH, Paine MF (2015) Milk thistle constituents inhibit raloxifene intestinal glucuronidation: a potential clinically relevant natural product–drug interaction. Drug Metab Dispos 43(9):1353–1359. https://doi.org/10.1124/dmd.115.065086

75. Alkharfy KM, Frye RF (2007) Effect of valerian, valerian/hops extracts, and valerenic acid on glucuronidation in vitro. Xenobiotica 37(2):113–123. https://doi.org/10.1080/00498250601050420
76. Vietri M, Pietrabissa A, Mosca F, Spisni R, Pacifici GM (2003) Curcumin is a potent inhibitor of phenol sulfotransferase (SULT1A1) in human liver and extrahepatic tissues. Xenobiotica 33(4):357–363. https://doi.org/10.1080/0049825031000065197
77. De Santi C, Pietrabissa A, Mosca F, Rane A, Pacifici GM (2002) Inhibition of phenol sulfotransferase (SULT1A1) by quercetin in human adult and foetal livers. Xenobiotica 32(5):363–368. https://doi.org/10.1080/00498250110119108
78. Hammerness P, Basch E, Ulbricht C et al (2003) St. John's Wort: a systematic review of adverse effects and drug interactions for the consultation psychiatrist. Psychosomatics 44(4):271–282. https://doi.org/10.1176/appi.psy.44.4.271
79. Zhu X, Jia YH (2014) Inhibition of catechol-o-methyltransferase (COMT) by myricetin, dihydromyricetin, and myricitrin. Pharmazie 69(3):183–186
80. European Scientific Cooperative on Phytotherapy (ESCOP) (2020) Table of herb-drug interactions based on the monographs of ESCOP. https://escop.com/interactions/. Accessed 31 Oct 2020
81. Silveira PF, Bandeira MAM, Arrais PSD (2008) Farmacovigilância e reações adversas às plantas medicinais e fitoterápicos: uma realidade. Rev Bras de Farmacogn 18(4):618–626. https://doi.org/10.1590/S0102-695X2008000400021
82. Ekor M (2013) The growing use of herbal medicines: issues relating to adverse reactions and challenges in monitoring safety. Front Pharmacol 4:177. https://doi.org/10.3389/fphar.2013.00177
83. Eisenberg DM, Davis RB, Ettner SL et al (1998) Trends in alternative medicine use in the United States, 1990–1997: results of a follow-up national survey. JAMA 280(18):1569–1575. https://doi.org/10.1001/jama.280.18.1569
84. Clarke TC, Black LI, Stussman BJ, Barnes PM, Nahin RL (2015) Trends in the use of complementary health approaches among adults: United States, 2002–2012. Natl Health Stat Rep 79:1–16
85. Tindle HA, Davis RB, Phillips RS, Eisenberg DM (2005) Trends in use of complementary and alternative medicine by US adults: 1997–2002. Altern Ther Health Med 11(1):42–49
86. Foster DF, Phillips RS, Hamel MB, Eisenberg DM (2000) Alternative medicine use in older Americans. J Am Geriatr Soc 48(12):1560–1565. https://doi.org/10.1111/j.1532-5415.2000.tb03864.x
87. Stussman BJ, Nahin RR, Barnes PM, Ward BW (2020) U.S. physician recommendations to their patients about the use of complementary health approaches. J Altern Complement Med 26(1):25–33. https://doi.org/10.1089/acm.2019.0303
88. Nahin RL, Barnes PM, Stussman BJ (2016) Expenditures on complementary health approaches: United States, 2012. Natl Health Stat Rep 95:1–11
89. Connor J, Buring JE, Eisenberg DM, Osypiuk K, Davis RB, Wayne PM (2020) Patient disclosure of complementary and integrative health approaches in an academic health center. Glob Adv Health Med 9:2164956120912730. https://doi.org/10.1177/2164956120912730
90. Graham RE, Ahn AC, Davis RB, O'Connor BB, Eisenberg DM, Phillips RS (2005) Use of complementary and alternative medical therapies among racial and ethnic minority adults: results from the 2002 National Health Interview Survey. J Natl Med Assoc 97(4):535–545
91. Black LI, Clarke TC, Barnes PM, Stussman BJ, Nahin RL (2015) Use of complementary health approaches among children aged 4–17 years in the United States: National Health Interview Survey, 2007–2012. Natl Health Stat Report 78:1–19
92. Dwyer J, Nahin RL, Rogers GT et al (2013) Prevalence and predictors of children's dietary supplement use: the 2007 national health interview survey. Am J Clin Nutr 97(6):1331–1337. https://doi.org/10.3945/ajcn.112.052373

93. Nahin RL, Dahlhamer JM, Taylor BL et al (2007) Health behaviors and risk factors in those who use complementary and alternative medicine. BMC Public Health 7:217. https://doi.org/10.1186/1471-2458-7-217
94. Honda K, Jacobson JS (2005) Use of complementary and alternative medicine among United States adults: the influences of personality, coping strategies, and social support. Prev Med 40(1):46–53. https://doi.org/10.1016/j.ypmed.2004.05.001
95. Afari N, Eisenberg DM, Herrell R et al (2000) Use of alternative treatments by chronic fatigue syndrome discordant twins. Integr Med 2(2):97–103. https://doi.org/10.1016/s1096-2190(99)00017-7
96. Yeh GY, Eisenberg DM, Davis RB, Phillips RS (2002) Use of complementary and alternative medicine among persons with diabetes mellitus: results of a national survey. Am J Public Health 92(10):1648–1652. https://doi.org/10.2105/ajph.92.10.1648
97. Morgan TK, Williamson M, Pirotta M, Stewart K, Myers SP, Barnes J (2012) A national census of medicines use: a 24-hour snapshot of Australians aged 50 years and older. Med J Aust 196(1):50–53. https://doi.org/10.5694/mja11.10698
98. Skalli S, Soulaymani Bencheikh R (2012) Safety monitoring of herb-drug interactions: a component of pharmacovigilance. Drug Saf 35(10):785–791. https://doi.org/10.1007/bf03261975
99. Haller CA, Dyer JE, Ko R, Olson KR (2002) Making a diagnosis of herbal-related toxic hepatitis. West J Med 176(1):39–44. https://doi.org/10.1136/ewjm.176.1.39
100. Asif M (2012) A brief study of toxic effects of some medicinal herbs on kidney. Adv Biomed Res 1:44. https://doi.org/10.4103/2277-9175.100144
101. Afolabi SO, Akindele AJ, Awodele O, Anunobi CC, Adeyemi OO (2012) A 90 day chronic toxicity study of Nigerian herbal preparation DAS-77 in rats. BMC Complement Altern Med 12:79. https://doi.org/10.1186/1472-6882-12-79
102. Mazzari ALDA, Prieto JM (2014) Monitoramento de interações farmacocinéticas entre plantas medicinais e fitoterápicos e os medicamentos convencionais pelo sistema de farmacovigilância brasileiro. Inf Dent 26(3):193–198. https://doi.org/10.14450/2318-9312.v26.e3.a2014.pp193-198
103. Graham RE, Gandhi TK, Borus J et al (2008) Risk of concurrent use of prescription drugs with herbal and dietary supplements in ambulatory care. In: Henriksen K, Battles JB, Keyes MA, Grady ML (eds) Advances in patient safety: new directions and alternative approaches, Technology and medication safety, vol 4. Agency for Healthcare Research and Quality (US), Rockville, MD
104. Bush TM, Rayburn KS, Holloway SW et al (2007) Adverse interactions between herbal and dietary substances and prescription medications: a clinical survey. Altern Ther Health Med 13(2):30–35
105. Agbabiaka TB, Spencer NH, Khanom S, Goodman C (2018) Prevalence of drug-herb and drug–supplement interactions in older adults: a cross-sectional survey. Br J Gen Pract 68(675):e711–e717. https://doi.org/10.3399/bjgp18X699101
106. Ezuruike U, Prieto JM (2016) Assessment of potential herb-drug interactions among nigerian adults with type-2 diabetes. Front Pharmacol 7:248. https://doi.org/10.3389/fphar.2016.00248
107. Leite PM, Castilho RO, Ribeiro AL, Martins MA (2016) Consumption of medicinal plants by patients with heart diseases at a pharmacist-managed anticoagulation clinic in Brazil. Int J Clin Pharm 38(2):223–227. https://doi.org/10.1007/s11096-016-0270-0
108. Aykan DA, Aykan AC (2019) Factors associated with the concomitant use of cardiovascular drugs and dietary herbal products: a cross-sectional study. J Cardiovasc Pharmacol Ther 24(2):146–152. https://doi.org/10.1177/1074248418794938
109. Martindale: the complete drug reference (2020) Royal Pharmaceutical Society. https://about.medicinescomplete.com/adr-checker/
110. Stockley's Herbal Medicines Interactions. https://about.medicinescomplete.com/publication/stockleys-herbal-medicines-interactions-2/
111. Drug Interactions Checker. https://www.drugs.com/drug_interactions.html

112. Drug Interaction Checker. https://reference.medscape.com/drug-interactionchecker
113. Drug Intraction Checker. https://www.webmd.com/interaction-checker/default.htm
114. Drug Interaction Checker. https://www.rxlist.com/drug-interaction-checker.htm
115. Watson Health Micromedex. https://www.micromedexsolutions.com/home/dispatch
116. Lexicomp. https://online.lexi.com/lco/action/login
117. Facts & Comparisons. https://online.lexi.com/lco/action/login
118. PEPID Knowledge Base (PKB). https://www.pepid.com/
119. Shakeel F, Fang F, Kidwell KM, Marcath LA, Hertz DL (2020) Comparison of eight screening tools to detect interactions between herbal supplements and oncology agents. J Oncol Pharm Pract 26(8):1843–1849. https://doi.org/10.1177/1078155220905009
120. World Health Organization (2000) General guidelines for methodologies on research and evaluation of traditional medicine. World Health Organization, Geneva
121. Rates SMK (2001) Promoção do uso racional de fitoterápicos: uma abordagem no ensino de Farmacognosia. Rev Bras Farmacogn 11(2):57–69. https://doi.org/10.1590/S0102-695X2001000200001
122. Balbino EE, Dias MF (2010) Farmacovigilância: um passo em direção ao uso racional de plantas medicinais e fitoterápicos. Rev Bras Farm 20(6):992–1000. https://doi.org/10.1590/S0102-695X2010005000031
123. World Health Organization (2004) Guidelines on safety monitoring of herbal medicines in pharmacovigilance systems. World Health Organization, Geneva. http://nla.gov.au/nla.cat-vn3411552
124. World Health Organization (2005) National policy on traditional medicine and regulation of herbal medicines: report of a WHO global survey. World Health Organization, Geneva
125. SUS (2006) Política Nacional de Práticas Integrativas e Complementares no SUS. http://189.28.128.100/dab/docs/publicacoes/geral/pnpic.pdf. Accessed 2 Nov 2016
126. Barnes J (2003) Pharmacovigilance of herbal medicines: a UK perspective. Drug Saf 26(12):829–851. https://doi.org/10.2165/00002018-200326120-00001
127. Shaw D, Graeme L, Pierre D, Elizabeth W, Kelvin C (2012) Pharmacovigilance of herbal medicine. J Ethnopharmacol 140(3):513–518. https://doi.org/10.1016/j.jep.2012.01.051

Chapter 6
Safety and Pharmacovigilance of Herbal Medicines in Pregnancy

Sally Stephens

6.1 Safety and Pharmacovigilance of Herbal Medicines in Pregnancy

6.1.1 Therapeutic Drug Use in Pregnancy

Therapeutic drug use in pregnancy is common, with many women accessing non-prescription ('over-the-counter') medicines to treat pregnancy-related symptoms, including constipation, heartburn and urinary tract infections. Some pregnant women will also be prescribed medication to treat more serious conditions, such as depression, epilepsy, and asthma. It is estimated that about 80% of women take at least one non-prescription medicine or prescription medicine throughout the course of pregnancy [1].

Many physiological changes occur during the (up to) 40 weeks of pregnancy and have the ability to alter the absorption, distribution and elimination of conventional and herbal medicines taken by pregnant women. These changes include increased cardiac output, the rate of liver metabolism, plasma volume, glomerular filtration rate, and extent of fat stores, as well as changes in gastrointestinal function. There is a misconception that there is a placental barrier providing protection to the foetus, but almost all drugs are able to pass through the placenta freely.

Some therapeutic drugs are known to be teratogenic to the developing foetus, increasing the risk of several adverse pregnancy outcomes, including miscarriage, birth defects, low birth weight and neurodevelopmental delay. Early pregnancy, during the first trimester, is the most sensitive time for a teratogenic insult to occur, but the foetus is potentially at risk of teratogenic effects of maternal drug exposure for

S. Stephens (✉)
UK Teratology Information Service, Regional Drug and Therapeutics Centre,
Newcastle-Upon-Tyne, UK
e-mail: sally.stephens7@nhs.net

J. Barnes (ed.), *Pharmacovigilance for Herbal and Traditional Medicines*,
https://doi.org/10.1007/978-3-031-07275-8_6

the full duration of pregnancy, and possibly even before conception. Up to 50% of pregnancies are unplanned, so exposure to medicines during conception and early pregnancy is not rare.

6.1.2 Herbal Medicine Use in Pregnancy

Pregnant women and their healthcare providers are often confused and/or concerned about the effects of taking therapeutic drugs in pregnancy. Safety data are often lacking, or conflicting, and women have historically been advised to avoid pharmaceutical drugs (conventional medicines) if possible because of the, often unknown, risks to the foetus. It is then perhaps unsurprising that some women may choose to use herbal medicines (or other 'natural' health products), believing that they are 'natural' and a 'safer' option during pregnancy, or in the preconception period, than are pharmaceutical medications [2]. Many other women in low- and middle-income countries often rely on herbal and other 'traditional' medicines as the only way to treat medical conditions and diseases [3].

6.1.3 Prevalence of Use of Herbal Medicines in Pregnancy

The prevalence of use of herbal medicines by women who are pregnant appears to be substantial. A multinational study, carried out in 23 countries, which included questionnaire data from 9459 women who were pregnant, or had a child under the age of one year, reported the prevalence of herbal medicine use during pregnancy to be 28.9% with most women using them to treat respiratory illness and nausea. The most commonly used herbs reported in this study were ginger, cranberry, valerian, and raspberry. The prevalence of use ranged from 9.4% to 82.3%, which may in part be explained by the use of different study methodologies and cultural and regional differences. Russia (69%), Poland (49.8%) and Australia (43.8%) reported the highest use of herbal medicines during pregnancy [4].

In developed countries, studies have shown that women who use herbal medicines in pregnancy are more likely to be middle-aged, have high levels of education and income, and be primiparous [5–7]. One study also reported that, among its sample, women who used herbal medicines were less likely to be smokers and more likely to be married [6].

6.2 Safety Concerns Associated with Use of Herbal Medicines During Pregnancy

6.2.1 Adulteration and Contamination of Herbal Medicines

The high risk of contamination or adulteration of herbal medicines with heavy metals, pesticides, and pharmaceutical medications [8, 9] is of particular concern with respect to use of these products by pregnant women.

The lack of regulation and 'product standards' for herbal medicines in some countries is problematic. Products may not be checked for the quantity or quality of active ingredients, for contaminants, may not provide patient information, and may not comply with internationally accepted pharmaceutical industry standards for assuring the quality of medicinal products. Consequently, products may be of poor and variable quality and have been found to contain high levels of bacterial pathogens [10], pesticides [11], and heavy metals [12]. In rare instances, even registered herbal medicines have been found to have serious quality problems; for example, in the UK, in 2016, St John's wort tablets were found to contain toxic pyrrolizidine alkaloids above the threshold recommended by the Committee on Herbal Medicinal Products, a European expert body. The contamination was thought to be from accidental collection of local weeds during harvesting [13].

Reports of contaminated herbal medicines used in pregnancy have associated use of oral Ayurvedic medicines from India with adverse pregnancy outcomes, when taken by several pregnant women in the US and one in Australia [14, 15]. Six asymptomatic pregnant women (with blood lead concentrations between 16 and 64 μg/dL) used ten Ayurvedic products which were found to have a high lead content (as high as 2.4%), as well as traces of arsenic and mercury. Two of the women miscarried before 20 weeks' gestation; both women had taken the product to promote fertility. It is unknown whether the pregnancy outcomes were related to reproductive issues or to the contaminated medicines [14]. Renal abnormalities (agenesis of one and absence of the other kidney), pulmonary hypoplasia and anhydramnios were reported in an infant exposed *in utero* to Ayurvedic medicines in the first trimester of pregnancy. Investigations showed very high maternal blood lead concentrations (67 μg/dL) and no genetic link to the abnormalities [15].

As pregnancy outcome data from women using contaminated medicines in pregnancy are scarce, it is difficult to establish causal associations. However, evidence from occupational or environmental exposure to heavy metals suggest that elevated concentrations of lead and arsenic in pregnant women have been associated with pregnancy loss, impaired intrauterine growth, and preterm labour [16–22]. Exposure to lead in pregnancy has also been associated with impaired postnatal neurodevelopment in the offspring [23–25].

Pregnant women should be made aware of the potential risks of contaminants and be advised not to use unregulated medicines prior to and during pregnancy.

6.2.2 Herbal Medicine Interactions with Conventional Medicines

Given that a large proportion of pregnant women use conventional medications, and the high prevalence of use of herbal medicines during pregnancy, a substantial number of pregnant women are likely to be using both conventional and herbal medicines, with the risk of potential herbal medicine-pharmaceutical drug interactions.

A survey of 889 women in North East Scotland, which collected data on medicines use in pregnancy, identified several potential moderate to severe herb-drug interactions in around 12% of the study cohort. Using the Natural Medicines Comprehensive Database [26] to assess the potential for herbal and natural product interaction with prescribed medicines, the survey identified 34 potential herb-drug interactions among 23 participants. Ginger (*Zingiber officinale*) was noted to have the potential to cause interactions with concurrent prescription medicines, including one major interaction with nifedipine, and three moderate interactions with metformin, insulin, and aspirin. Ondansetron and chamomile (type not specified) were also reported as having the potential to cause a minor interaction [27].

St John's wort (*Hypericum perforatum* L.) herb extract is well publicised as interacting with many conventional medicines: by inducing certain cytochrome P450 enzymes, St John's wort extracts lower serum concentrations of certain medicines below the therapeutic range. Their use has been reported to reduce the efficacy of several antihypertensive, anticonvulsant, immunosuppressant, and antipsychotic medicines [28], which in pregnancy could have devastating effects, including death, for both the mother and baby.

It is advisable for pregnant women to consult with their doctor or pharmacist before they use herbal products in combination with prescribed or non-prescription medication. Authoritative information sources on herbal medicines interactions, written for healthcare professionals, are available [29].

6.2.3 Issues Relating to Formulations, Routes of Exposure, Dose, and Dosage

As with conventional medicines, the dose, duration of use, and route of administration of herbal preparations are important to consider where such products are used or considered for use in pregnancy. Culinary use of small quantities of herbs would not be expected to increase the risk of foetal harm, but use of high doses of herbal substances, concentrated extracts, and/or prolonged use should be avoided as general precautions.

Essential oils used in small amounts at low concentrations in commercially produced shampoo and soap products are not thought to be in quantities that would cause foetal harm when used appropriately. However, this knowledge is based on unpublished experience. Oral ingestion of essential oils confers maternal and, therefore, foetal toxicity and should be avoided in pregnancy.

Tinctures—alcohol extracts of herbal substances—should generally be avoided in pregnancy as alcohol is a known teratogen. However, inadvertent exposure to small quantities during pregnancy would not be considered to increase the risk of adverse pregnancy outcomes.

6.2.4 The Availability of Information on the Use of Herbal Medicines in Pregnancy

Online sources and books provide a wealth of information on herbal use for women who are, or intending to become, pregnant, but the majority of this is historical, empirical, and observational with little pharmacologic and animal safety data. Much of the information is of poor quality, often exaggerating the perceived benefits and trivialising, or ignoring, the potential harms. Reputable resources are available for health professionals to help provide some assessment of risk for exposed pregnancies [26, 29–30]; Medicines Complete [29, 30], and many countries commission Teratology Information Services to provide advice to pregnant women and/or their healthcare providers about drug and chemical exposures in pregnancy, including herbal medicines. Contact is usually via telephone but some services also provide online written information (Box 6.1). However, as with conventional medicines, providing a risk assessment can be difficult as there are substantial gaps in knowledge, data are often conflicting, dose and duration of use are poorly defined, and information is typically written in a generic fashion for a particular herbal substance rather than relating to a specific product.

Box 6.1 Authoritative Resources on Herbal Medicine Use in Pregnancy for Women and/or Their Healthcare Providers

If available, healthcare providers and women can contact their local Teratology Information Service for patient specific risk assessments where exposure to herbal medicine has occurred.
https://www.entis-org.eu/centers. [31]
https://mothertobaby.org/locations/. [32]
Authoritative Information resources on the use of some herbal medicines in pregnancy:
www.medicinesinpregnancy.org [33]
http://naturaldatabase.therapeuticresearch.com/ [26]
Williamson E, Driver S, Baxter K, Preston CL (eds), Stockley's Herbal Medicines Interactions. [online] London: Pharmaceutical Press. http://www.medicinescomplete.com [29]
Pharmaceutical Press Editorial. Herbal Medicines. [online] London: Pharmaceutical Press. http://www.medicinescomplete.com [30]

6.3 Herbal Medicines Commonly Used to Treat Pregnancy-Related Conditions

Herbal medicines cited in the literature as most frequently being used during pregnancy vary between studies, and herbal substances are typically described only using common names, which are not precise, or not defined at all. Herbal medicines that are commonly reported across studies from Western countries include ginger root, 'chamomile' tea, cranberry, and echinacea [5, 7, 34–36]. The recommendations for using these herbals during pregnancy are conflicting. In all cases, further large pharmacoepidemiological studies are required to accurately assess the likelihood of adverse health outcomes for mothers and babies associated with use during pregnancy.

*6.3.1 Ginger Root (*Zingiber officinale *Roscoe)*

Ginger root (*Zingiber officinale*) has been studied in pregnancy for reducing nausea and vomiting, but with many of the studies only investigating efficacy, not foetal safety.

The available literature consists of 14 randomised controlled trials (RCTs) that include a total of 617 women who were exposed to ginger root preparations at fewer than 20 weeks' gestation [37]. Prospective cohort studies contribute a further 1366 gestational ginger root exposures, with first trimester exposure confirmed in 593 of these women. A population-based, case-control study assessed associations between *in utero* ginger root exposure and specific congenital malformations in the infant. Ginger was consumed as either a powder, essence, extract or fresh and the dose ranged from 500 mg/d to 2.5 g/d. Data from four of the RCTs described above have also been meta-analysed. Collectively, these studies do not suggest an increase in risk of adverse foetal outcomes with exposure to ginger root preparations [37].

Two systematic reviews using overlapping data from 18 studies compared the antiemetic effects of ginger with placebo, vitamin B6, metoclopramide, and dimenhydrinate in pregnant women. Both reviews concluded ginger supplementation significantly relieved nausea compared with placebo, but there were no significant effects on vomiting [37, 38].

Several conventional pharmaceuticals can be recommended for use in pregnancy to treat symptoms of nausea and vomiting. Women with excessive nausea and vomiting (hyperemesis gravidarum) may need to seek hospital treatment for fluid and electrolyte replacement and treatment with a pharmaceutical antiemetic.

*6.3.2 Chamomile (*Chamaemelum nobile *(L.) All)*

The quality of data relating to the use of chamomile (*Chamaemelum nobile*) in pregnancy is poor, with the available cohort studies collecting exposure details retrospectively, after the participants' pregnancies had ended. The types of chamomile preparations were not explicitly reported, described as an oral exposure most commonly by infusion in one study [39], and as oral or topical in another [40]. Common indications for use were anxiety, digestive problems, and stretch marks [40]. The data come from three Italian studies, which all compared pregnancies of daily consumers of chamomile to those of non-users, finding an association with use of chamomile and preterm labour [6, 39, 40] and threatened miscarriage [40]. In one study, all participants were classed as 'healthy', and in the other two, the analyses were part of larger studies, making it difficult to extrapolate the influence potential maternal risk factors may have had on the outcomes.

Due to the limited available data, it is recommended to avoid chamomile-containing products during pregnancy.

6.3.3 *Cranberry (*Vaccinium macrocarpon *Aiton)*

Cranberries are used for the treatment and prevention of urinary tract infections (UTIs), a common complaint in pregnancy due to hormonal changes. Research on the safety of cranberries during pregnancy is based on one cohort study (including women <16 weeks pregnant) and one randomised, placebo-controlled pilot study (most exposed in 'early pregnancy'), showing no increased risk of maternal or foetal outcomes in >1000 pregnancies [41, 42] following the consumption of cranberry juice.

Data regarding the efficacy of cranberry for the prevention of UTIs are derived from studies using different cranberry preparations (juice, capsules, and tablets) and provide no clear consensus of clinical benefit [43]. Untreated UTIs can cause serious adverse outcomes, and conventional pharmaceutical treatment should not be withheld on account of pregnancy.

6.3.4 *Echinacea (*Echinacea *spp.:* E. angustifolia *DC.,* E. pallida *(Nutt.) Nutt.,* E. purpurea *(L.) Moench)*

Echinacea species are often used by patients to 'strengthen the immune system' and to treat colds and other upper respiratory tract infections, urinary tract infections, and slow-healing wounds.

Two studies have investigated the use of 'echinacea' preparations in pregnancy. One study explored pregnancy outcomes among 206 women who had used an echinacea preparation for 5 to 7 days. Most women used either *E. angustifolia* or *E. purpurea* (not further specified) as capsules/tablets (58%) or tincture (38%). The dose ranged from 250-1000 mg/d or 2-10 tincture drops. Of these, 112 women (54%) were exposed in the first trimester and 17 (8%) in all three trimesters. These women were compared with non-users of echinacea matched for maternal age, alcohol intake, and smoking status. No differences in the rates of gestational age, birth weight, foetal distress, or major congenital malformations were reported among the echinacea group, compared with the control group [44]. However, this study included only very small numbers of participants and does not provide conclusive evidence on the likelihood of adverse outcomes in the foetus following exposure to echinacea *in utero*.

The second study from the Norwegian Mother and Child cohort study reported 363 women using echinacea in pregnancy within a cohort of 68,522. Details regarding the dose, echinacea species, preparation type, or stage of pregnancy at exposure were not reported. Users of echinacea were older and less likely to smoke but did not have an increased risk of having a child with a malformation or an adverse pregnancy outcome when compared to non-users [45].

The European Medicines Agency (EMA) does not recommend the use of echinacea in pregnancy due to the limited safety data available [46].

6.3.5 Other Herbs with Known Adverse Effects in Pregnancy

Single studies have reported adverse pregnancy outcomes following the use of almond oil and liquorice *(Glycyrrhiza glabra* L.*)*.

Regular users of almond oil ($n = 123$) had a significantly higher risk of a preterm delivery than non-users even when other risks factors for preterm birth, such as age, smoking, twin pregnancy, and recreational drug use, were controlled for [39]. Most of the women in this study applied almond oil to their abdomen to reduce the risk of stretch marks during their pregnancy.

Women who took up to 2104 mg/day of liquorice (*Glycyrrhiza glabra*) between the fourth day and 25th week of gestation had a marginally increased risk of stillbirth than those not exposed. Other adverse outcomes were not reported [47]. When comparing pregnancy outcomes in fourteen women who regularly use liquorice during pregnancy with 238 pregnancies in non-users, a higher frequency of threatening miscarriages (35.7%) and preterm labour (16.7%) was seen in the liquorice-exposed group [40].

Although most herbal medicines have not been formally investigated for their embryotoxicity, teratogenic, and abortifacient potential, many are not recommended for use during pregnancy due to the theoretical risks from their known constituents and pharmacological effects. Examples include black cohosh (*Actaea racemosa* L.), blue cohosh (*Caulophyllum thalictroides* (L.) Michx.), and motherwort (*Leonurus cardiaca* L.), all of which are not recommended as they have traditionally been used to stimulate menstruation or provoke abortion by acting on the smooth muscle of the uterus [48–50].

6.4 Safety Monitoring Systems for Medicines Used in Pregnancy

Safety information for pregnant women and their clinicians is frequently reliant on observational studies where data have been voluntarily reported by women who have often been inadvertently or unavoidably exposed to medicines in pregnancy.

Study designs are usually cohort or case-control investigations. Difficulties arising with this type of data collection are due to, but not limited to, selection bias with regard to those pregnancies that are successfully followed up, small sample sizes, and inaccurate or missing data on exposures and dose, timings, duration and indication.

Many spontaneous reporting systems from national healthcare registries and marketing authorisation holders encourage reporting of suspected adverse reactions associated with prescribed and herbal medicines, which can provide useful data for safety signal detection. However, adverse event reporting can create bias, where adverse pregnancy outcomes are overrepresented with no data to determine the frequency of risk within an exposed population. Teratology Information Services (TIS) provide routine collection of data from pregnant women, or their healthcare providers, who contact the service for advice regarding maternal use of medicines. Although this method allows prospective data collection, methodological issues remain, including difficulties collecting adequate numbers of exposed pregnancies for rarely used medicines, and sample bias, as enquiries to TISs are often biased towards high-risk pregnancies. Further, not all countries have these voluntary reporting systems in place and those that do find it very labour intensive to track and follow up pregnancies to collect outcome data, and busy clinicians find the burden of reporting pregnancies too great.

6.4.1 *Reporting of Adverse Reactions Associated with Herbal Medicine Use in Pregnancy*

Between 2006 and 2014, the UK 'Yellow Card Scheme' (the national system for voluntary reporting of suspected adverse reactions associated with medicines, including herbal medicines) received around 60 reports involving herbal medicines each year; 40% of these were submitted by the general public (e.g. patients or their carers) rather than by health professionals [51]. The number of pregnancy-specific reports was not provided.

Healthcare providers are rarely informed by their patients about herbal medicine use in pregnancy [52], and health professionals may omit to ask their patients about this. Teratology Information Services are also not commonly asked for advice about the use of herbal medicines by either patients or their healthcare providers. The Berlin Teratology Information Service (Embryotox) reported that only 6% of enquiries to the service were regarding 'alternative medications' [53]. For the UK Teratology Information Service (UKTIS), which has been in operation and collecting surveillance data, including those concerning herbal medicines, since 1984, fewer than 1% of all enquiries made to the service are in relation to herbal products. Enquiries regarding the use of peppermint oil, St John's wort and 'senna' are the most commonly recorded herbal medicine exposures in the database.

6.4.2 *Safety Monitoring Systems for the Collection of Herbal Medicine Exposure Data in Pregnancy*

In order to increase reporting of medicines in pregnancy in countries that may not have national reporting systems or registries in place, and to improve some of the issues around missing maternal and infant details, collection of exposure data on non-prescription medicines, and to reduce the burden on clinicians, web-based applications are being developed.

The UKTIS has created a bespoke, online, patient-oriented pregnancy recording system, *BUMPS* (Best Use of Medicines in Pregnancy), where women are invited to set up an account and input information regarding all exposures in pregnancy. There is a dedicated section to record use of herbal medicines, including the proprietary and ingredient names, dose, dose frequency, duration of use, route of administration, and stage of pregnancy of exposure (see Boxes 6.2a and 6.2b). The system is designed to be updated throughout pregnancy and to record the outcome(s) once the pregnancy has ended. Women are requested to report ongoing pregnancies in early pregnancy, preferably before any prenatal screening or knowledge of the pregnancy outcome has occurred. Although the data are analysed separately, the system also permits the reporting of previous pregnancies. Where a liveborn child is recorded in the system, neurodevelopmental milestones are requested annually from 6 months of age. Engagement with women in this way provides an opportunity to stay in contact throughout their child's life to improve detection of longer-term outcomes, including childhood illnesses, and development milestones.

Box 6.2a Extracts from the Best Use of Medicines in Pregnancy (BUMPS) questionnaire: Medicines and Vaccines

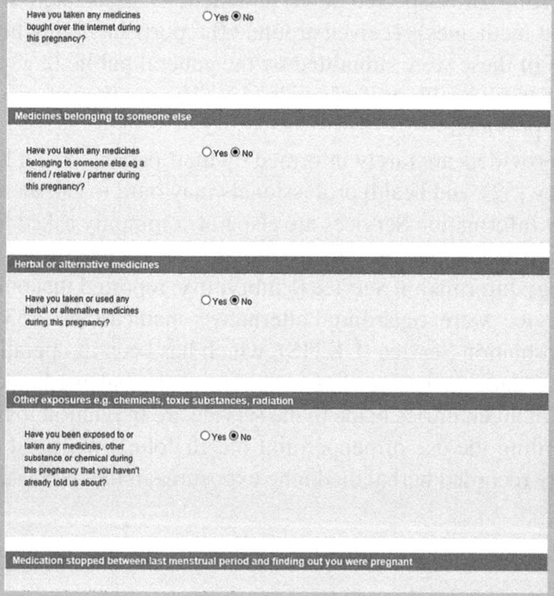
Have you taken any medicines bought over the internet during this pregnancy? ○ Yes ◉ No

Medicines belonging to someone else

Have you taken any medicines belonging to someone else eg a friend / relative / partner during this pregnancy? ○ Yes ◉ No

Herbal or alternative medicines

Have you taken or used any herbal or alternative medicines during this pregnancy? ○ Yes ◉ No

Other exposures e.g. chemicals, toxic substances, radiation

Have you been exposed to or taken any medicines, other substance or chemical during this pregnancy that you haven't already told us about? ○ Yes ◉ No

Medication stopped between last menstrual period and finding out you were pregnant

Box 6.2b Extract from 'Medicines Used in the 3 Months Before or During Pregnancy' Questionnaire

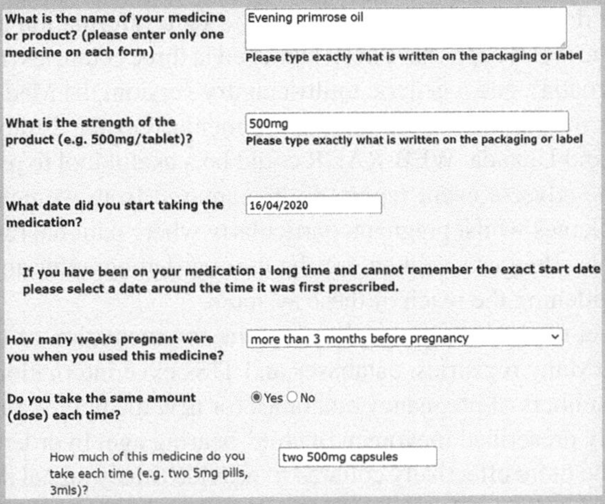

What is the name of your medicine or product? (please enter only one medicine on each form) — Evening primrose oil
Please type exactly what is written on the packaging or label

What is the strength of the product (e.g. 500mg/tablet)? — 500mg
Please type exactly what is written on the packaging or label

What date did you start taking the medication? — 16/04/2020

If you have been on your medication a long time and cannot remember the exact start date please select a date around the time it was first prescribed.

How many weeks pregnant were you when you used this medicine? — more than 3 months before pregnancy

Do you take the same amount (dose) each time? — Yes ◉ No ○

How much of this medicine do you take each time (e.g. two 5mg pills, 3mls)? — two 500mg capsules

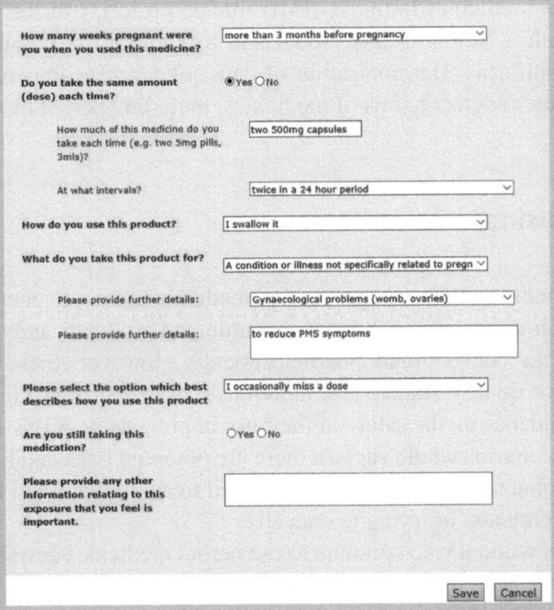

How many weeks pregnant were you when you used this medicine? — more than 3 months before pregnancy

Do you take the same amount (dose) each time? — Yes ◉ No ○

How much of this medicine do you take each time (e.g. two 5mg pills, 3mls)? — two 500mg capsules

At what intervals? — twice in a 24 hour period

How do you use this product? — I swallow it

What do you take this product for? — A condition or illness not specifically related to pregn

Please provide further details: — Gynaecological problems (womb, ovaries)

Please provide further details: — to reduce PMS symptoms

Please select the option which best describes how you use this product — I occasionally miss a dose

Are you still taking this medication? — Yes ○ No ○

Please provide any other information relating to this exposure that you feel is important.

Save Cancel

Targeted promotion of the system is required, highlighting the opportunity and importance of women informing UKTIS of their use of herbal products in pregnancy. Only 1.5% of women who have reported to the system to date have reported these types of exposures. The most common products to be reported have been essential oils, used topically or accidentally ingested.

Other monitoring systems have been designed to provide easier ways to capture data. One such development is WEB-RADR [54]: recognising adverse drug reactions, a mobile application enabling patients and healthcare providers to report suspected adverse drug reactions and receive up-to-date information and news alerts. Country-specific mobile apps have been launched in three countries (the Netherlands, the UK, and Croatia), and a generic multi-country version, the Med Safety app, has been launched in Burkina Faso, Zambia, Armenia, Ghana, Ethiopia, Botswana, Cote d'Ivoire, and Uganda. WEB-RADR could be a useful tool to provide a way for herbal medicine adverse event reports to be captured from women who regularly use herbal medicines whilst pregnant, particularly where national reporting systems are not available. Pregnant women are the correct demographic to be targeted by social media widening the reach of these systems.

Despite novel methods of data collection, the fragmentation of this information is problematic. Many registries, databases and TISs exist internationally, often collecting small numbers of pregnancy outcomes for new medicines, or medicines that are not routinely prescribed to women of child-bearing age. In order to address how these data can be more effectively collated to provide timely signal detection, a five-year Innovative Medicines Initiative (IMI)-funded project, ConcePTION, is underway to establish a common data model and to explore the possibility of secure data-sharing platforms. Harmonisation of data collection is imperative to achieve timely responses to detect harms of medicines, including herbal medicines.

6.5 Conclusions

The available published data suggest herbal medicine use in pregnancy is associated with possible teratogenicity, risk from contaminated products, and the potential for interactions with conventional pharmaceuticals. However, these findings often come from poor quality studies and individual case reports making it difficult to provide any guidance on the safety of their use in pregnancy. A risk-benefit analysis for almost all scenarios would suggest there are potential risks and little evidence of any benefit. Women should therefore be advised to avoid the use of herbal products when they are pregnant or trying to conceive.

For pregnant women who continue to use herbal medicines despite the warnings, or because of the lack of evidence of harm, data collection is vital to improve the quantity and quality of information provided to women and their healthcare providers about the products they are using.

Reporting systems designed for conventional pharmaceutical medicines can be utilized and have been adapted to collect adverse event reports following herbal medicines use also. Online data systems in the form of pregnancy registries, databases and apps provide an opportunity to collect data from women internationally about the herbal products they use in pregnancy and their birth outcomes. It remains to be seen if women will be willing to use these systems in enough numbers that data can be collected on a large enough scale, to provide meaningful guidance on the use of individual herbal products. Education and promotion are the next steps to encourage reporting and providing robust evidence-based safety data for pregnant women in the future.

References

1. Lupattelli A, Spigset O, Twigg MJ et al (2014) Medication use in pregnancy: a cross-sectional, multinational web-based study. BMJ Open 4(2):e004365. https://doi.org/10.1136/bmjopen-2013-004365
2. Pallivalapila AR, Stewart D, Shetty A, Pande B, Singh R, McLay JS (2015) Use of complementary and alternative medicines during the third trimester. Obstet Gynecol 125(1):204–211. https://doi.org/10.1097/AOG.0000000000000596
3. Towns AM, van Andel T (2016) Wild plants, pregnancy, and the food-medicine continuum in the southern regions of Ghana and Benin. J Ethnopharmacol 179:375–382. https://doi.org/10.1016/j.jep.2016.01.005
4. Kennedy DA, Lupattelli A, Koren G, Nordeng H (2013) Herbal medicine use in pregnancy: results of a multinational study. BMC Complement Altern Med 13:355. https://doi.org/10.1186/1472-6882-13-355
5. Broussard CS, Louik C, Honein MA, Mitchell AA (2010) Herbal use before and during pregnancy. Am J Obstet Gynecol 202(5):443.e1–443.e6. https://doi.org/10.1016/j.ajog.2009.10.865
6. Trabace L, Tucci P, Ciuffreda L et al (2015) "Natural" relief of pregnancy-related symptoms and neonatal outcomes: above all do no harm. J Ethnopharmacol 174:396–402. https://doi.org/10.1016/j.jep.2015.08.046
7. Forster DA, Denning A, Wills G, Bolger M, McCarthy E (2006) Herbal medicine use during pregnancy in a group of Australian women. BMC Pregnancy Childbirth 6:21. https://doi.org/10.1186/1471-2393-6-21
8. Saper RB, Kales SN, Paquin J et al (2004) Heavy metal content of ayurvedic herbal medicine products. JAMA 292(23):2868–2873. https://doi.org/10.1001/jama.292.23.2868
9. Chan K (2003) Some aspects of toxic contaminants in herbal medicines. Chemosphere 52(9):1361–1371. https://doi.org/10.1016/S0045-6535(03)00471-5
10. de Sousa Lima CM, Fujishima MAT, de Paula LB, Mastroianni PC, de Sousa FFO, da Silva JO (2020) Microbial contamination in herbal medicines: a serious health hazard to elderly consumers. BMC Complement Med Ther 20(1):17. https://doi.org/10.1186/s12906-019-2723-1
11. Luo L, Dong L, Huang Q et al (2021) Detection and risk assessments of multi-pesticides in 1771 cultivated herbal medicines by LC/MS-MS and GC/MS-MS. Chemosphere 262:127477. https://doi.org/10.1016/j.chemosphere.2020.127477
12. Luo L, Wang B, Jiang J et al (2020) Heavy metal contaminant in herbal medicines: determination, comprehensive risk assessments, and solutions. Front Pharmacol 11:595335. https://doi.org/10.3389/fphar.2020.595335
13. Medicines and Healthcare products Regulatory Agency (2016) Press Release: Precautionary recall-six batches of St John's Wort Tablets. https://www.gov.uk/government/news/precautionary-recall-six-batches-of-st-johns-wort-tablets. Accessed 1 Nov 2020
14. Center for Disease Control and Prevention (2012) Lead poisoning in pregnant women who used Ayurvedic medications from India – New York City, 2011–2012. Morb Mortal Wkly Rep 61(33):641–646
15. Wong A, Dargan P, Koutsogiannis Z, Sokol J, Ramkrishna J, Greene SL (2015) Chronic Ayurvedic medicine use associated with major and fatal congenital abnormalities. Med J Aust 203(11):443–444. https://doi.org/10.5694/mja15.00636
16. Quansah R, Armah FA, Essumang DK et al (2015) Association of arsenic with adverse pregnancy outcomes/infant mortality: a systematic review and meta-analysis. Environ Health Perspect 123(5):412–421. https://doi.org/10.1289/ehp.1307894

17. Wang H, Li J, Zhang X et al (2018) Maternal serum arsenic level during pregnancy is positively associated with adverse pregnant outcomes in a Chinese population. Toxicol Appl Pharmacol 356:114–119. https://doi.org/10.1016/j.taap.2018.07.030
18. Borja-Aburto VH, Hertz-Picciotto I, Rojas Lopez M, Farias P, Rios C, Blanco J (1999) Blood lead levels measured prospectively and risk of spontaneous abortion. Am J Epidemiol 150(6):590–597. https://doi.org/10.1093/oxfordjournals.aje.a010057
19. González-Cossío T, Peterson KE, Sanín LH et al (1997) Decrease in birth weight in relation to maternal bone-lead burden. Pediatrics 100(5):856–862. https://doi.org/10.1542/peds.100.5.856
20. Nishioka E, Yokoyama K, Matsukawa T et al (2014) Evidence that birth weight is decreased by maternal lead levels below 5 μg/dl in male newborns. Reprod Toxicol 47:21–26. https://doi.org/10.1016/j.reprotox.2014.05.007
21. Jelliffe-Pawlowski LL, Miles SQ, Courtney JG, Materna B, Charlton V (2006) Effect of magnitude and timing of maternal pregnancy blood lead (Pb) levels on birth outcomes. J Perinatol 26(3):154–162. https://doi.org/10.1038/sj.jp.7211453
22. Rothenberg SJ, Schnaas L, Perroni E, Hernández RM, Martínez S, Hernández C (1999) Pre- and postnatal lead effect on head circumference: a case for critical periods. Neurotoxicol Teratol 21(1):1–11. https://doi.org/10.1016/s0892-0362(98)00034-8
23. Hu H, Téllez-Rojo MM, Bellinger D et al (2006) Fetal lead exposure at each stage of pregnancy as a predictor of infant mental development. Environ Health Perspect 114(11):1730–1735. https://doi.org/10.1289/ehp.9067
24. Schnaas L, Rothenberg SJ, Flores M-F et al (2006) Reduced intellectual development in children with prenatal lead exposure. Environ Health Perspect 114(5):791–797. https://doi.org/10.1289/ehp.8552
25. Liu J, Chen Y, Gao D, Jing J, Hu Q (2014) Prenatal and postnatal lead exposure and cognitive development of infants followed over the first three years of life: a prospective birth study in the Pearl River Delta region, China. Neurotoxicology 44:326–334. https://doi.org/10.1016/j.neuro.2014.07.001
26. The Natural Medicines Comprehensive Database. http://naturalmedicines.therapeuticresearch.com. Accessed 1 Nov 2020
27. McLay JS, Izzati N, Pallivalapila AR et al (2017) Pregnancy, prescription medicines and the potential risk of herb-drug interactions: a cross-sectional survey. BMC Complement Altern Med 17(1):543. https://doi.org/10.1186/s12906-017-2052-1
28. British National Formulary. BMJ Group and Pharmaceutical Press. http://www.medicinescomplete.com. Accessed 1 Nov 2020
29. Williamson E, Driver S, Baxter K, Preston CL (eds) Stockley's herbal medicines interactions. Pharmaceutical Press, London. https://about.medicinescomplete.com/publication/stockleys-herbal-medicines-interactions-2/
30. Pharmaceutical Press Editorial. Herbal Medicines. Pharmaceutical Press, London. https://about.medicinescomplete.com/publication/herbal-medicines/
31. European Network of Teratology Information Services (ENTIS). https://www.entis-org.eu/centers. Accessed 1 Nov 2020
32. MotherToBaby, a service of the Organisation of Teratology Information Specialists (OTIS), USA. https://mothertobaby.org/affiliates/. Accessed 1 Nov 2020
33. Best Use of Medicines in Pregnancy, a service of the UK Teratology Information Service (UKTIS). www.medicinesinpregnancy.org. Accessed 1 Nov 2020
34. Kennedy DA, Lupattelli A, Koren G, Nordeng H (2016) Safety classification of herbal medicines used in pregnancy in a multinational study. BMC Complement Altern Med 16:102. https://doi.org/10.1186/s12906-016-1079-z
35. Nordeng H, Bayne K, Havnen GC, Paulsen BS (2011) Use of herbal drugs during pregnancy among 600 Norwegian women in relation to concurrent use of conventional drugs and pregnancy outcome. Complement Ther Clin Pract 17(3):147–151. https://doi.org/10.1016/j.ctcp.2010.09.002
36. Holst L, Wright D, Haavik S, Nordeng H (2011) Safety and efficacy of herbal remedies in obstetrics-review and clinical implications. Midwifery 27(1):80–86. https://doi.org/10.1016/j.midw.2009.05.010

37. Stanisiere J, Mousset P-Y, Lafay S (2018) How safe is ginger rhizome for decreasing nausea and vomiting in women during early pregnancy? Foods 7(4):50. https://doi.org/10.3390/foods7040050
38. Hu Y, Amoah AN, Zhang H et al (2020) Effect of ginger in the treatment of nausea and vomiting compared with vitamin B6 and placebo during pregnancy: a meta-analysis. J Matern Neonatal Med 1–10. https://doi.org/10.1080/14767058.2020.1712714
39. Facchinetti F, Pedrielli G, Benoni G et al (2012) Herbal supplements in pregnancy: unexpected results from a multicentre study. Hum Reprod 27(11):3161–3167. https://doi.org/10.1093/humrep/des303
40. Cuzzolin L, Francini-Pesenti F, Verlato G, Joppi M, Baldelli P, Benoni G (2010) Use of herbal products among 392 Italian pregnant women: focus on pregnancy outcome. Pharmacoepidemiol Drug Saf 19(11):1151–1158. https://doi.org/10.1002/pds.2040
41. Heitmann K, Nordeng H, Holst L (2013) Pregnancy outcome after use of cranberry in pregnancy—the Norwegian Mother and Child Cohort Study. BMC Complement Altern Med 13:345. https://doi.org/10.1186/1472-6882-13-345
42. Wing DA, Rumney PJ, Preslicka CW, Chung JH (2008) Daily cranberry juice for the prevention of asymptomatic bacteriuria in pregnancy: a randomized, controlled pilot study. J Urol 180(4):1367–1372. https://doi.org/10.1016/j.juro.2008.06.016
43. de Rossi P, Cimerman S, Truzzi JC et al (2020) Joint report of SBI (Brazilian Society of Infectious Diseases), FEBRASGO (Brazilian Federation of Gynecology and Obstetrics Associations), SBU (Brazilian Society of Urology) and SBPC/ML (Brazilian Society of Clinical Pathology/Laboratory Medicine): recommendations for the clinical management of lower urinary tract infections in pregnant and non-pregnant women. Braz J Infect Dis 24(2):110–119. https://doi.org/10.1016/j.bjid.2020.04.002
44. Gallo M, Sarkar M, Au W et al (2000) Pregnancy outcome following gestational exposure to echinacea: a prospective controlled study. Arch Intern Med 160(20):3141–3143. https://doi.org/10.1001/archinte.160.20.3141
45. Heitmann K, Havnen GC, Holst L, Nordeng H (2016) Pregnancy outcomes after prenatal exposure to echinacea: the Norwegian Mother and Child Cohort Study. Eur J Clin Pharmacol 72(5):623–630. https://doi.org/10.1007/s00228-016-2021-5
46. European Medicines Agency (2015) European Union herbal monograph on Echinacea purpurea (L.) Moench, herba recens. https://www.ema.europa.eu/en/documents/herbal-monograph/final-european union herbal monograph echinacea-purpurea-l-moench-herba-recens_en.pdf
47. Choi JS, Han JY, Ahn HK et al (2013) Fetal and neonatal outcomes in women reporting ingestion of licorice (Glycyrrhiza uralensis) during pregnancy. Planta Med 79(2):97–101. https://doi.org/10.1055/s-0032-1328102
48. Dugoua J-J, Perri D, Seely D, Mills E, Koren G (2008) Safety and efficacy of blue cohosh (Caulophyllum thalictroides) during pregnancy and lactation. Can J Clin Pharmacol 15(1):e66–e73
49. Dugoua J-J, Seely D, Perri D, Koren G, Mills E (2006) Safety and efficacy of black cohosh (Cimicifuga racemosa) during pregnancy and lactation. Can J Clin Pharmacol 13(3):e257–e261
50. Liu J, Peng C, Zhou Q-M, Guo L, Liu Z-H, Xiong L (2018) Alkaloids and flavonoid glycosides from the aerial parts of Leonurus japonicus and their opposite effects on uterine smooth muscle. Phytochemistry 145:128–136. https://doi.org/10.1016/j.phytochem.2017.11.003
51. Safety, regulation and herbal medicines: a review of the evidence. A report prepared by the UHMAC (HMAC) for the HM and PWG (2014) https://assets.publishing.service.gov.uk/government/uploads/system/uploads/attachment_data/file/545681/HMAC_-_HerbalsafetyOctober2014Final.pdf#:~:text=
52. Gardiner P, Jarrett K, Filippelli A, Pecci C, Mauch M, Jack B (2013) Herb use, vitamin use, and diet in low-income, postpartum women. J Midwifery Womens Health 58(2):150–157. https://doi.org/10.1111/j.1542-2011.2012.00240.x
53. Philipps W (2017) Herbal medicines in pregnancy – pharmacovigilance and risk communication. https://www.bfarm.de/SharedDocs/Downloads/DE/Service/Termine-und-Veranstaltungen/dialogveranstaltungen/dialog_2017/170914/15_Philipps.pdf?__blob=publicationFile&v=3. Accessed 1 Nov 2020
54. WEB-RADR: Recognising adverse drug reactions. https://web-radr.eu/. Accessed 1 Nov 2020

Chapter 7
Advances in Methods and Techniques in Pharmacovigilance for Herbal and Traditional Medicines and Other Natural Health Products

Joanne Barnes

7.1 Introduction

Pharmacovigilance is '*the science and activities relating to the detection, assessment, understanding and prevention of adverse effects or any other drug-related problem*' [1]; it is an essential public health function [2]. Initially, pharmacovigilance was focused on post-marketing surveillance. The first 'early-warning' systems for (mainly) doctors to report their concerns about harms associated with the use of medicines were introduced in the early 1960s; subsequently, for many years, pharmacovigilance relied heavily on monitoring submitted case reports of suspected adverse drug reactions (ADRs) associated with the use of medicines to detect signals of harm(s). Over the last two decades, progress in pharmacovigilance and signal detection for *conventional* medicines has been—and continues to be—swift [3], and contemporary pharmacovigilance involves the pre-marketing phase of medicines development, as well as post-marketing [4], and extends to a broad range of activities, including monitoring the production and quality of medicines [5].

In contrast, while some of these activities are also undertaken for herbal medicines (HMs), pharmacovigilance for these products, and for traditional medicines (TMs) and other natural health products (NHPs), is evolving more slowly. Clearly, pharmacovigilance for HMs/TMs/NHPs stands to benefit from general developments in pharmacovigilance methods, techniques and practices; however, these types of products bring unique challenges to pharmacovigilance that need to be considered. Comprehensive accounts of the methods and techniques used in pharmacovigilance and signal detection for herbal medicines, and the challenges that herbal medicines present for these, have been published previously (see de Smet [6] and Barnes [7]). This chapter reviews methodological advances in herbal

J. Barnes (✉)
School of Pharmacy, Faculty of Medical and Health Sciences, University of Auckland, Auckland, New Zealand
e-mail: j.barnes@auckland.ac.nz

J. Barnes (ed.), *Pharmacovigilance for Herbal and Traditional Medicines*,
https://doi.org/10.1007/978-3-031-07275-8_7

medicines' pharmacovigilance over the last 15–20 years, discusses the challenges in applying developments in pharmacovigilance in general—such as the interest in 'big data' and real-world approaches—to pharmacovigilance for herbal medicines, and considers possible future directions in pharmacovigilance for herbal medicines. The chapter focusses on herbal medicines, but much of the content is also relevant to other traditional medicines and NHPs, so, at times, reference is made to this broader group as HTMs/NHPs, or 'traditional and complementary medicines' (T&CMs), as described by the World Health Organization [8].

7.1.1 The Need for Pharmacovigilance for Herbal Medicines

The need for continuous monitoring of the safety/harms profiles of conventional medicinal products is rarely questioned, but for HTMs/NHPs understanding and acceptance of the importance of pharmacovigilance for these products is less entrenched, at least among some stakeholders. The concept and practice of pharmacovigilance extend to all medicinal-type products, including HTMs/NHPs, as all products have the potential to result in ADRs, including lack of efficacy. For HTMs/NHPs, the high prevalence of use of these products, and the ways in which they are accessed and used, further underpin the need for safety monitoring.

Many HTMs/NHPs are chemically rich complex mixtures, and some of their constituents are known to be toxic; therefore, it is to be expected that some HTMs/NHPs have been associated with ADRs, including serious ADRs, as well as interactions with certain conventional medicines [9]. It is beyond the scope of this chapter to summarise the scientific literature on ADRs associated with HTMs/NHPs, and readers are encouraged to consult authoritative reference texts for summaries of this information (for example, see Herbal Medicines [10], the Natural Medicines Database [11]) and information provided in Chaps. 3–6 of this book).

The use of HTMs/NHPs for health maintenance and well-being, and for the treatment and prevention of symptoms and acute and chronic disease, is extensive and ubiquitous worldwide. The specific HTMs/NHPs used vary between regions and countries, but there is strong evidence that the use of these products/preparations is a popular healthcare choice among patients and consumers across low-, middle- and high-income nations [8]. In some countries, the use of traditional medicines is the only accessible and affordable primary healthcare available to many people [8]. Generally speaking, many people use HTMs/NHPs as self-treatment without involvement of a statutory-registered health professional, although some users may disclose use to their health professionals, particularly if prompted. Some users access HTMs/NHPs through consultations with traditional-medicine or natural-health practitioners. The prevalence and patterns of use of HTMs/NHPs, and of consultations with traditional-medicine and natural-health practitioners who use HTMs/NHPs in their practice, is discussed in Chap. 2 of this book.

Other factors supporting the need for pharmacovigilance for HTMs/NHPs arise from the ways in which these products/preparations are perceived. Many users of

HTMs/NHPs believe these products to be safe, usually because such products are considered to be of natural origin, and users may overlook, or be unaware that, like conventional medicines, HTMs/NHPs can have adverse effects [12, 13]. Thus, some consumers of HTMs/NHPs may be unconvinced, or unaware of, the need for safety monitoring for these products. Likewise, traditional-medicine or natural-health practitioners may focus on the benefits of HTMs/NHPs and be less familiar with ADRs associated with these products/preparations; they may attribute ADRs to other factors and/or may consider adverse reactions a necessary, desirable or unavoidable part of treatment [14–16]. In addition, there may be limited understanding or awareness among traditional-medicine/natural-health practitioners as to why ADRs should be reported and monitored [16, 17].

7.1.2 Advocacy and Drivers for Change in Pharmacovigilance for Herbal Medicines

Advocacy and drivers for improved pharmacovigilance for HTMs/NHPs have come from many stakeholders. These include medicines' regulators, pharmacovigilance professionals, academics, professional societies for pharmacovigilance [7, 12, 18, 19] and others, including representatives of and consultants to the HTMs/NHPs industry [20, 21]. Significantly, the importance of considering and including HTMs/NHPs in pharmacovigilance initiatives is recognised at the international level by health agencies and organisations. In 2004, the WHO published guidelines on pharmacovigilance for herbal medicines [22] and, in 2017/18, a briefing note for policy makers and traditional-medicine practitioners [23]. More recently, the Council for International Organizations of Medical Sciences (CIOMS)—a non-governmental, non-profit organisation established by WHO and the United Nations Educational, Scientific and Cultural Organization (UNESCO)—included a section on liver adverse reactions associated with the use of 'herbal and dietary supplements' in its consensus report on drug-induced liver injury, along with a call for this complex issue to be addressed through a range of pharmacovigilance initiatives [24].

7.1.2.1 Regulatory Pharmacovigilance for Herbal Medicines as a Driver for Change

A key driver in the development of pharmacovigilance activities for herbal medicines, at least with respect to activities and obligations of manufacturers/sponsors, is the introduction of regulatory systems for HTMs/NHPs. Over the last 20 years, several countries, including Australia, Canada, the UK and member states of the European Union (EU), among others, have implemented regulations for herbal medicines, traditional medicines, other (sub)categories of NHPs, or for the broad category of NHPs, or 'complementary medicines'. These regulations usually place obligations on manufacturers to undertake pharmacovigilance activities for their products that are authorised, approved, listed or registered (there are differences in terms and regulatory status for HTMs/NHPs in different countries) under these regulations.

For example, the introduction of the Traditional Herbal Medicinal Products Directive in the EU requires sponsors of traditional herbal medicines registered under Traditional Herbal Medicinal Products Registration schemes (set up in EU member states under this Directive) to undertake pharmacovigilance activities outlined in Directive 2001/83/EC (which relates to medicinal products for human use in the European Community). This means that manufacturers (holders of registrations) of traditional herbal medicines registered under the scheme have the same obligations regarding pharmacovigilance as do marketing authorisation holders for conventional medicines, with the following exception: traditional-use registration (TUR) holders are not required to submit periodic safety update reports (PSURs) for their TUR products, except when this is a condition of a marketing authorisation or requested by a competent authority [25]. Elsewhere, several countries have similar (but not identical) requirements. In Australia, sponsors of medicines registered or listed on the Australian Register of Therapeutic Goods are required to comply with the Australian Therapeutic Goods Administration's guidance on pharmacovigilance responsibilities of medicine sponsors [26, 27]. Pharmacovigilance requirements for herbal medicines are summarised for several other countries in Chaps. 16–27 of this book.

While the introduction of regulatory requirements for pharmacovigilance clearly impacts manufacturers/sponsors of regulated HTMs/NHPs, the effect on other stakeholders, particularly consumers of HTMs/NHPs, traditional-medicine/natural-health practitioners, and health professionals who interact with users of these products, is not known. Differences in perceived, hypothetical ADR reporting behaviour for conventional non-prescription medicines and herbal medicines have been described among users of these products and some types of health professionals [28, 29], but it is not clear whether these differences are due to HTMs/NHPs being considered 'unconventional' products, unregulated products, or both. Investigation of the impact of regulation of HTMs/NHPs on reporting behaviour among consumers, health practitioners and other stakeholders is warranted.

7.1.2.2 Traditional-Medicine/Natural-Health Practitioner-Focused Drivers for Change

Traditional-medicine and natural-health practitioners who administer, sell or supply HTMs/NHPs to their patients clearly have important roles in identifying and reporting suspected ADRs, and in encouraging patients to do the same. There are traditional-medicine and natural-health practitioners who identify and report suspected ADRs, at least in the context of research studies [17, 30]. Also, in the UK, several traditional-medicine and natural-health practitioner organisations demonstrated early leadership in pharmacovigilance for HTMs/NHPs through introducing reporting schemes for their members to report suspected ADRs associated with the use of treatments administered or supplied by them [7, 31–33]. These schemes, which are largely dormant now, fed the ADR reports they received to the UK MHRA for inclusion in its ADR database.

Statutory regulation of traditional-medicine/natural-health practitioners is an important driver for change with respect to pharmacovigilance for HTMs/NHPs. This is because, in many countries, statutory regulation would be expected to involve the development of codes of ethics and practice standards for traditional-medicine/natural-health practitioners that would likely include reference to the participation of the health practitioner in adverse event reporting schemes, as is the expectation for other statutory-regulated health practitioners, such as doctors and pharmacists. For example, in Australia, where there is statutory regulation of Chinese medicine practitioners, the Chinese Medicine Board of Australia (the national board that works with the Australian Health Practitioner Regulation Agency to implement the regulatory framework) has published Guidelines for Safe Chinese Herbal Medicine Practice. These guidelines state: '*It is the professional responsibility of all Chinese medicine practitioners to report suspected adverse events to the Therapeutic Goods Administration (TGA)*' and that reporting of adverse events contributes to knowledge and improved health outcomes for patients [34]. How practitioners interpret these statements, and how this impacts their identification and reporting of suspected ADRs, are not known. Information on how many reports are received by the TGA from Chinese medicine practitioners in Australia is not (publicly) available; based on published reporting statistics for the year 2017 [35], numbers are likely to be very low.

With some notable exceptions, few countries at present have statutory regulation for even some categories of traditional-medicine/natural-health practitioners. Without this, embedding pharmacovigilance activities into routine clinical practice for traditional-medicine/natural-health practitioners is challenging.

7.2 Passive Surveillance in Pharmacovigilance for Herbal Medicines

Passive surveillance in pharmacovigilance is the analysis of unsolicited spontaneous reports of suspected ADRs submitted to national pharmacovigilance centres (NPCs). Spontaneous reporting schemes for health professionals to describe suspected ADRs were first established following, and in response to, the thalidomide tragedy of the late 1950s and early 1960s when over 10,000 babies worldwide were born with phocomelia due to *in-utero* exposure to thalidomide. Many of the national spontaneous ADR reporting schemes established in response to the thalidomide tragedy have now (in 2021) been operational for over 50 years; however, the history of spontaneous reporting of suspected ADRs associated with herbal medicines does not necessarily have the same long history.

In the early years, even decades, of their operation, spontaneous ADR reporting schemes in at least some countries involved reporting for authorised medicinal products only; this is still the case in some countries today [36]. Many HTMs/NHPs were, or remain, unauthorised products, so, effectively, these products were/are not formally monitored by spontaneous reporting schemes. Even in countries with (some) authorised herbal medicines, the inclusion of these products in the national spontaneous ADR reporting scheme may not have been well-publicised. For example, in the UK, for several decades, the inclusion of licensed herbal medicines in the national spontaneous reporting scheme was not promoted until the scheme was extended to include reporting for *unlicensed* herbal remedies in 1996 [7, 37]. Further, at that time in the UK, only doctors, dentists and coroners were formally recognised as reporters of ADRs; it was not until the scheme was extended to include ADR reporting by hospital and, particularly, community pharmacists (in 1997 and 1999, respectively) that greater attention was given to stimulating ADR reporting for herbal medicines. This followed a one-year pilot scheme, in which community pharmacists were found to submit a greater proportion of reports for herbal medicines than did general practitioners (although actual numbers of reports submitted for herbal medicines were very low). Community pharmacists were then asked to focus on reporting suspected ADRs for licensed and unlicensed herbal products and other non-prescription medicines [7, 38]. In the Netherlands, interviews with various stakeholder organisations involved with the vigilance process for 'non-registered healthcare products' (such as 'nutritional products'/food supplements) identified a lack of clear roles and responsibilities, and no underpinning legal basis, with respect to their handling of reports they receive describing suspected ADRs associated with these products [39]. The authors concluded that opportunities exist to improve public protection in relation to these types of products through improvements in transparency and communication about signals of safety concerns associated with non-registered healthcare products, as well as consistency and collaboration at the regional level among EU member states and medicines regulatory agencies.

Globally, there is evidence that an increasing number of countries undertakes pharmacovigilance for at least some types of HTMs/NHPs. In 2005 and 2012, 59 (of 141) and 84 (of 170) WHO member states, respectively, reported that their country had a 'market surveillance system' for herbal medicines [40]. However, several countries have not yet integrated herbal medicines, or 'traditional and complementary medicines' (T&CMs, as referred to by WHO), into their pharmacovigilance systems, or do not have a clearly designated pharmacovigilance system [41]. It is not clear in these cases whether this means that reports of ADRs associated with herbal medicines and T&CMs are simply not submitted by health professionals (and others), or whether they are not processed, if received. In some of these countries, ADR reports for T&CMs may be reported through other systems: for example, it may be mandatory for marketing authorisation holders to report adverse events or defective products to a regulatory committee concerned with traditional medicines [41].

7.2.1 *Developments in Spontaneous Reporting for Herbal Medicines*

Several initiatives aimed at stimulating reporting of ADRs in general (i.e. for all medicines) have been introduced in different countries over the last 20 years or more. These include expansion of recognised reporter groups (in certain countries) to allow ADR reporting by non-physician healthcare practitioners (such as nurses, midwives), as well as the introduction of technical interventions, such as electronic reporting, through the use of on-line reporting forms and reporting applications ('apps').

7.2.1.1 Direct Patient Reporting of ADRs Associated with Herbal and Traditional Medicines

One of the most significant developments in spontaneous reporting has been the introduction in several countries of direct reporting of suspected ADRs by users of medicines, i.e. patients, or consumers. Medicines' users make a valuable contribution to pharmacovigilance for conventional medicines, particularly with respect to providing a patient perspective on experiences of ADRs [42], as well as information on medicines and ADRs different to that reported by healthcare professionals [43]. As HTMs/NHPs are used widely, and usually on a self-care/self-selection basis, the introduction of direct patient reporting for ADRs could be particularly important in the context of HTMs.

There has, however, been only limited investigation of the contribution that patient/consumer reports make to pharmacovigilance for HTMs/NHPs. An analysis of data from the UK 'yellow card scheme' (YCS) for spontaneous reporting of ADRs found that patients submitted significantly more reports of ADRs associated with 'herbal and complementary/alternative medicines (CAMs)' (as well as certain other medicine-group types, classified by Anatomical-Therapeutic-Chemical (ATC) code) than did health professionals [43]. Qualitative analysis of reports from the same dataset revealed some examples of patient reports describing 'extreme symptoms' following use of certain herbal medicines. However, absolute numbers of reports of ADRs associated with 'herbals/CAMs' submitted by patients were low [43].

Limited reporting of ADRs associated with HTMs/NHPs may, in part, be due to low awareness of spontaneous reporting schemes among the general public. A study exploring awareness of the UK Yellow Card Scheme (YCS) among a sample of individuals recruited through the HealthWise Wales online platform found that fewer than one-third of respondents was aware that the YCS applied to herbal and homoeopathic medicines [44]. The study also found that, before receiving an educational intervention about the YCS, fewer than 20% of respondents (a third of whom were health professionals) said they knew how to report an ADR to the YCS.

One of the drivers for implementing direct patient reporting of suspected ADRs was the unique information that patient reports can bring to emerging signals of drug safety concerns. Since 2017, a new feature implemented into the UK YCS online reporting form for direct patient reports is the ability for reporters to select ADR terms from a 'patient-friendly' list. The list was compiled from free-text descriptions of ADRs reported most frequently by patients/consumers and corresponded (at the time) to about 1400 'lower level terms' in the Medical Dictionary for Regulatory Activities (MedDRA) [45, 46]. A study exploring the use of this list by patient/consumer reporters to the UK YCS found that most users described ADRs in their own (free text) words rather than by selecting terms from the 'patient-friendly' terms list. Such lists (including where they are translated into other languages), if they are to be useful for pharmacovigilance professionals and, equally important, relevant for medicines' users, need to evolve to better reflect terms/descriptions of ADRs used by patients [45]. Further, in the context of HTMs/NHPs, it may be important for patients (and HTM/NHP practitioners) to be able to report suspected ADRs using terms used in traditional medicine systems [see Sect. 7.4.2.1 in this chapter], as well as to use terms describing ADRs relating to cultural, spiritual and other dimensions of health that are important aspects of well-being for many people [47]. Other aspects that may be considered relevant by patients and consumers include the social/psychobiosocial impact of ADRs [43, 48, 49] and the severity of ADRs, as experienced by patients [50]. Thus, pharmacovigilance systems in the future may need to be able to accommodate a more pluralistic approach to health and healthcare that is taken by many individuals [51].

At the international level, an analysis of patient/consumer reporting of ADRs associated with HTMs was undertaken in VigiBase using data up to February 2018. Over 50,000 reports for HTMs were identified using the UMC-assigned ATC code 'V90' ('unspecified herbal and traditional medicine', which, in VigiBase, is assigned to every product given an herbal ATC code, or if no ATC (herbal or chemical) is suitable for products with at least one herbal ingredient—see Chap. 9 for further explanation). Of these, 26.7% listed 'consumer' as a reporter, and 51.5% of these listed 'consumer' as the sole reporter [52]. HTMs and ADRs reported most frequently were generally similar for consumer and health professional reports, although consumer-only reports (and not health-professional reports) included ADR terms relating to lack of efficacy (of HTMs) and drug dependence.

Alongside the implementation of direct patient reporting of suspected ADRs has been the introduction in several countries of online public access to (anonymised) brief information about reported suspected ADRs, provided through a searchable (usually by generic drug name, 'active ingredient' or similar) interface on competent authority websites. Drivers for this have often been the need to increase transparency and public access to official information. For example, in New Zealand, Medsafe (the New Zealand competent authority for regulation of medicines and medical devices) launched its 'Suspected Medicine Adverse Reaction Search' function in 2012 to allow public access to data on reported suspected ADRs. (It should be noted that access is to a subset of ADR reports in the database: reports relating to very rare ADRs, and/or those involving drugs rarely reported are excluded to protect patient privacy). Similar systems have been introduced in Australia [53] and Canada [54], among others.

In the UK, the MHRA provides public access to interactive Drug Analysis Profiles (iDAPs) for all licensed drugs, including licensed herbal medicines, and certain other products, including some NHPs (e.g. glucosamine, halibut liver oil) for which it has received reports of suspected ADRs [55]. The iDAPS allow users to filter reports, including by patient characteristics (age group, sex), reporter type (health professional, patient/consumer), seriousness (fatal, other serious, non-serious), year received, MedDRA system organ class, and to view numbers of reports within SOCs down to the preferred-term level. The drug search functionality is limited to selecting the generic name of the drug/substance of interest from an alphabetised list. For herbal medicines, these generic names may be vernacular (e.g. 'hazel', 'deadnettle, white'), genus (e.g. 'harpagophytum', 'hypericum'), or species names (e.g. 'Ginkgo biloba'). The iDAPS for these herbal substances do list proprietary names of single- and multiple-constituent products containing the herbal ingredient of interest and which have been listed in reports of suspected ADRs, but it is not possible to filter reports for a specific proprietary product (on the public-facing system).

At the regional and international levels, Eudravigilance (the European database of suspected ADR reports) provides an interface for public access to its data, which can be viewed for products and substances at a 'line-listing' level [56], and the WHO provides public access to brief information from reports held in VigiBase (the WHO global database of individual case safety reports (ICSRs), maintained by the Uppsala Monitoring Centre (UMC) in Sweden, an independent centre for drug safety and scientific research, on behalf of WHO) through VigiAccess™ [57]. For all these systems, how this type of information is found, used, and interpreted by patients/consumers and health practitioners, and its impact on their behaviour in identifying and reporting suspected ADRs associated with HTMs/NHPs, deserves evaluation. Such research could provide useful insights that could help improve ADR reporting for these products.

7.2.1.2 Promotional Campaigns for Spontaneous Reporting Schemes

Alongside the extension of spontaneous reporting schemes to allow direct patient reporting of ADRs—and central to the success of this development—there has been regional and international investment in the production of promotional materials aimed at raising awareness among the general public and among health professionals of the importance of spontaneous reporting and how to report suspected ADRs. These materials and strategies have included the use of social media campaigns and infographics. Over the years 2013–2017, the UK MHRA led a Europe-wide SCOPE (Strengthening Collaborations to Operate Pharmacovigilance in Europe) Joint Action project, funded by the European Commission and national competent authorities (NCAs). The project aimed to share pharmacovigilance expertise and best practice among EU member states and to develop practical tools for use in pharmacovigilance activities [58]. One of the outputs of the work programme was an ADR reporting 'awareness' toolkit and an EU-wide social media campaign distributed by 21 NCAs [58, 59]. The terminology and animations used in these awareness campaigns to date have comprised the word 'medicines', and images of injections and solid-dose forms, such as manufacturer packs of tablets; how these are interpreted by users of HTMs/NHPs, and what impact the campaigns have on users' behaviour towards identifying and reporting suspected ADRs associated with these products, warrant investigation.

At the international level, the UMC has developed a campaign webpage aimed at informing users of medicines (i.e. patients) about ADRs and encouraging patients to report ADRs they experience to a health professional [60]. The impact that these initiatives have on reporting of suspected ADRs associated with HTMs/NHPs is not known. Other strategies introduced by UMC include an annual Medication Safety week ('MedSafetyWeek') run in conjunction with medicines regulatory authorities around the world and aimed at encouraging reporting of suspected ADRs [61]. Also at the international level, World Patient Safety Day (WPSD) was established in 2019 with the objective of enhancing global understanding, engagement and action on healthcare safety matters [62]. The first WPSD, in 2019, focused on adverse events in healthcare, including preventable harms arising from ADRs [63].

7.2.1.3 Modified Spontaneous Reporting Forms

One development that has clear implications for reporting of ADRs associated with HTMs/NHPs is modifications made to existing ADR reporting forms with the intention of improving the likelihood and quality of collecting data relating to HTMs/NHPs. The limitations of existing reporting forms for collecting precise information relating to implicated and concurrently used HTMs have been considered previously (see Barnes [7] and WHO [22]). In its 2004 guidelines on pharmacovigilance for herbal medicines, the WHO indicated that national pharmacovigilance centres should modify their national reporting forms to facilitate the reporting of suspected ADRs associated with herbal medicines [22]. In conjunction with this recommendation, in 2004, the WHO proposed a model ADR reporting form intended to be applicable to all medicines (including herbal medicines) and vaccines, and which contained additional questions to be completed for herbal medicines listed as being used by the patient concerned [22].

This development is an important step in recognising and raising awareness of this issue, at least with respect to herbal medicines, but falls short of using a broader term that would include all traditional medicines and other NHPs, as well as identifying that such preparations might be self-prepared, or accessed through traditional medicine or natural-health practitioners. While it is not desirable to have long, complex reporting forms, or bespoke forms for different types of products, the lack of certainty around the exact HTMs/NHPs substance(s) involved in ADR reports is a key issue affecting the quality of the information provided by these ADR reports and, ultimately, their usefulness for pharmacovigilance purposes. In the Republic of Korea, a reporting form designed specifically for Korean 'folk medicine' includes several data fields for collecting information on aspects of traditional medicines. These include harvesting time, type of material (e.g. fresh/dried), and type of preparation (e.g. decoction/poultice/tincture), along with prompts for the classification of the material (e.g. plant/animal/fungi/mineral), and its scientific and vernacular names [64].

In several other countries, ADR reporting forms have been refreshed to include, or expand, terms specifically relating to herbal medicines (and other types of NHPs) to prompt the reporter to consider these products at the time of completing the report. For example, the UK MHRA form for ADR reporting by health professionals now includes the term '*including complementary remedies*' [65], and its public-facing reporting form encourages patients/consumers to '*report suspected side effects to any medicines, vaccines, herbal medicines and homeopathic remedies*' as well as referring to (somewhat inconsistently) '*herbal remedies*' elsewhere in the form [66]. In New Zealand, the Centre for Adverse Reactions Monitoring invites reports for 'natural health products', and guidance/forms prompt reporters to include 'over-the-counter', 'alternative medicines' and 'nutritional suppl' use [67, 68]. There is scope for other modifications to ADR reporting forms, supporting documents and processes to improve consistency and further emphasise their applicability to HTMs/NHPs, to allow better capture of product and ingredient information, including using photographs and, perhaps, product barcodes, and submission of retained samples.

7.2.1.4 Bespoke Spontaneous Reporting Schemes for Herbal Medicines

While many countries include ADR reports for HTMs/NHPs in their national database, in some instances dedicated spontaneous reporting schemes for these products have been established. For example, in Italy, an online 'phytovigilance' scheme, coordinated by the Italian National Institute of Health, was introduced in 2002 to stimulate and collect spontaneous reports of suspected ADRs associated with food supplements, herbal products, and compounded preparations containing herbal ingredients [69] . This system—discussed in detail in Chap. 17 of this book—runs separately to the medicines pharmacovigilance system (which also collects reports of ADRs associated with *authorised* or *registered* herbal medicinal products). While having separate pharmacovigilance systems for specific types of products is not usually considered desirable, the apparent success of this scheme in stimulating reporting and identifying signals of safety concerns associated with herbal products may mean this approach warrants further consideration. Such schemes could be more appealing to reporters, particularly patients/consumers. Research exploring this theme could contribute to better understanding of patient reporting for HTMs/NHPs.

7.2.2 *Numbers of Spontaneous Reports of Suspected ADRs Associated with Herbal Medicines*

Despite the initiatives described above, for the most part, numbers of reports of suspected ADRs associated with HTMs/NHPs submitted to many national pharmacovigilance centres remain low and usually comprise only a small proportion of the total number of ADR reports received (see Chaps. 16–27 of this book). For example, an analysis of spontaneous reports of ADRs associated with herbal medicinal products and other 'natural remedies' received by the Swedish Medical Products Agency for the period 2007–2015 inclusive identified 116 reports describing a total of 259 suspected ADRs [70]. Elsewhere, an analysis of almost 75,000 ADR reports received by the Malaysian Centre for ADR Monitoring database over a 15-year period included 930 reports (1.2% of the total number of reports received for the period) for 'CAM products', comprising traditional medicines (including traditional Chinese, Malay and Ayurvedic medicines) and 'health supplements' with at least one CAM ingredient as the suspected agent [71]. In Singapore, for the period 1998–2009, 627 reports of suspected ADRs associated with the use of 'CAM products', including Chinese patent medicines and 'health supplements', were identified in the pharmacovigilance database of the Health Services Authority; this number represented 3.8% of the total number of ADR reports for this period [72]. In India, for the period 2011–2013, only 39 reports of suspected ADRs associated with 'herbal products' were received by the India Pharmacovigilance Program; these reports involved single-ingredient products (e.g. 'turmeric'), as well as multi-ingredient Ayurvedic remedies [73].

The introduction of mandatory reporting of suspected adverse events has resulted in increased reporting of serious adverse events to national adverse event reporting schemes where this has been made mandatory. For example, in the USA, for dietary supplements (which, in the USA, are defined as including products that contain vitamin, mineral, herbs/botanical, amino acids and any concentrate, metabolite, constituent, extract, or combination of any of these ingredients), since 2007, manufacturers have been mandated to submit reports of *serious* adverse events to the Food and Drug Administration (FDA) Center for Food Safety and Applied Nutrition (CFSAN) Adverse Event Reporting System (CAERS) within 15 business days of first receiving notification. An analysis of adverse event reports associated with dietary supplements held in the CAERS database for the period 2004–2013 included 15,430 reports that specified at least one suspected dietary supplement product [74]. Of the reports examined, around two-thirds were mandatory submissions, and the remainder was voluntary submissions. Also in the USA, while not an analysis of reports submitted to the national adverse event reporting system, an analysis of adverse event reports received by two large dietary supplement companies from March 2014 to August 2016 involved 41,121 adverse event reports, of which 203 (0.48%) related to serious adverse events [75]. While the proportion of reported adverse events meeting the definition for a serious adverse event was low, non-serious adverse events can still reflect substantial harm, including impact on quality of life, from the patient's perspective. Also, non-serious adverse events may be important in the detection of signals of more serious concerns [75].

One country that is an exception to low numbers of reports of ADRs associated with HTMs/NHPs being received by NPCs is China (see Chap. 23 of this book). In China, the use of traditional Chinese medicines (TCM) both as manufactured, formulated products and as simple preparations of crude Chinese herbal drugs (from plants and other natural resources) is ubiquitous, and access occurs as part of the main health system. In China, around 10–15% of reports received by the National Centre are for suspected ADRs associated with TCM drugs, mostly manufactured products [76]; ADRs associated with TCM injections account for over 50% of all reported ADRs for TCMs [77].

Pharmacovigilance activities for HTMs/NHPs also occur at the international level. Many countries are members of, and contribute their national spontaneous report data to, the WHO Programme for International Drug Monitoring (PIDM). The PIDM was established with ten member countries in 1968 after the thalidomide tragedy of the 1950s and 1960s; since then, the number of member countries has grown to 148 (at August 2021). PIDM member countries send their national spontaneous reports of suspected ADRs (i.e. ICSRs), in an E2B compatible format, at least quarterly, to the UMC. The reports are stored in VigiBase, the World Health Organization's (WHO) global database of ICSRs, maintained by the UMC on behalf of WHO [78]. The total number of ICSRs in the database is now over 25 million (at August 2021).

A descriptive analysis of VigiBase for the period 1968–1997, when there were 55 member countries of the WHO PIDM, found that around 0.5% of the two million ICSRs held in the database (at the time) involved herbal substances [18]. Many of the newer member countries of the WHO PIDM have a strong tradition of, and, in many cases, a reliance on (at least for primary healthcare) the use of HTMs but have only a very short history of organised pharmacovigilance. A new descriptive analysis of VigiBase reports involving herbal medicines for the period 1968–2019 inclusive, and which includes ICSRs contributed by these newer members of the WHO PIDM, is reported in Chap. 16.

In addition to reports of ADRs associated with HTMs being received by NPCs, in some countries, reports of suspected ADRs associated with these products/preparations are received by poisons control centres, in addition to, or instead of, a report being made to an NPC. Thus, the databases in poisons centres can also be a source of pharmacovigilance information for HTMs/NHPs [79–82].

7.2.3 Challenges in Spontaneous Reporting for Herbal and Traditional Medicines

The strengths and limitations of spontaneous reporting as a method for pharmacovigilance are well documented [83], including in the context of herbal and traditional medicines (see de Smet [6], Barnes [7] and Shaw et al. [84]), and are summarised in Box 7.1.

Box 7.1 Summary of Strengths and Limitations of Spontaneous Reporting Schemes for Suspected Adverse Drug Reactions

Strengths	Limitations
• Provide continuous safety monitoring for all healthcare products, including those not authorised as medicines, such as many herbal and traditional medicines and other natural health products • Can provide early warnings of undocumented safety concerns • Relative to other methods are inexpensive to run	• Under-identification and under-reporting of suspected adverse drug reactions • Poor quality of data provided; forms often do not cater well for herbal and traditional medicines and other natural health products • Spontaneous report data cannot be used to estimate the frequency of suspected ADRs because they do not collect denominator data, i.e. information on the number of individuals exposed to a medicine/healthcare product of interest

Among these issues, perhaps the most challenging to research and to resolve are the perennial problems of under-identification and under-reporting of suspected ADRs to spontaneous reporting systems. These issues occur with almost all medicinal and healthcare products [85] and, for several reasons, may be even more substantial for HTMs/NHPs [7]. As discussed in Sect. 7.1.1 of this chapter, many users of HTMs/NHPs believe these products do not cause adverse effects [13]; this could mean that users holding these beliefs may not recognise ADRs if they occur or may not associate experiences of ADRs with the use of HTMs/NHPs. Patients' motivations for reporting suspected ADRs differ and include altruism and reporting for personal reasons [43, 86, 87]. Some users of HTMs/NHPs do recognise ADRs and associate them with the use of these products but are reluctant to discuss these experiences with health professionals [29, 88]. Users may prefer to disclose these ADRs to health-food-store staff, or to HTMs/NHPs' manufacturers, or not to report at all [89], even for ADRs they perceive to be serious [88]. ADRs reported to those sectors may go unnotified to official channels. For example, health-food-store staff are unlikely to be familiar with ADR reporting systems and may have other barriers or motivators for reporting [90]. Where reports of suspected ADRs reach manufacturers of HTMs/NHPs, those manufacturers are unlikely to be mandated to report suspected ADRs for unauthorised HTMs/NHPs to a relevant competent authority.

Health professionals may also have a passive approach towards identifying and reporting suspected ADRs associated with HTMs/NHPs. Community pharmacists are the most accessible health professionals, and most community pharmacies sell HTMs/NHPs, so pharmacists are well placed to engage with ADR reporting for these products [91]. However, even though community pharmacists are professionally responsible for overseeing sales of HTMs/NHPs in pharmacies in which they practise, in many instances, these transactions are undertaken by non-pharmacist staff members, who may not be aware of, or trained in, identifying and reporting ADRs. Community pharmacists' limited interactions with purchasers of these products may mean that users of HTMs/NHPs do not inform pharmacists about experiences of ADRs [92]. Where users of these products do report suspected ADRs to pharmacists, or to other health professionals, those reports are not necessarily reported on by health professionals to the national spontaneous reporting scheme [28, 92–94]. Reasons for pharmacists under-identification and under-reporting of suspected ADRs associated with HTMs/NHPs include: perceptions that HTMs/NHPs are 'low-risk products; personal professional limitations (e.g. knowledge, training, confidence, competence) with respect to HTMs/NHPs, perceived or actual [91, 94–97]; lack of awareness that spontaneous reporting schemes apply to HTMs/NHPs [28]; lack of awareness about ADRs, including recognising ADRs, and how and what to report [98, 99], particularly for HTMs/NHPs [89, 93, 94, 100]. Reasons for lack of engagement with ADR reporting in general—described by Inman [101] and others—may also apply in the context of HTMs/NHPs [92].

7.2.4 Published Reports of Suspected ADRs Associated with Herbal Medicines

Published case reports, or case series, of ADRs associated with HTMs/NHPs deserve comment, as they represent another manner in which health professionals and others may report, or raise awareness of, their concerns about harms associated with these products. Published individual case reports can be useful in identifying signals and developing hypotheses about safety issues that may require further, more formal investigation. However, publication itself can be perceived to lend credibility to the reports, or imply a causal relationship, that is not necessarily justified [102].

Published case reports/series, including those involving HTMs/NHPs, are of variable quality [103]. There are numerous examples of published case reports of ADRs associated with HTMs/NHPs that omit, or their authors were unable to ascertain, key pieces of information. Typically, details on the implicated/suspected HTM/NHP and exposure to it are insufficient and/or poorly reported. This includes, for example:

- Evidence of exposure to the product may be limited.
- The product/preparation may be described using the common name(s) for HTM/NHP ingredient(s), or, in the case of more crude HTM preparations, there may be no name or indication of the ingredient(s).
- Information on the plant/animal part(s) used, type of extract and so forth may be lacking.

- Labels and/or samples of the preparation may not have been obtained or be unavailable.
- Pharmaceutical analysis of product samples for identification of ingredients and presence/absence of adulterants may not be undertaken.

These issues may, of course, also apply to ADR reports submitted to spontaneous reporting schemes and, in either case, this may have consequences. The lack of, or limited, information may lead to the 'wrong' ingredients being implicated, or HTM/NHP ingredients being implicated instead of adulterants/contaminants. There may also be notoriety bias in favour of, or against, HTM/NHP ingredients/products, including in cases where conventional medicines are being taken alongside HTMs/NHPs. It is also possible, with respect to ADRs associated with conventional medicines, that a reporter (e.g. health professional) may be unaware of a patient's HTMs/NHPs use, and so these products are not considered, or excluded, as a possible cause of the ADR(s).

Guidelines—endorsed by the International Society for Pharmacoepidemiology (ISPE) and the International Society of Pharmacovigilance (ISoP)—aimed at improving the quality of reporting of ADRs, including those associated with HTMs, include reference to information items to consider when writing and submitting reports for publication [102]. Other guidelines, such as the Consolidated Standards for Reporting Trials (CONSORT) extensions for describing herbal medicine interventions [104, 105] and Chinese herbal medicine formulae [106] in randomised trials, also provide useful information for authors that can be applied when preparing for publication case reports of suspected ADRs associated with HTMs/NHPs.

An assessment of the quality of case reports of ADRs associated with herbal medicines published in major scientific medical databases found that the quality of reports had improved over time: the proportion of high-quality reports had increased, and that of low-quality reports had decreased, for reports published between 2006–08 compared with those published during 1986–88 [103]. Whether the publication of the ISPE/ISoP (and other) reporting guidelines had any influence on this apparent improvement in the quality of published case reports of ADRs associated with herbal medicines is not known. To be truly impactful, such guidelines would need to be actively promoted, adopted by journal editors, and applied and enforced in the peer-review process.

An important element to consider with respect to published reports is that they may (or may not) be based on spontaneous reports submitted to a national pharmacovigilance centre. Therefore, whether published reports duplicate, or add to, reports held by formal spontaneous reporting schemes should be explored. The ISPE/ISoP guidelines state that '*authors of adverse event case reports should have reported the case to the appropriate regulatory authority and, if possible, provide the report number to help identify duplicates that might also be included in reports submitted by the authority*' [102]. However, the inclusion of unique identifying numbers in publications for reports that were also submitted to a regulatory authority/national pharmacovigilance centre as spontaneous reports may not be possible for reasons of patient confidentiality and privacy.

7.3 Active Surveillance Methods in Pharmacovigilance for Herbal Medicines

Active surveillance methods aim to identify the number of adverse events occurring during or after treatment with a particular medicine (or other healthcare product) using a pre-determined process, which would usually involve follow-up with patients prescribed the medicine and/or their health professionals, over time. Active surveillance is usually used to explore the adverse event profile for a single medicine or, in some instances, a distinct 'class' of medicines. The latter does not have a clear and consistent equivalent in the context of HTMs/NHPs, in large part due to the chemical complexity and diversity of these types of products.

7.3.1 Active Surveillance Studies Involving Herbal and Traditional Medicines and Other Natural Health Products

Active surveillance methods have been applied to exploring the adverse event profile of specific herbal and traditional medicines. These include numerous formal post-marketing surveillance/observational studies carried out in Germany for specific, authorised herbal medicinal products (HMPs) [107]. Some of these studies achieved patient cohorts involving 10,000 or more individuals. For example, an observational study involving a *Ginkgo biloba* L. leaf extract (LI-1370) included 10,815 patients treated for 3 months, of whom 1.7% were reported to experience 'side effects' [107]. In a separate study, in which 11,296 patients with depressive disorders were treated with an *Hypericum perforatum* L. herb extract (Laif 600), 0.2% and < 0.1% of patients experienced adverse events and 'side effects', respectively [108]. These studies typically recruited cohorts through primary-care physicians in Germany at a time when HMPs were prescribed by physicians, and were reimbursed (which is no longer the case); whether it is possible to enrol such large cohorts today, without this formal process for access in place, is not clear. Further, even these large-scale observational studies involving specific manufactured HMPs have several methodological limitations: such studies typically are conducted to collect data on beneficial as well as adverse health outcomes, and are often used for marketing purposes; the terms used (e.g. adverse events, side effects) are sometimes poorly defined, and methods used for collecting these data are not always described [108]. Further, the absence of any reports of serious adverse events in these studies raises concerns about the reliability of the data, since at least some adverse events would be expected to occur, if only by chance.

Several other observational studies have collected data on adverse events following use of specific categories or types of herbal and traditional medicinal products. A prospective, observational study in Germany—the Evaluation of Anthroposophic Medicine (EVAMED) study—explored the frequency of ADRs associated with prescriptions of anthroposophic medicines. (*Anthroposophic medicines are single- or multiple-ingredient preparations of plant, animal, mineral and, sometimes, isolated chemical compounds. They may be prepared using standard or, in some cases, bespoke procedures. Finished products, which can take numerous pharmaceutical dose forms, including injections, may contain ingredients at standard concentrations and, sometimes, homoeopathic potencies*). In the study, data on prescriptions written for conventional or 'complementary' medicines and reports of serious suspected ADRs were collected from 38 general practitioners and specialists (who were all members of the German National Association of Anthroposophic Physicians) using a web-based system; physicians were remunerated 15 euro for each ADR report submitted [109]. In total, 44,642 patients received a total of 311,731 prescriptions corresponding to 1722 different anthroposophic medicines. Overall, 95 patients experienced a total of 100 ADRs assessed as having a certain, probable or possible relationship with anthroposophic medicine treatment and relating to 83 different anthroposophic medicinal products.

Another analysis using data collected in the EVAMED study focused on products containing extracts obtained from several plants in the Asteraceae family, which have been associated with allergic reactions and cross-sensitivity [110]. Over a two-year period, 50,115 patients with different medical conditions were evaluated, of whom over one-third (n = 18,830) received 25,652 prescriptions for 42,378 preparations containing one or more of the herbal ingredients being monitored. Some preparations were homoeopathic preparations of the ingredients (although not necessarily high dilutions), and some were external use preparations. Among the 18,830 patients who received preparations containing herbal ingredients derived from plants in the Asteraceae family, no serious ADRs were reported. Among a subset of 6961 patients for whom non-serious ADRs were also collected, 11 such ADRs occurred (0.16% of patients); of these, all but one were in relation to the use of homoeopathic preparations at 1:10 or 1:100 dilutions.

A similar approach was piloted in the UK in 2004 in a study involving qualified herbal-practitioner members of the UK Register of Chinese Herbal Medicine (RCHM), a self-regulating professional association representing practitioners of Chinese herbal medicine (CHM). All 549 RCHM members were invited to participate in the study, of whom 13% ($n = 71$) enrolled [30]. Enrolled practitioners were asked to invite ten consecutive patients (regardless of presenting complaint and CHM treatment prescribed) to participate in the study; study activities comprised baseline (patient characteristics, medical history) and 4-week follow-up questionnaires collecting data on exposure (duration and dose form of CHM treatment) and outcomes (contact with health system, AEs experienced) directly from patient participants. Of the 194 eligible patients returning a baseline questionnaire, 144 (74%) completed a 4-week follow-up; of these, 20 patients of 14 practitioners reported a total of 32 AEs, all non-serious, and most frequently diarrhoea, nausea and fatigue [30]. However, the study did not collect true AE data (i.e. all clinical events, regardless of association with treatment) as patient participants were asked to report—from a checklist—those AEs they associated with their use of CHM, and no AE data were collected from CHM practitioners for their patients enrolled in the study.

As with some similar studies, this approach experienced low participation rates for both practitioners and patients, despite including all patients regardless of presenting condition and the CHM treatment received from their practitioner. With these broad inclusion criteria, the study provided preliminary descriptive data on self-reported AEs during or following practitioner-initiated treatment with CHM. However, to provide estimates of the risk of harms associated with specific CHM ingredients, a modified approach, similar to that described below (see Sect. 7.3.2) would be required. A longer follow-up period, as well as including ways of capturing AEs identified or reported to other health providers, would also be important.

An active surveillance study conducted in Taiwan comprised a multi-centre, prospective observational study involving women experiencing menopausal symptoms who were then treated with a multi-ingredient traditional Chinese herbal medicine preparation three times daily for 12 weeks; outcome data relating to adverse events were collected from case report forms, laboratory results and directly from participants [111]. Although described as collecting data on adverse events, it is not clear if this exploratory study truly collected adverse *event* data as participants were provided with a checklist of possible adverse events, rather than being asked to record all health events during or following treatment. The study enrolled 134 participants, among whom 203 adverse events were recorded; of these, 23 participants withdrew from the study, seven due to adverse events (no further information provided). It was not stated how participants were recruited and, clearly, a much larger cohort would be required for robust pharmacoepidemiological analyses.

In Canada, an active surveillance method involving selected community pharmacies was developed to explore the frequency of adverse reactions resulting from drug interactions between NHPs and conventional medicines [112]. The method was subsequently modified and extended to other healthcare settings, including mental health clinics [113, 114]. This body of work is discussed in Chap. 12 of this book.

7.3.2 Intensive Monitoring Methods Involving Herbal and Traditional Medicines and Other Natural Health Products

'Prescription-event' ('drug-event', 'intensive', or 'cohort') monitoring (PEM) methods, used in pharmacovigilance for conventional prescription medicines, are prospective, observational (i.e. non-interventional) cohort studies (usually) involving patients prescribed a specific medicine; information on 'clinical events' (a much broader concept than that of 'suspected ADRs') the patient experiences during, or after, the use of the monitored medicine is collected from questionnaires (usually) sent to patients' GPs [115]. These 'intensive monitoring' studies typically aim to recruit around 10,000 patients who have been prescribed the medicine of interest.

Such studies are usually used to explore the adverse event profile of newly marketed, prescription-only medicines, but have also been used to monitor (prescription-only) medicines meeting other criteria. Detailed accounts of intensive/prescription-event monitoring methods used in NZ and the UK have been published [115]. Intensive monitoring methods have made an important contribution to pharmacovigilance for conventional prescription-only medicines and continue to do so. In the UK, the methodology has developed to comprise modified prescription-event monitoring (M-PEM) [116] and, more recently, specialist cohort event monitoring (SCEM) studies—observational studies that monitor a cohort of patients prescribed a medicine in secondary care (hospital) settings [117]. In NZ, PEM studies were undertaken by the Intensive Medicines Monitoring Programme (IMMP) until its closure in 2014 due to funding issues [118]. Medicines were previously selected for monitoring in IMMP studies based on Medicines Assessment Advisory Committee (MAAC) criteria, which considered factors such as the expected extent and length of exposure, and whether a medicine belonged to a new class of medicines [115].

Applying intensive monitoring methods to monitoring the safety profile of HTMs/NHPs presents several challenges. HTMs/NHPs (in most countries, with notable exceptions of China, Republic of Korea, Japan and several other Asian countries) are not usually prescribed by health professionals, and so identifying and recruiting a patient cohort in the same manner as for conventional (prescription-only) medicines is not possible. Even in countries where HTMs are prescribed by a health professional, or by traditional-medicine or natural-health practitioners, the HTM 'prescription' (or 'formula' for some TM systems) typically comprises multiple ingredients, which may change over a course of treatment, so clearly defining exposure to a single substance of interest is challenging. In addition, most HTMs have a long history of use (although many, of course, are now prepared in ways, and used for indications, that are very different to those of their traditional uses) [119] and are not 'newly marketed', so there may be preconceptions and documented information about their safety profile [7].

Despite these challenges, several innovations in the last 20 years have begun to explore the possibility of applying intensive monitoring/modified PEM methods to monitoring the safety of HTMs. A modified PEM method, developed by the UK Drug Safety Research Unit (the organisation that undertakes PEM, M-PEM and SCEM studies in the UK) in collaboration with the UK National Institute of Medical Herbalists and others, proposed to identify and recruit patients treated with a specific herbal medicine by herbalists, and to collect adverse event (outcome) data from herbalists (and patients' GPs where consent was provided) [120]. The herbal medicine selected for the study was St John's wort (*Hypericum perforatum* L.); however, the proposed study did not achieve the required funding to be operationalised, and so was unable to examine the feasibility of this method, including whether a cohort of sufficient size could be recruited in an acceptable time frame.

Other modified PEM methods applied to HTMs/NHPs have involved using a community-pharmacy-based approach to recruit 'medicines-purchaser' cohorts. Similar approaches have been applied effectively to surveillance of specific *conventional non-prescription* medicines, but not without methodological issues, including difficulties with recruiting pharmacies and medicines users, and the lack of reliable and comprehensive data on exposures [121–124]. Most of these issues also apply to HTMs/NHPs as (typically) non-prescription products [125], alongside other challenges unique to applying these study designs to monitoring HTMs [47].

Studies in the UK and NZ have piloted 'purchaser-event' monitoring for herbal medicines and other NHPs. In the UK, one study piloted a 'purchaser-cohort' monitoring method for products containing *Ginkgo biloba* ('ginkgo') leaf extract purchased in community pharmacies, with the intention of collecting exposure and outcome (all clinical events during or following use of the product) data directly from ginkgo-product purchasers enrolled into the study [126]. A similar study, conducted in NZ, evaluated a web-based, 'purchase-event' intensive monitoring method that aimed to recruit purchasers of ginkgo products in pharmacies, and to collect exposure and outcome data directly from ginkgo-purchaser participants using web-based questionnaires [47]. A subsequent study in NZ recruited purchasers of *any* NHPs in pharmacies as a pragmatic solution to low participation rates in the previous study [47]. However, very few community pharmacists and NHPs purchasers participated in these pharmacy-based, purchase-event monitoring studies and key feasibility issues could not be explored [47, 126]. These issues included, for example, the amount, quality and completeness of exposure and outcome data collected directly from NHP purchasers through web-based questionnaires, and NHP-purchaser provision of key patient-identifier information (such as a patients' name/address, date of birth and, National Health Index (NHI) number, which most NZ citizens/residents have), that (theoretically) would allow data linkage with national health datasets [47].

Reasons for low participation among pharmacists and consumers in pharmacy-based intensive monitoring studies have been considered elsewhere (see Bond and Hannaford [125], Barnes [7] and Barnes et al. [47]). One of the barriers to participation in these studies may have been that, as research studies, individual formal informed consent procedures were required [47, 127]. (In some instances (which will vary from country to country), individual consent may not (at least previously) have been required for observational cohort studies, for example, if conditions were met for a waiver of the need to seek individual informed consent). To address these and other methodological issues, engagement with NHPs users and other stakeholders to explore social, behavioural and human-factors aspects of engagement with this method is needed [128]. This engagement should include consideration of extensions of intensive monitoring methods, for example, to other settings in which NHPs are accessed (such as in health-food stores and through natural-health/traditional medicine practitioners).

As web-based, purchase-event, pharmacy-based, intensive monitoring studies appear impractical at present, other methods of recruiting HTMs users are needed, perhaps through direct enrolment of users of HTMs/NHPs. Such studies may need to collect data on beneficial outcomes, including using patient-reported and person-centred outcome measures, in addition to collecting data on harms [47]. Other methodological issues abound, such as obtaining sufficient information about each enrolled individual's medical history so that bias and confounding can be considered. In some countries, such as NZ, it would be possible to obtain some of this information (e.g. hospital admissions, mortality data) from data-linkage studies with national health datasets, with linkage undertaken using a patient's key identifiers, provided these could be collected. Other challenges include how to select and recruit an appropriate comparison group in order that robust data analyses can be undertaken, and whether future studies based on this approach should recruit existing, or only new, users of the monitored HTM/NHP [47, 125].

7.3.3 Registries

Patient registries are health databases listing individuals possessing one or more defined characteristics, such as having a particular disease/condition, or procedure ('disease registries', e.g. cancer registries, hip replacement registries), or having experienced a particular exposure, e.g. to a specific drug/medicine ('exposure registries', e.g. lenalidomide exposure registry). Pregnancy registries are a particular type of registry that focuses on collecting data on exposures to medicines during and after pregnancy, and exploring health outcomes, including effects on children born to women enrolled in the register. These organised systems use observational study methods to collect data to explore health outcomes. Registries can be powerful tools for pharmacovigilance purposes, particularly for exploring safety aspects for which there is limited information available for new medicines approvals [129, 130]. However, registries are underutilised in general [131] possibly because many are established by pharmaceutical companies, which has implications for data access, and are particularly underutilised for HTMs/NHPs, likely for several reasons.

Ideally, disease registries would include collection of data on all medicines and other health products to which a patient is exposed, including HTMs/NHPs. The extent to which existing disease and pregnancy registries collect data on exposures to HTMs/NHPs, if at all, is not known. Where disease registries do collect data on HTMs/NHPs exposures, the quality and completeness of these data should be included in data quality checks and data should be coded in a systematic manner. As mentioned elsewhere in this chapter, accurately establishing exposures in relation to HTMs/NHPs is problematic, and challenges in coding and classifying these data abound; these issues are discussed in Chaps. 8 and 9 of this book.

While exposure registries could be a useful method of exploring the harms profile of specific HTMs/NHPs, to date, these databases have almost exclusively been established for conventional medicines, or medical devices. Typically, these registries are introduced by manufacturers, or sponsors, of medicinal products, sometimes because this may be imposed by a medicines regulator at the time of approval of a new medicine [129]. As the regulatory frameworks (where they exist) for HTMs/NHPs usually provide for registrations for 'low-risk', non-prescription products, it seems unlikely that a competent authority would impose a requirement for a registry for a specific HTM/NHP at the point of registration/approval. Manufacturers or sponsors of HTMs/NHPs could, of course, take the initiative to establish registries for their products, but may not have the resource, expertise, and/or incentive to do so. If they were to take this approach, manufacturers would need to be aware that, in some jurisdictions, regulatory obligations for post-authorisation safety studies (PASS) apply to product registries initiated by manufacturers/sponsors of products where patient eligibility is determined by exposure to a defined medicinal product [132].

One context in which establishing exposure registries for certain types of HTMs/NHPs is more feasible is where these products are an accepted, integrated part of the health system and even included on lists of funded and/or essential medicines [133]. For example, in China, a registry study has been conducted to explore the harms profile of certain TCM preparations formulated and administered by injection. The registry involves patients from inpatient and outpatient departments of 25 hospitals who received one or more of ten TCM injections during the period 2012–2015 [134]. In China, a stronger focus on evaluating the harms profile of TCM injections has been identified as one of several priority areas with respect to progressing regulatory science for TCMs [135].

Provided the data they hold can be accessed, registries may have potential for evaluating health outcomes associated with the use of HTMs/NHPs since they may use a range of patient-reported outcome measures among the data elements collected. This may be particularly important in the context of HTMs/NHPs, since such products/preparations are often selected by the user, or recommended by traditional-medicine/natural-health practitioners, for their reputed effects on general health and well-being, and for spiritual reasons, which are not well-served by observer/health-professional-reported health outcomes assessments.

7.4 'Real-World' Health Data

'Real-world' data are defined as '*data relating to patient health status and/or the delivery of health care routinely collected from a variety of sources*' [136], including routinely collected healthcare data (such as electronic medical/pharmacy/health records), administrative healthcare databases (such as health insurance claims and billing data), event (births, deaths), disease and product registries, health surveys, observational datasets and networks of observational datasets, as well as data collected through, for example, mobile technologies and social media [137]. In recent years, the term 'big data' has been used increasingly to describe these data sources, and there is substantial interest in big data analytics in healthcare and its potential for improving health outcomes and reducing costs [138]. The potential for big data to bring about transformative change has also been discussed in the context of pharmacovigilance, but, at present, there is good reason to be cautious about the contribution that big data can make to improving the health, safety and well-being of individual patients and the public [139].

In order to understand the contribution that real-world data on HTMs/NHPs exposures and outcomes could make to pharmacovigilance for these products, there is a need for initiatives taking the first steps towards realising routine data collection for HTMs/NHPs and, ultimately, data linkage with, or inclusion in, large, observational datasets. Challenges that require resolution include achieving comprehensive and consistent collection of exposure data given the numerous ways in which HTMs/NHPs can be accessed (including from pharmacies, health-food stores, supermarkets, online, from traditional-medicine/natural-health practitioners), and, as with other methods, robust coding and classification of HTMs/NHPs exposures.

7.4.1 Electronic Medical/Health Record Data

The potential for using structured/coded and unstructured data (e.g. in clinical narratives) held in patients' electronic health records and other 'real-world' data sources has been discussed and explored in the context of conventional medicinal products and remains an active area of research endeavour [140]. Use of these routinely collected data on exposure and health outcomes is an important pharmacovigilance tool for identifying safety concerns associated with medicines; such data are also used for exploring associations between medicines and health benefits. However, there are multiple challenges inherent in the use of electronic health records for pharmacovigilance (and other) purposes for HTMs/NHPs, particularly issues relating to data capture: HTMs/NHPs are usually accessed without the involvement of a registered health professional, HTMs/NHPs' users often do not disclose use to their health professionals, and health professionals rarely record patients' HTMs/NHPs' use and associated health outcomes on patients' health records. Given the widespread and largely undocumented use of HTMs/NHPs, it is important from public health, health economic and, not least, patient perspectives that steps are taken towards achieving routine capture of data on HTMs/NHPs' exposures and health outcomes.

There are some indicators internationally of some progress in this context. The Observational Health Data Sciences and Informatics (OHDSI) collaboration has realised an ambitious vision to create and implement open-source data analysis capability to numerous health databases with the ultimate goal of improving patient health outcomes [141, 142]. The potential that OHDSI may hold for health outcomes research in the context of conventional medicine(s) has been debated [139]. While, in theory, the OHDSI network could be used to undertake collaborative observational studies for HTMs/NHPs, it is not clear whether exposure data for HTMs/NHPs are of sufficient quantity and quality to undertake such studies: there may be methodological issues unique to these HTMs/NHPs that require resolution in order to apply this approach to this unique category of products. For example, it is likely that issues with coding and classification of HTMs/NHPs, described in Chaps. 8 and 9 in this book, will also apply to drug coding terminologies employed in OHDSI and to data available in contributing datasets.

On a much smaller scale, the EVAMED study in Germany, described earlier in this chapter, was able to extract anonymised patient data automatically from electronic medical records held by anthroposophical physician outpatient practices [109]. Physicians documented ADRs using an electronic case report form linked to the physicians' existing EMRs [143]. Identifying the products concerned was possible due to the existence of, and access to, a Federal German Pharmacist Associations database, which contained information on ingredients, pharmaceutical form, among other items, of all medicinal drugs and substances available at the time [110].

Another approach is the use of an integrated data infrastructure that allows for 'whole-of-population' data analyses. NZ is one of only a small number of countries that has such a system; in NZ, this is known as the Integrated Data Infrastructure (IDI). This resource holds longitudinal data—linkable at the individual level—for an 'ever-resident' population (including, e.g. students and temporary workers), with data sourced from government health and other administrative sources, the 2013 census, and several social questionnaire surveys from samples of the population [144]. However, as with other routinely collected data resources, this data resource is not yet useful for exploring associations between the use of HTMs/NHPs and health outcomes: the data include records of (conventional) pharmaceutical medicines dispensed to patients, but there are no records relating to non-prescription-medicine use (including HTMs/NHPs) nor of healthcare accessed outside the organised system, such as visits to traditional-medicine or natural-health practitioners.

7.4.2 *Patient Experience Data*

One approach to capturing data on exposures to non-prescription medicines (including HTMs/NHPs) may lie, in part, in data collection directly from HTMs/NHPs' users, i.e. harvesting patient experience data. Patient experience data have been defined by the USA FDA (see Box 7.2).

Box 7.2 Definition and Description of Patient Experience Data

Patient experience data are defined as including data that *'are collected by any persons (including patients, family members and caregivers of patients, patient advocacy organisations, disease research foundations, researchers, and drug manufacturers); are intended to provide information about patients' experiences with a disease or condition, including the impact (including physical and psychosocial impacts) of such disease or condition, or a related therapy, on patients' lives; and patient preferences with respect to treatment of such disease or condition'.*

Patient experience data include information that *'captures patients' experiences, perspectives, needs, and priorities related to (but not limited to): (1) the symptoms of their condition and its natural history; (2) the impact of the conditions on their functioning and quality of life; (3) their experience with treatments; (4) input on which outcomes are important to them; (5) patient preferences for outcomes and treatments, and (6) the relative importance of any issue as defined by patients'* [145]

A subset of patient experience data is 'patient-generated data', where patients themselves contribute the data, and control the sharing of it; this is not without challenges relating to data collection, linkage, patients' privacy, and the time and commitment required of patients to contribute [146]. 'PatientsLikeMe' (PLM) is a USA-based, online, patient network and health-information-sharing website that collects self-reported exposure (including for HTMs/NHPs) and outcomes data from registered patient users [147]. Patient users can list health conditions and symptoms, treatments used, laboratory test results, health measures, consultations with health professionals, and a daily 'holistic' self-rating score. Users can also record 'side effects', including their severity, and whether any treatment(s) was/were stopped due to 'side effects'. Treatments that patients can list include prescription and non-prescription medicines, surgeries, physical therapies, procedures, exercises, and diets/nutrition, among others. Over 2500 'supplements' (including vitamins, minerals, herbal medicines, dietary supplements, and 'Chinese herbs') are listed, as well as over 200 'complementary and alternative medicines'. Most of the latter are actually CAM therapies/techniques (such as acupuncture, meditation) although some are herbal substances, types of mushroom, essential oils, or proprietary herbal products. Each different description of a product or ingredient appears to be listed as a separate entity, and identifying all 'treatments' or products containing a particular ingredient does not seem possible. Thus, at present, these data have recognised limitations, including fundamental issues with the recording and coding of HTMs/NHPs' exposures, that preclude their use for epidemiological studies and for signal detection purposes. Nevertheless, this dataset is a rich resource of lived patient experiences with healthcare, including the use of and responses to a wide range of HTMs/NHPs. Studies exploring these data—as well as the development and evaluation of similar systems—could bring further insights to understanding the patient perspective on healthcare use.

7.4.2.1 Mobile Applications and Social Media

Several innovations in pharmacovigilance internationally have explored the collection of adverse event data for medicines using web-based mobile applications ('apps'), and mining data shared on social media websites. The realm of the internet, including social media, may present an opportunity for pharmacovigilance for HTMs/NHPs, since it is not limited to the collection and sharing of information on prescription medicines only, as are many other resources: the internet provides users of *any* medicinal/healthcare products, as well as healthcare practitioners, the freedom to share information on their experiences.

Experiences with implementing mobile applications for reporting of suspected ADRs have generally been positive, and uptake and use of this method is expected to increase, alongside the increasing use of smartphones and as medicines' users become more 'app-literate' [148, 149]. For example, the Innovative Medicines Initiative Web-Recognising Adverse Drug Reactions (IMI WEB-RADR) project included the evaluation of mobile app technology for instantaneous reporting of suspected ADRs by patients and healthcare practitioners. The WEB-RADR app also allowed two-way communication—by disseminating personalised medicinal product safety alerts and other information about user-selected medicines—from national pharmacovigilance centres/competent authorities directly to healthcare professionals and the public. It also provides a means for users to search aggregated information on numbers and certain characteristics of ADR reports received by the competent authority [149]. These latter features, along with other considerations, such as protection of data privacy, appeared to be important with respect to achieving user engagement with such apps. As with other systems, how these tools and their features are used and perceived by different users [150], including users of HTMs/NHPs, who may have different views on safety of medicines, and/or hold different perceptions of HTMs/NHPs product safety alerts issued by a competent authority, deserves investigation.

The use of social media to discuss, retrieve and share health information is vast, yet research exploring methods of harvesting and analysing social media data in the context of identifying signals of medicines' safety concerns has only recently begun to gather pace [151]. Studies have shown that adverse events are identifiable in social media [151], but that broad-ranging statistical signal detection using social media is not of value when compared with existing pharmacovigilance approaches [152], at least in the context of conventional medicines. However, the 'worthiness' or added value that mining social media (such as Facebook and Twitter) might bring for HTMs/NHPs, for which existing pharmacovigilance methods and tools have substantial limitations, warrants investigation in its own right. For example, the need for 'custom dictionaries' to reflect colloquial words and terms that may be used by patients has been recognised for conventional medicines [153]; there are additional nuances in this context that require consideration for HTMs/NHPs, along with the well-documented challenges in coding and classifying these types of products/preparations.

For example, many HTMs/NHPs are used in traditional medicine systems that typically use concepts of health, wellness and illness that are not aligned with those of western medicine. These systems have their own extensive vocabularies and describe diagnoses, indications for use and effects of traditional medicines in terms that have no equivalent concepts in western medicine. For example, practitioners and users may refer to, for example, 'heart Qi deficiency' (in traditional Chinese medicine) and 'excess pita' (in Ayurvedic medicine) as explanations for illness. These terms/descriptions may be subject to gross misinterpretation if a western medicine interpretation is attempted [51]. The International Classification of Diseases 11th revision includes, for the first time, a supplementary chapter on traditional medicine conditions, which provides a harmonised classification system for traditional medicine (health) conditions relating to the traditional Chinese, Japanese, and Korean medicine systems; this system originated from the World Health Organization's International Standard Terminologies on Traditional Medicine in the Western Pacific Region [154–156]. The hierarchical system includes at least nine different categories of 'patterns', including 'Organ system patterns', which are further classified into liver-, heart-, spleen-, lung-, and kidney-system patterns. Some examples of 'lowest level terms' from these classifications are 'kidney yin and yang deficiency pattern', 'fire-heat factor pattern', 'lung heat transmitting into the intestine pattern', and 'liver fire flaming upward pattern' [154]. The complexities of developing adverse event search algorithms incorporating these types of terms, and colloquial forms of them, are somewhat obvious but yet to be explored formally.

Ultimately, the special considerations inherent in social media in pharmacovigilance for HTMs/NHPs need to be incorporated into regulatory guidance on this. Principles for the use of social media in pharmacovigilance [157], while developed in the context of conventional medicines, are relevant for manufacturers/sponsors of HTMS/NHPs; MAHs of authorised or registered HTMs/NHPs need to be mindful of these.

7.5 Traditional Observational Study Designs

Methods for case-control and cohort study designs used to examine risks of harm(s) associated with the use of medicinal products, including when conducted using computerised health record databases, are well established, and their strengths and limitations are well documented. The principles of these methods also apply to HTMs, although this category of products brings additional challenges, as discussed by de Smet [6] and Barnes [7]. There are particular issues in defining and establishing exposure to the HTMs/NHPs of interest: there are multiple manufacturers' products and other types of preparations containing the same ingredients, but which will have variations in the profile of their chemical constituents. Also, as mentioned elsewhere in this chapter, these types of products typically are not prescribed, and rarely recorded on patients' electronic medical records, even where patients disclose the use of these products to their healthcare professionals [91, 158].

Several case-control and cohort studies exploring risks of harms associated with certain HTMs/NHPs have been conducted, although these studies typically have included only small numbers of participants and have other important methodological limitations. A review of all studies using traditional pharmacoepidemiological (observational) study designs, including case-control and cohort designs, to assess harms (and/or benefits) associated with specific HTMs/NHPs is beyond the scope of this chapter. Summaries of many of these studies that are available for these products can be found in authoritative reference sources, such as 'Herbal Medicines' [10] and the Natural Medicines Database [11].

7.6 Randomised Clinical Trials

Randomised clinical trials (RCTs) have well-documented strengths and limitations with respect to their use as a method for evaluating the safety profiles of medicines [159], and these apply equally to herbal medicines and related products. Common methodological limitations with respect to RCTs involving herbal medicines (and related healthcare products) interventions include small sample sizes and inadequate collection (and reporting) of data relating to harms. Some RCTs do provide greater certainty about the precise contents of herbal products to which study participants have been exposed, particularly if evidence of botanical authentication of raw ingredient(s) and pharmaceutical analysis of finished product(s) is undertaken and if details are published. In 2006, an herbal medicines extension to the Consolidated Standards of Reporting Trials (CONSORT) statement (2001) was published aimed at improving descriptions of herbal medicine interventions tested in RCTs [104, 105]. A similar extension was published in 2017 for reporting of RCTs involving Chinese herbal medicine formulae [106]. Authors of trials of herbal medicine interventions are encouraged to follow these herbal-specific CONSORT extensions, as well as general CONSORT guidelines on reporting RCTs [160] and harms [161]. However, comprehensive adoption, application and adherence to these guidelines across publications describing clinical trials involving herbal and traditional medicines is yet to occur. Ultimately, improving the quality of reporting of herbal medicine interventions in trials—so that the information can be better used in the context of herbal medicines pharmacovigilance—may require journal editors to insist on the use of these authoritative guidelines.

Many systematic reviews (SRs) and meta-analyses (MAs), including Cochrane SRs, of RCTs testing herbal medicine interventions are now available. Their usefulness in detecting signals, however, is limited at present to reporting relative frequencies of adverse events (although data collection in the original trials may not necessarily have been undertaken and/or reported optimally). Possibly the most comprehensive and robust analysis of adverse event/suspected ADR data from RCTs involving an herbal medicine is presented in the Cochrane SR/MA for 'St. John's wort' (SJW; *Hypericum perforatum* L.) in major depression. The review, which included 29 trials involving a total of 5489 participants, presents forest plots for numbers of patients reporting adverse events, discontinuing treatment/withdrawing from a study due to adverse events, and study withdrawals, for SJW-containing products versus placebo, and for SJW versus standard antidepressant medicines, including sub-analyses for SJW versus 'older' antidepressants, and versus the 'newer' selective serotonergic reuptake inhibitors (SSRIs) [162]. The review also reported a sub-analysis of trials grouped by each specific manufacturer's product/extract for an efficacy outcome. While this was not reported for harms data, it represents a considerable advance in terms of recognition of the variability that may exist in different manufacturers' products. Future presentation of adverse event data in this way is essential to allow for a 'grouping' (where analyses include all products containing a specific herbal ingredient) or a 'splitting' approach (where analyses are undertaken at the level of each specific manufacturer's product), or both, with respect to signal detection.

7.7 Signal Detection in Pharmacovigilance for Herbal Medicines

Signal detection in medicines' safety surveillance began with manually quantifying, or counting, voluntary spontaneous reports of suspected ADRs submitted by (usually) doctors [159]. It has since advanced considerably to include routine use of a range of different data sources, research methods, and sophisticated statistical data mining techniques, such as disproportionality analysis, to identify and quantify signals of drug safety concerns [163–165].

There is evidence in the context of conventional medicines that this approach—the collection, assessment and quantitative analysis of ICSRs—remains an effective one for medicines safety signal detection [3]. However, few countries have applied these statistical signal detection techniques to the identification of safety signals for HTMs/NHPs. This is due, in part, to low numbers of reports for specific manufacturers' products, or for specific herbal substances, although this may change as numbers of reports increase [7]. Thus, at present, at the national level, signal detection for HTMs/NHPs typically relies on more rudimentary indicators, such as an increase in the number of reports for specific herbal substances (or products containing specific ingredients), or another change in the reporting pattern for the herbal-product-suspected ADR combination; this is then followed by manual clinical review and causality assessment for the ICSRs concerned. Case causality assessment remains an essential activity in signal detection and is discussed in the context of herbal medicines in Chaps. 10 and 11 of this book. This approach has resulted in the identification of several safety concerns associated with certain herbal medicines. These include, for example, hepatotoxicity associated with products containing extracts of black cohosh (*Actaea racemosa* L.; synonym: *Cimicifuga racemosa* L.) root/rhizome [166, 167], green tea (*Camellia sinensis* (L.) Kuntze) [168], and, in New Zealand, *Artemisia annua* L. herb [169].

7.7.1 *Application of Statistical Methods for Signal Detection in Pharmacovigilance for Herbal Medicines*

Despite the challenges, some countries have begun to investigate using measures of disproportionality in pharmacovigilance for herbal and traditional medicines. For example, an analysis of ICSRs in the Thai national pharmacovigilance database for the period 2002 (when reports for Thai traditional medicines (TTMs) were first accepted) to 2013 applied statistical signal detection techniques for reports involving TTMs [170]. In total, 502 reports contained TTMs–ADR events, relating to 97 different types of ADR and 58 different TTMs. These data were used to calculate reporting odds ratios (RORs) for specific TTM-ADR pairs, against the background of all drug–ADR pairs in the database. This approach resulted in several significant associations for serious ADRs associated with specific TTMs being identified, although these were still based on very small numbers of reports.

RORs have also been used as a measure of disproportionality to investigate a suspicion of a drug-herb interaction from spontaneous reporting system data collected during 2003–2014 by the China Guangdong Provincial Centre of ADR Monitoring [171]. The suspected interaction was a possible increased risk of anaphylaxis with co-administration of benzylpenicillin and Qingkailing injection. Qingkailing injection is a well-known and frequently used TCM preparation of several herbal, animal and mineral ingredients, as well as several isolated chemical compounds, given intravenously for numerous different medical conditions in adults and children. In China (and some other countries) the combined use of conventional medicines and TCMs is an accepted, expected and routine healthcare approach. The unique issues relating to the potential for using spontaneous report data to detect and further explore signals of suspected ADRs associated with herbal and traditional medicines, including drug-herb and herb-herb interactions, merits further attention.

At the international level, statistical methods for signal detection comprising measures of disproportionality are routinely used in VigiBase for all substances, including HTMs/NHPs, listed on ICSRs as suspected or interacting medicines. In brief, the UMC method for disproportionality analysis calculates 'information component' (IC) values as an indicator of disproportionality in the numbers of observed and expected reports for a specific 'drug'-ADR combination [164]. Signals are identified from VigiBase data through routine monitoring and through 'sprint runs', some of which may have a particular focus (e.g. identifying signals from reports submitted by patients). UMC has also introduced the use of 'vigiRank', a predictive model that uses disproportionate reporting, as well as indicators of the quality and content of ICSRs, as a way of prioritising drug-ADR combinations for manual assessment according to their strength of evidence [172, 173]. Once identified and prioritised, signals are assessed by UMC staff and UMC signal reviewers: the clinical evaluation of signals remains an essential component of the signal detection process. Signals are published in VigiLyze (the UMC's signal detection and management system that is available free of charge to national pharmacovigilance centres in all member countries of the WHO Programme for International Drug Monitoring) and, later, in the WHO Pharmaceuticals Newsletter. Box 7.3 summarises the process, analysis and outcome of a comprehensive signal assessment that was undertaken for the herbal medicine *Ginkgo biloba* L. and the adverse reaction group cardiac arrhythmias [174].

Box 7.3 *Ginkgo biloba* L. and Cardiac Arrhythmias (Extracted and Summarised from Barnes and van Hunsel [174])

Background Ginkgo (*Ginkgo biloba* L.; Ginkgoaceae) has been used in medicine for around 5000 years. Traditional Chinese medicine (TCM) uses the seeds (kernel/nuts) and leaves of ginkgo trees. The chemical constituents of the leaves and seeds of *G. biloba* (Gb) are different, although both contain ginkgolic acids. Standardised concentrated extracts and other formulations of Gb leaves are marketed worldwide, and used for a range of health conditions.

In 2016, an analysis (unpublished) undertaken by the Netherlands Pharmacovigilance Centre and the Uppsala Monitoring Centre (UMC) concluded that the VigiBase* cases ($n = 123$) and literature reports available suggested a signal relating to *Ginkgo biloba* and cardiac arrhythmias.

Methods This new analysis, undertaken at the request of the UMC, involved a data extract (dataset: 11 September 2019) from VigiBase for *Ginkgo biloba* (substance) and Cardiac arrhythmias (MedDRA Standardised MedDRA Query (SMQ) broad). The search included only single-ingredient Gb products; notably, the Cardiac arrhythmias (MedDRA SMQ broad) query includes the preferred term syncope and loss of consciousness, but not dizziness. The observed number of reports for the SMQ was roughly as expected. Case narratives, if present, were not translated except where the authors had some knowledge of the language in which the narrative was written (French, German).

Findings This analysis considered 162 reports, which came from 18 countries; no reports were from China, which is unusual given GB's long history of use in TCM. For all reports where Gb was the sole suspect drug ($n = 92$), there were 46 cases with dechallenge information; of those, 39 had a positive dechallenge. There were 25 reports with a high completeness score (≥ 0.75) and Gb was the sole suspect drug for 20 reports; dechallenge information was given for 14 of these cases, all of which provided some documentation of positive dechallenge. For most of this subset of 14 reports, the specified time to onset of the reactions was within days. Pre-existing cardiac arrhythmias may cause various symptoms, including tinnitus (the reason for the use of ginkgo in 18 of the 162 reports); thus, confounding by indication cannot be excluded.

A mechanism by which Gb could induce cardiac arrhythmias is not clear; however, the number, nature and diversity (geographical origin, range of products implicated) of the reports and published cases indicate a signal [174].

*VigiBase is the World Health Organization's (WHO) global database of individual case safety reports, maintained by the Uppsala Monitoring Centre on behalf of WHO.

The underlying assumptions made in disproportionality analysis and the importance of considering the effects of different 'backgrounds' (i.e. comparison sets used in (dis)proportionality analysis) has been discussed with respect to conventional medicines [175]. Given the unique reporting and other biases presented by herbal medicines and similar products, this issue merits nuanced consideration in respect of this category of products [7]. In 2006, when VigiBase contained around three million ICSRs, the UMC undertook some exploratory data mining for signals associated with certain herbal substances with the analyses conducted against the 'background' (comparator) of all ICSRs for herbal products (rather than the usual 'background' of all reports in the database) [176]. These investigations indicated that further work on appropriate comparators for herbal substances is warranted. Selecting different comparators for herbal medicines (and related products) is not necessarily straightforward: it requires a simple and accurate way of selecting all ICSRs involving, for example, herbal substances in the database. Then, decisions would need to be made as to which reports are included in the 'herbal' background; for example, whether or not to include ICSRs listing multi-ingredient products that contain both herbal and non-herbal substances (including 'conventional' medicines), and reports listing conventional medicines as suspected ingredients.

7.7.2 Challenges in Statistical Methods of Signal Detection for Herbal Medicines

The development of the contemporary science of signal detection has occurred almost exclusively in the context of conventional medicines, initially those comprising a single chemical entity and, more recently, with consideration of the particular challenges presented by new biologic entities (biological therapies, or 'biotherapeutics') and 'biosimilars' used as medicines [177]. Some of the challenges with these contemporary medicinal products, particularly those relating to the production of these highly complex, heterogeneous mixtures, are similar to those identified for herbal medicines. Herbal medicines contain multiple (usually hundreds, sometimes thousands) of different chemical constituents, many of which, even at low concentrations, may have pharmacological and/or toxicological effects. The profile of these constituents is not constant and varies qualitatively (present or absent) and quantitatively (concentration present) in different batches of raw and processed herbal (and other traditional) materials and products. The final profile of constituents varies depending on numerous factors, including the growing and harvesting conditions, processing and extraction methods used, and so forth [7]. Further, the numerous different manufacturers' products containing the same herbal ingredient(s) will have differences in their precise chemical composition; products may contain multiple ingredients, herbal or otherwise, and product formulations and labelled ingredients (whether present or not) may change. Similarly, preparations of herbal ingredients compounded by traditional-medicine and natural-health practitioners will also have substantial differences in their chemical composition. It is beyond the scope of this chapter to discuss in detail the quality of herbal medicines and related products, and readers are guided to authoritative reference sources (such as Heinrich et al. [178]) for overviews on this topic. Clearly, quality has fundamental importance for the safety of the product used by the patient/consumer and has numerous implications for pharmacovigilance [5].

A key question in pharmacovigilance for HTMs/NHPs is, therefore, at what 'level' should signal detection and assessment occur for individual substances? In considering this, it may be useful to reflect upon the view that data mining needs to make the most of the data available, and that signals need to be sufficiently specific to be actionable [176]. For HTMs/NHPs, in different contexts, it might be appropriate to consider the data at different 'levels'. For example, for a particular herbal substance (e.g. *Ginkgo biloba* L. leaf extract) there may be instances where a safety concern (e.g. product quality issue) relates to a specific manufacturer's product only. If data mining is undertaken at the (aggregated) herbal substance level, then a safety concern associated with a sole specific manufacturer's product might only be identified through manual review of reports, and, even then, only if detailed product information was reported. In other contexts (e.g. toxicity suspected to be associated with a specific chemical constituent, or group of constituents, in the herbal ingredient/substance), it could be necessary to consider all reports involving products containing herbal ingredients with those constituents. This approach is similar in some respects to investigating a 'class effect' with conventional medicines. With respect to the latter scenario, the challenges that herbal medicines present rapidly become evident: for many herbal substances, their profile of chemical constituents has not been fully (or, in some cases, not even partially) documented, so identifying herbal ingredients containing constituents of interest is deeply problematic.

Of course, the purpose of pharmacovigilance is to detect signals of previously unrecognised ADRs (or other reasons for harms) and, therefore, prior knowledge of the types of safety concerns that may exist in the reports in the database is not available. Routine data mining is undertaken at a particular 'level'; for herbal medicines, this is (usually) the herbal substance, sometimes further specified by plant part (e.g. root, leaf). Extending the considerations described above suggests that routine signal detection for herbal medicines should, perhaps, include several 'levels' of analysis, although this would have resource implications. The development of standardised medical queries (SMQs) provided by MedDRA (Medical Dictionary for Regulatory Affairs), and other novel term groupings for outcomes, has been a useful development in improving signal detection practices [159]. The feasibility of taking a similar approach and developing 'standardised herbal queries' ('SHQs') in relation to herbal medicines exposures may deserve evaluation. For example, SHQs could be developed based on grouping herbal substances containing the same chemical constituent, or group of constituents (e.g. aristolochic acids, unsaturated pyrrolizidine alkaloids), and studies undertaken to determine whether implementing these would confer any benefits in signal detection for herbal medicines. The question then becomes: which constituent(s) should be monitored? The concept behind SHQs is also similar to that of the Anatomical Therapeutic Chemical (ATC) classification system, which classifies conventional medicines based on the body system/organ on which they act, and their therapeutic, pharmacological and chemical properties. An Herbal-ATC system has been developed using a similar approach, but is not without limitations.

The Herbal-ATC system, the complexities of nomenclature for plants and plant ingredients in herbal medicines, and the implications of this for pharmacovigilance, not least coding and classifying herbal products and their ingredients, are discussed comprehensively in Chaps. 8 and 9 of this book. In short, the many ways in herbal medicines and herbal ingredients are described—such as with common names (e.g.

'echinacea'), Latinised 'pharmaceutical names', scientific names (with or without including the botanical authority), synonyms of scientific names, proprietary product names, and formula names—can lead to substantial ambiguity about the product/material of interest.

Pharmacovigilance methods other than spontaneous reporting, such as intensive monitoring methods, and randomised clinical trials, use different approaches to analysing data to identify potential signals. For example, prescription-event/intensive monitoring studies calculate incidence densities and reporting rates, among other outputs; data from randomised clinical trials of medicines are used to compare frequencies of adverse events between interventions (and/or placebo). With the exception of randomised trials, application of these study designs to pharmacovigilance for HTMs/NHPs is relatively limited.

Several methods have been proposed for detecting signals of safety concerns associated with medicinal products in the context of data mining using healthcare administrative databases. These methods, which include disproportionality analysis, traditional pharmacoepidemiologic study designs, sequence symmetry analysis, and supervised machine learning, each have strengths and limitations, and as yet there is no consensus about the most suitable approach to use [179]. As healthcare administrative databases typically contain data on prescription medicine use only, these signal detection methods cannot yet be applied in this context to safety surveillance for HTMs/NHPs.

An exception to this may be in countries where certain traditional medicine formulations/products are regulated and prescribed (and, in some instances, reimbursed) as part of the healthcare system. In this context, it may be possible to undertake hypothesis-driven studies relating to specific traditional medicines in particular settings [133]. For example, in China, the China Hospital Pharmacovigilance system collects electronic medical records data from 300 hospitals [135, 180]. However, even in these contexts, routine surveillance of electronic medical records and healthcare administrative databases for signal detection for HTMs/NHPs remains a vision. In realising a vision of routine collection of data on all medicinal product exposures, i.e. including all non-prescription medicines, and HTMs/NHPs, a consequence for signal detection is likely be to an increase in the number of false-positive signals and, therefore, an even greater need for ways of distinguishing these [181–183].

Several recommendations for improving signal detection practices were made in 2016 following an Innovative Medicines Initiative project in Europe [159]. While the recommendations made are not limited to particular types of medicines or healthcare products, and several different databases and data sources were used in the project, the evidence-base supporting the recommendations appears to be founded on studies involving conventional pharmaceutical products. Some of the recommendations may require nuanced interpretation in the context of herbal medicines and related products, and some of the outputs, such as the creation of a structured database of the ADR information in sect. 4.8 (Undesirable effects) of the summary of product characteristics document for all European centrally authorised medicinal products, coded using MedDRA [159], have little relevance at this point for most herbal products.

Ultimately, whether or not statistical methods and/or manual methods are used for signal detection, information on safety concerns identified needs to be communicated to stakeholders. Risk communication in the context of herbal medicines is discussed in Chap. 15 of this book.

7.8 Conclusion

Pharmacovigilance for HTMs/NHPs continues to rely almost exclusively on the analysis of unsolicited spontaneous reports of suspected ADRs submitted to national pharmacovigilance centres to identify signals of safety concerns associated with these products. Despite this reliance on spontaneous reporting, numbers of reports involving HTMs/NHPs received by national pharmacovigilance centres remain low in most countries (with some notable exceptions, e.g. China). Pharmacovigilance for HTMs/NHPs has probably benefitted from keen progress and general developments in pharmacovigilance methods, techniques and practices, such as the implementation in recent years of direct patient reporting. However, pharmacovigilance specifically in the context of HTMs/NHPs is evolving far more slowly; it may be years (or decades) before there is the collective will and resource for quantum leaps in pharmacovigilance for these types of products and preparations.

There is, however, evidence of increasing activity in pharmacovigilance for HTMs/NHPs at national and international levels, alongside other signs of progress in the science and practice of pharmacovigilance for this category of products. Specific advances in pharmacovigilance for HTMs/NHPs over the last 15–20 years include the application of active surveillance methods, including intensive monitoring, to obtain more comprehensive information on adverse event profiles of HTMs/NHPs. However, these targeted initiatives have taken place largely in a research context, and some have encountered substantial methodological limitations. These approaches require further development, deeper and wider engagement with stakeholder groups (including HTMs/NHPs' users, traditional-medicine/natural-health practitioners, the HTMs/NHPs industry, and health professionals), as well as adequate resourcing, to allow them to evolve into tangible methods for safety surveillance for this category of products. Further, for these and other methods (such as registries) to be relevant for users of HTMs/NHPs, they may need to incorporate patient-reported/-centred outcome measures relating to the benefits and harms of these products, including those relating to cultural, spiritual and other aspects of well-being, and other outcomes important to patients/consumers. Methods and tools may also need to integrate the extensive vocabularies used in traditional medicine systems that describe diagnoses, indications for use and effects of traditional medicines in terms that have no equivalent concepts in western medicine. In essence, pharmacovigilance systems in the future may need to be able to accommodate a more pluralistic approach to health.

Other areas of future focus in pharmacovigilance for HTMs/NHPs may include initiatives aimed at realising routine capture of data on HTMs/NHPs' exposures and health outcomes and, ultimately, data linkage with, or inclusion in, large, observational datasets. For example, these could include developing ways of capturing electronically exposures to these types of products used in the primary care setting, including in self-treatment, either through including these products in existing electronic medical/health record systems, and/or through bespoke registries using data-linkage approaches to connect these with other datasets. In addition, the potential of the internet, including social media, and other real-world and big data approaches, for detecting signals of safety concerns associated with HTMs/NHPs has not yet been explored.

Many important challenges in pharmacovigilance for HTMs/NHPs remain. Particular issues include collecting information that is sufficiently detailed at the product/preparation level, coding and classifying that information using comprehensive, internationally recognised product and substance dictionaries, and determining at what level (e.g. plant species; specific herbal drug substance; specific manufacturer's product) to apply statistical signal detection techniques to data relating to adverse reactions associated with HTMs/NHPs.

Key drivers for improvements and change in pharmacovigilance for HTMs/NHPs will continue to be high-profile safety concerns associated with HTMs/NHPs, and the introduction and enforcement of regulatory requirements relating to pharmacovigilance activities to be undertaken by manufacturers/sponsors of HTMs/NHPs. Similarly, statutory regulation of traditional-medicine/natural-health practitioners, along with expectations that registered practitioners should meet practice standards with respect to identifying and reporting suspected ADRs, would also contribute to driving progress in pharmacovigilance for HTMs/NHPs. Beyond these (largely) national developments, the continuing recognition of the importance of pharmacovigilance for HTMs/NHPs by international health agencies and organisations (such as the WHO and CIOMS) as well as pharmacovigilance service and research centres, such as the Uppsala Monitoring Centre, will also be pivotal to strengthening pharmacovigilance for HTMs/NHPs.

As pharmacovigilance for HTMs/NHPs evolves and matures, the (sub)discipline may develop into a key strand of regulatory science dedicated to designing, testing, implementing, evaluating and communicating pharmacovigilance strategies, and researching their outcomes and impact, in the context of HTMs/NHPs. This should include exploring initiatives through the different lens(es) relevant to this complex category of products and preparations, most importantly, those of users of HTMs/NHPs, and of health practitioners who administer, sell, supply, and/or recommend HTMs/NHPs to patients or consumers, or who are otherwise responsible for their healthcare. Ultimately, the greatest impact potentially could come from changes in patient/consumer perceptions and behaviour with respect to HTMs/NHPs, their (patients'/consumers') appetite for advocating for stronger vigilance for these products, and their willingness to be part of the solution.

Acknowledgement The author thanks Dr. Mira Harrison-Woolrych and Dr. Aleksandra Milosavljevic for helpful comments on this chapter, and E Lyn Lee for administrative assistance.

References

1. Regulation and Prequalification. World Health Organization. https://www.who.int/teams/regulation-prequalification/regulation-and-safety/pharmacovigilance/health-professionals-info/pidm. Accessed 7 Sept 2021
2. European Medicines Agency (2014) Guidelines on good pharmacovigilance practices (GVP). Introductory cover note, last updated with final definition annex revision 2. EMA/781168/2013, London
3. Arlett P, Straus S, Rasi G (2020) Pharmacovigilance 2030: invited commentary for the January 2020 "futures" edition of clinical pharmacology and therapeutics. Clin Pharmacol Ther 107(1):89–91. https://doi.org/10.1002/cpt.1689
4. Shakir SA, Layton D (2002) Causal association in pharmacovigilance and pharmacoepidemiology: thoughts on the application of the Austin Bradford-Hill criteria. Drug Saf 25(6):467–471. https://doi.org/10.2165/00002018-200225060-00012
5. Sardella M, Belcher G, Lungu C et al (2021) Monitoring the manufacturing and quality of medicines: a fundamental task of pharmacovigilance. Ther Adv Drug Saf 12:1–17. https://doi.org/10.1177/20420986211038436
6. de Smet PAGM (1997) An introduction to herbal pharmacovigilance. In: De Smet PAGM, Keller K, Hansel R, Chandler R (eds) Adverse effects of herbal drugs, vol 3. Springer-Verlag, Berlin, pp 1–13
7. Barnes J (2003) Pharmacovigilance of herbal medicines: a UK perspective. Drug Saf 26(12):829–851. https://doi.org/10.2165/00002018-200326120-00001
8. World Health Organization (2013) WHO traditional medicine strategy: 2014–2023. World Health Organization. https://www.who.int/medicines/publications/traditional/trm_strategy14_23/en/. Accessed 16 Mar 2020
9. Barnes J (2012) Adverse drug reactions and pharmacovigilance of herbal medicines. In: Talbot J, Aronson J (eds) Stephens' detection and evaluation of adverse drug reactions, 6th edn. Wiley, Chichester, pp 645–683
10. Herbal Medicines. Medicines complete. Pharmaceutical Press. https://about.medicinescomplete.com/publication/herbal-medicines/. Accessed 16 Mar 2021
11. Natural Medicines Database. https://naturalmedicines.therapeuticresearch.com/. Accessed 16 Mar 2021
12. Hartigan-Go K (2002) Developing a pharmacovigilance system in the Philippines, a country of diverse culture and strong traditional medicine background. Toxicology 181–182:103–107. https://doi.org/10.1016/s0300-483x(02)00263-9
13. Giveon SM, Liberman N, Klang S et al (2004) Are people who use "natural drugs" aware of their potentially harmful side effects and reporting to family physician? Patient Educ Couns 53(1):5–11. https://doi.org/10.1016/s0738-3991(03)00241-6
14. Dodoo ANO, Appiah-Danquah A, Gyansa-Lutterodt M et al (2006) Safety monitoring of herbal medicines in Ghana: challenges and opportunities. In: Abstracts of the Pharmacovigilance of herbal medicines: current state and future directions, London, UK, 26–28 April 2006
15. Dodoo A, Appiah-Danquah A (2006) Safety of herbal medicines: the practitioners' view. In: Abstracts of the pharmacovigilance of herbal medicines: current state and future directions, London, UK, 26–28 April 2006
16. Teng L, Shaw D, Barnes J (2008) Practice of traditional Chinese herbal medicine shops in central London. Phytochem Lett 1(2):94–98. https://doi.org/10.1016/j.phytol.2008.04.001
17. Bensoussan A, Myers SP, Carlton AL (2000) Risks associated with the practice of traditional Chinese medicine: an Australian study. Arch Fam Med 9(10):1071–1078. https://doi.org/10.1001/archfami.9.10.1071
18. Farah MH, Edwards R, Lindquist M et al (2000) International monitoring of adverse health effects associated with herbal medicines. Pharmacoepidemiol Drug Saf 9(2):105–112. https://doi.org/10.1002/(SICI)1099-1557(200003/04)9:2<105::AID-PDS486>3.0.CO;2-2

19. Barnes J (2020) The International Society of Pharmacovigilance (ISoP) special interest group on herbal and traditional medicines: towards progress in pharmacovigilance for herbal and traditional medicines and other "natural health" products. Drug Saf 43(7):619–622. https://doi.org/10.1007/s40264-020-00937-0
20. Schmitz SM, Lopez HL, MacKay D (2014) Nutravigilance: principles and practices to enhance adverse event reporting in the dietary supplement and natural products industry. Int J Food Sci Nutr 65(2):129–134. https://doi.org/10.3109/09637486.2013.836743
21. Mills SY (2006) The ESCOP perpsective on ADRs and ADR reporting. In: Abstracts of the Pharmacovigilance of herbal medicines: current state and future directions, London, UK, 26–28 April 2006
22. World Health Organization (2004) WHO guidelines on safety monitoring of herbal medicines in pharmacovigilance systems. World Health Organization. https://apps.who.int/medicinedocs/documents/s7148e/s7148e.pdf. Accessed 16 Mar 2020
23. World Health Organization Regional Office for South-East Asia (2018) Pharmacovigilance for traditional medicine products: why and how? World Health Organization Regional Office for South-East Asia. https://apps.who.int/iris/handle/10665/259854. Accessed 24 Aug 2021
24. Council for International Organizations of Medical Sciences (CIOMS) (2020) Drug-induced liver injury (DILI): Current status and future directions for drug development and the post-market setting. A consensus by a CIOMS Working Group. Council for International Organizations of Medical Sciences (CIOMS), Geneva
25. European Medicines Agency (2017) Traditional herbal medicine products and simplified registrations for homeopathic medicinal products: pharmacovigilance requirements and EudraVigilance access. Note for clarification. EMA/80556/2017. Inspections, Human Medicines, Pharmacovigilance and Committees Division, London
26. Larter C (2019) Pharmacovigilance and complementary medicines. Regulatory requirements. Therapeutic Goods Administration, Department of Health, Australia
27. Therapeutic Goods Administration (2021) Pharmacovigilance responsibilities of medicine sponsors. Australian recommendations and requirements Version 2.2. Therapeutic Goods Administration, Department of Health, Australia
28. Barnes J (2001) An examination of the role of the pharmacist in the safe, effective and appropriate use of complementary medicines. Dissertation. University of London, London
29. Barnes J, Mills SY, Abbot NC et al (1998) Different standards for reporting ADRs to herbal remedies and conventional OTC medicines: face-to-face interviews with 515 users of herbal remedies. Br J Clin Pharmacol 45(5):496–500. https://doi.org/10.1046/j.1365-2125.1998.00715.x
30. MacPherson H, Liu B (2005) The safety of Chinese herbal medicine: a pilot study for a national survey. J Altern Complement Med 11(4):617–626. https://doi.org/10.1089/acm.2005.11.617
31. Broughton AL (2006) Adverse event reporting by herbal practitioners: The National Institute of Medical Herbalists yellow card reporting scheme. In: Abstracts of the Pharmacovigilance of herbal medicines: current state and future directions, London, UK, 26–28 April 2006
32. Booker AJ (2006) The register of Chinese herbal medicine card scheme. In: Abstracts of the Pharmacovigilance of herbal medicines: current state and future directions, London, UK, 26–28 April 2006
33. Howie JA (2006) Adverse events and aromatherapy practice: a UK initiative on reporting. In: Abstracts of the Pharmacovigilance of herbal medicines: current state and future directions, London, UK, 26–28 April 2006
34. Chinese Medicine Board of Australia (2015) Guidelines for safe Chinese herbal medicine practice, Melbourne
35. Therapeutic Goods Administration (2018) Medicines and vaccines postmarket vigilance. Statistics for 2017 Version 1.0. Therapeutic Goods Administration, Department of Health, Australia
36. Kim M, Woo Y, Han C-H (2021) Current status of the spontaneous reporting and classification/coding system for herbal and traditional medicine in pharmacovigilance. Integr Med Res 10(1):100467–100467. https://doi.org/10.1016/j.imr.2020.100467

37. Anonymous (1996) Extension of the yellow card scheme to unlicensed herbal remedies. Curr Prob Pharmacovig 22:10
38. Davis S, Coulson R (1999) Community pharmacist reporting of suspected ADRs: (1) the first year of the yellow card demonstration scheme. Pharm J 263:786–788
39. de Boer A, Geboers L, van de Koppel S et al (2021) Governance of nutrivigilance in the Netherlands: reporting adverse events of nonregistered products. Health Policy [first published online] https://doi.org/10.1016/j.healthpol.2022.05.011
40. World Health Organization (2019) WHO Global report on traditional and complementary medicine 2019. World Health Organization. https://www.who.int/traditional-complementary-integrative-medicine/WhoGlobalReportOnTraditionalAndComplementaryMedicine2019.pdf?ua=1. Accessed 24 Apr 2020
41. World Health Organization (2019) Pharmacovigilance and traditional and complementary medicine in South-East Asia: a situation review. World Health Organization. https://apps.who.int/iris/bitstream/handle/10665/325982/9789290227250-eng.pdf. Accessed 24 Apr 2020
42. Anderson C, Krska J, Murphy E et al (2011) The importance of direct patient reporting of suspected adverse drug reactions: a patient perspective. Br J Clin Pharmacol 72(5):806–822. https://doi.org/10.1111/j.1365-2125.2011.03990.x
43. Avery AJ, Anderson C, Bond CM et al (2011) Evaluation of patient reporting of adverse drug reactions to the UK 'Yellow Card Scheme': literature review, descriptive and qualitative analyses, and questionnaire surveys. Health Technol Assess 15(20):1–234. https://doi.org/10.3310/hta15200
44. Bracchi RC, Tseliou F, Copeland L et al (2021) Public awareness in Wales of the UK Yellow Card scheme for reporting suspected adverse drug reactions. Br J Clin Pharmacol 87(8):3344–3348. https://doi.org/10.1111/bcp.14726
45. de Vries ST, Harrison J, Revelle P et al (2019) Use of a patient-friendly terms list in the adverse drug reaction report form: a database study. Drug Saf 42(7):881–886. https://doi.org/10.1007/s40264-019-00800-x
46. MedDRA. Medical Dictionary for Regulatory Activities. https://www.meddra.org/. Accessed 7 Sept 2021
47. Barnes J, Sheridan JL, Dong CY et al (2020) Evaluation of a web-based, 'purchase event' intensive monitoring method for pharmacovigilance of natural health products: lessons and insights from development work in New Zealand. Drug Saf 43(10):981–998. https://doi.org/10.1007/s40264-020-00963-y
48. Castillon G, Salvo F, Moride Y (2019) The social impact of suspected adverse drug reactions: an analysis of the Canada vigilance spontaneous reporting database. Drug Saf 42(1):27–34. https://doi.org/10.1007/s40264-018-0713-8
49. Riera-Arnau J, Alvarado Aguirre LA, Garcia Doladé N et al (2021) Patients' contribution to drug safety in Catalonia: the interest of personal feelings on adverse drug reactions. Eur J Clin Pharmacol 77(4):637–642. https://doi.org/10.1007/s00228-020-03033-5
50. Rolfes L, Haaksman M, van Hunsel F et al (2020) Insight into the severity of adverse drug reactions as experienced by patients. Drug Saf 43(3):291–293. https://doi.org/10.1007/s40264-019-00890-7
51. Green G, Bradby H, Chan A et al (2006) "We are not completely Westernised": dual medical systems and pathways to health care among Chinese migrant women in England. Soc Sci Med 62(6):1498–1509. https://doi.org/10.1016/j.socscimed.2005.08.014
52. van Hunsel F, Skalli S, Barnes J (2018) Consumers' contributions to pharmacovigilance for herbal medicines: analyses of global reports in Vigibase. In: Abstracts of the 18th ISoP Annual Meeting "Pharmacovigilance without borders", Geneva
53. Database of Adverse Event Notifications (DAEN). Therapeutic Goods Administration, Department of Health, Australian Government. https://www.tga.gov.au/database-adverse-event-notifications-daen. Accessed 30 Jul 2021
54. Canada Vigilance Adverse Reaction Online Database. Health Canada. https://www.canada.ca/en/health-canada/services/drugs-health-products/medeffect-canada/adverse-reaction-database.html. Accessed 30 Jul 2021

55. Interactive Drug Analysis Profiles. Medicines and Healthcare products Regulatory Agency. https://yellowcard.mhra.gov.uk/iDAP/. Accessed 30 Jul 2021
56. Eudravigilance—the European database of suspected adverse drug reaction reports. European Medicines Agency. https://www.adrreports.eu/en/index.html. Accessed 30 Jul 2021
57. VigiAccess. Uppsala Monitoring Centre and World Health Organization. http://www.vigiaccess.org/. Accessed 30 Jul 2021
58. Loughlin L (2017) Strengthening collaborations for operating pharmacovigilance in Europe joint action. In: 11th stakeholder forum on the pharmacovigilance legislation. https://www.ema.europa.eu/en/documents/presentation/presentation-strengthening-collaborations-operating-pharmacovigilance-europe-joint-action-l-loughlin_en.pdf. Accessed 23 Apr 2021
59. MHRA Campaigns. https://yellowcard.mhra.gov.uk/campaigns/. Accessed 23 Apr 2021
60. Getting patients involved in medicines safety. Uppsala Monitoring Centre. https://www.who--umc.org/global-pharmacovigilance/pharmacovigilance-communications/taketell/. Accessed 23 Apr 2021
61. Raising pharmacovigilance awareness on social media. Uppsala Monitoring Centre. https://www.who-umc.org/global-pharmacovigilance/pharmacovigilance-communications/med-safetyweek/. Accessed 25 Aug 2021
62. Announcing World Patient Safety Day 2021. World Health Organization. https://cdn.who.int/media/docs/default-source/world-patient-safety-day/announcing-world-patient-safety-day-2021.pdf?sfvrsn=a5e55c4a_32. Accessed 25 Aug 2021
63. Patient Safety. World Health Organization. https://www.who.int/news-room/fact-sheets/detail/patient-safety. Accessed 25 Aug 2021
64. Park JH, Choi SM, Moon S et al (2017) Development of an adverse events reporting form for Korean folk medicine. Pharmacoepidemiol Drug Saf 26(5):498–508. https://doi.org/10.1002/pds.4077
65. Medicines and Healthcare products Regulatory Agency Healthcare professional yellow card reporting form. https://yellowcard.mhra.gov.uk/_assets/files/Healthcare-Professional-Yellow-Card-Reporting-Form-(July-2019).pdf. Accessed 23 Apr 2021
66. Medicines and Healthcare products Regulatory Agency Patient yellow card form. https://yellowcard.mhra.gov.uk/_assets/files/Patient-YC-form.pdf. Accessed 23 Apr 2021
67. New Zealand Pharmacovigilance Centre. https://nzphvc.otago.ac.nz/. Accessed 25 Aug 2021
68. CARM Reporting card. https://nzphvc.otago.ac.nz/wp-content/uploads/2013/12/Report.pdf. Accessed 25 Aug 2021
69. Menniti-Ippolito F, Mazzanti G, Santuccio C et al (2008) Surveillance of suspected adverse reactions to natural health products in Italy. Pharmacoepidemiol Drug Saf 17(6):626–635. https://doi.org/10.1002/pds.1566
70. Svedlund E, Larsson M, Hägerkvist R (2017) Spontaneously reported adverse reactions for herbal medicinal products and natural remedies in Sweden 2007–15: report from the Medical Products Agency. Drugs Real World Outcomes 4(2):119–125. https://doi.org/10.1007/s40801-017-0104-y
71. Rahman S, Aziz S (2020) Complementary and alternative medicine: Pharmacovigilance in Malaysia and predictors of serious adverse reactions. J Clin Pharm Ther 45(5):946–958. https://doi.org/10.1111/jcpt.13106
72. Patel DN, Low WL, Tan LL et al (2012) Adverse events associated with the use of complementary medicine and health supplements: an analysis of reports in the Singapore Pharmacovigilance database from 1998 to 2009. Clin Toxicol (Phila) 50(6):481–489. https://doi.org/10.3109/15563650.2012.700402
73. Kalaiselvan V, Saurabh A, Kumar R et al (2015) Adverse reactions to herbal products: an analysis ofspontaneous reports in the database of the pharmacovigilance programme of India. J Herbal Med 5(1):48–54. https://doi.org/10.1016/j.hermed.2015.01.006
74. Timbo BB, Chirtel SJ, Ihrie J et al (2018) Dietary supplement adverse event report data from the FDA Center for Food Safety and applied nutrition adverse event reporting system (CAERS), 2004–2013. Ann Pharmacother 52(5):431–438. https://doi.org/10.1177/1060028017744316

75. Schmitz SM, Hl L, Mackay D et al (2020) Serious adverse events reported with dietary supplement use in the United States: a 2.5 year experience. J Diet Suppl 17(2):227–248. https://doi.org/10.1080/19390211.2018.1513109
76. Zhang L, Yan J, Liu X et al (2012) Pharmacovigilance practice and risk control of Traditional Chinese Medicine drugs in China: current status and future perspective. J Ethnopharmacol 140(3):519–525. https://doi.org/10.1016/j.jep.2012.01.058
77. Li H, Wang S, Yue Z et al (2018) Traditional Chinese herbal injection: current status and future perspectives. Fitoterapia 129:249–256. https://doi.org/10.1016/j.fitote.2018.07.009
78. Uppsala Monitoring Centre WHO Programme for International Drug Monitoring. https://www.who-umc.org/global-pharmacovigilance/who-programme-for-international-drug-monitoring/ Accessed 27 Apr 2021
79. Shaw D, Leon C, Kolev S et al (1997) Traditional remedies and food supplements. A 5-year toxicological study (1991–1995). Drug Saf 17(5):342–356. https://doi.org/10.2165/00002018-199717050-00006
80. Dennehy CE, Tsourounis C, Horn AJ (2005) Dietary supplement-related adverse events reported to the California Poison Control System. Am J Health Syst Pharm 62(14):1476–1482. https://doi.org/10.2146/ajhp040412
81. Gardiner P, Sarma DN, Dog TL et al (2008) The state of dietary supplement adverse event reporting in the United States. Pharmacoepidemiol Drug Saf 17(10):962–970. https://doi.org/10.1002/pds.1627
82. Chan TY (2016) Herbal medicines induced anticholinergic poisoning in Hong Kong. Toxins (Basel) 8(3). https://doi.org/10.3390/toxins8030080
83. Mann R, Andrews EB (2014) Introduction: updated from second edition. In: Andrews EB, Moore N (eds) Mann's pharmacovigilance, 3rd edn. John Wiley & Sons, Chichester, pp 1–10
84. Shaw D, Graeme L, Pierre D et al (2012) Pharmacovigilance of herbal medicine. J Ethnopharmacol 140(3):513–518. https://doi.org/10.1016/j.jep.2012.01.051
85. Hazell L, Shakir SA (2006) Under-reporting of adverse drug reactions: a systematic review. Drug Saf 29(5):385–396. https://doi.org/10.2165/00002018-200629050-00003
86. Jacobs TG, Hilda Ampadu H, Hoekman J et al (2018) The contribution of Ghanaian patients to the reporting of adverse drug reactions: a quantitative and qualitative study. BMC Public Health 18(1):1384. https://doi.org/10.1186/s12889-018-6285-9
87. Hartigan-Go K (2015) Empowering consumers as contributors for health product safety: lessons from the Philippines. Drug Saf 38(4):329–335. https://doi.org/10.1007/s40264-015-0274-z
88. Walji R, Boon H, Barnes J et al (2010) Consumers of natural health products: natural-born pharmacovigilantes? BMC Complement Altern Med 10:8. https://doi.org/10.1186/1472-6882-10-8
89. Chiba T, Sato Y, Kobayashi E et al (2017) Behaviors of consumers, physicians and pharmacists in response to adverse events associated with dietary supplement use. Nutr J 16(1):18. https://doi.org/10.1186/s12937-017-0239-4
90. Walji R, Boon H, Barnes J et al (2009) Adverse event reporting for herbal medicines: a result of market forces. Healthc Policy 4(4):77–90
91. Barnes J, Abbot NC (2007) Professional practices and experiences with complementary medicines: a cross-sectional study involving community pharmacists in England. Int J Pharm Pract 15(3):167–175. https://doi.org/10.1211/ijpp.15.3.0003
92. Barnes J, Butler R (2020) Community pharmacists' views and experiences with ADR reporting for complementary medicines: a qualitative study in New Zealand. Drug Saf 43(11):1157–1170. https://doi.org/10.1007/s40264-020-00980-x
93. Green CF, Mottram DR, Raval D et al (1999) Community pharmacists' attitudes to adverse drug reaction reporting. Int J Pharm Pract 7(2):92–99. https://doi.org/10.1111/j.2042-7174.1999.tb00955.x
94. Walji R, Boon H, Barnes J et al (2011) Reporting natural health product related adverse drug reactions: is it the pharmacist's responsibility? Int J Pharm Pract 19(6):383–391. https://doi.org/10.1111/j.2042-7174.2011.00150.x

95. Kanjanarach T, Krass I, Cumming RG (2006) Exploratory study of factors influencing practice of pharmacists in Australia and Thailand with respect to dietary supplements and complementary medicines. Int J Pharm Pract 14(2):123–128. https://doi.org/10.1211/ijpp.14.2.0006
96. Charrois TL, Hill RL, Vu D et al (2007) Community identification of natural health product-drug interactions. Ann Pharmacother 41(7):1124–1129. https://doi.org/10.1345/aph.1H463
97. Kheir N, Gad HY, Abu-Yousef SE (2014) Pharmacists' knowledge and attitudes about natural health products: a mixed-methods study. Drug Healthc Patient Saf 6:7–14. https://doi.org/10.2147/DHPS.S57756
98. Elkalmi RM, Hassali MA, Ibrahim MIM et al (2011) A qualitative study exploring barriers and facilitators for reporting of adverse drug reactions (ADRs) among community pharmacists in Malaysia. J Pharm Health Serv Res 2(2):71–78. https://doi.org/10.1111/j.1759-8893.2011.00037.x
99. Li R, Curtain C, Bereznicki L et al (2018) Community pharmacists' knowledge and perspectives of reporting adverse drug reactions in Australia: a cross-sectional survey. Int J Clin Pharm 40(4):878–889. https://doi.org/10.1007/s11096-018-0700-2
100. Tabali M, Jeschke E, Bockelbrink A et al (2009) Educational intervention to improve physician reporting of adverse drug reactions (ADRs) in a primary care setting in complementary and alternative medicine. BMC Public Health 9:274. https://doi.org/10.1186/1471-2458-9-274
101. Inman WH (1996) Attitudes to adverse drug reaction reporting. Br J Clin Pharmacol 41(5):434–435
102. Kelly WN, Arellano FM, Barnes J et al (2007) Guidelines for submitting adverse event reports for publication. Drug Saf 30(5):367–373. https://doi.org/10.2165/00002018-200730050-00001
103. Hung SK, Hillier S, Ernst E (2011) Case reports of adverse effects of herbal medicinal products (HMPs): a quality assessment. Phytomedicine 18(5):335–343. https://doi.org/10.1016/j.phymed.2010.07.007
104. Gagnier JJ, Boon H, Rochon P et al (2006) Recommendations for reporting randomized controlled trials of herbal interventions: explanation and elaboration. J Clin Epidemiol 59(11):1134–1149. https://doi.org/10.1016/j.jclinepi.2005.12.020
105. Gagnier JJ, Boon H, Rochon P et al (2006) Reporting randomized, controlled trials of herbal interventions: an elaborated CONSORT statement. Ann Intern Med 144(5):364–367. https://doi.org/10.7326/0003-4819-144-5-200603070-00013
106. Cheng CW, Wu TX, Shang HC et al (2017) CONSORT extension for Chinese herbal medicine formulas 2017: recommendations, explanation, and elaboration (traditional Chinese version). Ann Intern Med 167(2):W7–W20. https://doi.org/10.7326/IsTranslatedFrom_M17-2977_1
107. Schulz V, Hänsel R, Tyler VE (2001) Rational phytotherapy: a physicians' guide to herbal medicine, 4th edn. Springer-Verlag, Berlin, Heidelberg
108. Linde K, Knüppel L (2005) Large-scale observational studies of hypericum extracts in patients with depressive disorders: a systematic review. Phytomedicine 12(1–2):148–157. https://doi.org/10.1016/j.phymed.2004.02.004
109. Hamre HJ, Glockmann A, Heckenbach K et al (2017) Use and safety of anthroposophic medicinal products: an analysis of 44,662 patients from the EvaMed pharmacovigilance network. Drugs Real World Outcomes 4(4):199–213. https://doi.org/10.1007/s40801-017-0118-5
110. Jeschke E, Ostermann T, Luke C et al (2009) Remedies containing Asteraceae extracts: a prospective observational study of prescribing patterns and adverse drug reactions in German primary care. Drug Saf 32(8):691–706. https://doi.org/10.2165/00002018-200932080-00007
111. Hsieh S-C, Lai J-N, Chen P-C et al (2006) Development of active safety surveillance system for traditional Chinese medicine: an empirical study in treating climacteric women. Pharmacoepidemiol Drug Saf 15(12):889–899. https://doi.org/10.1002/pds.1340
112. Vohra S, Cvijovic K, Boon H et al (2012) Study of natural health product adverse reactions (SONAR): active surveillance of adverse events following concurrent natural health product and prescription drug use in community pharmacies. PLoS One 7(9):e45196. https://doi.org/10.1371/journal.pone.0045196

113. Sparks E, Zorzela L, Necyk C et al (2020) Study of Natural products Adverse Reactions (SONAR) in children seen in mental health clinics: a cross-sectional study. BMJ Paediatr Open 4(1):e000674. https://doi.org/10.1136/bmjpo-2020-000674
114. Zorzela L, Khamba B, Sparks E et al (2021) Study of Natural Products Adverse Reactions (SONAR) in adults with mental health conditions: a cross-sectional study. Drug Saf 44(9):999–1006. https://doi.org/10.1007/s40264-021-01092-w
115. Harrison-Woolrych M, Coulter DM (2007) PEM in New Zealand. In: Mann R, Andrews EB (eds) Pharmacovigilance, 2nd edn. John Wiley & Sons, Chichester, pp 317–332
116. Layton D, Hazell L, Shakir SA (2011) Modified prescription-event monitoring studies: a tool for pharmacovigilance and risk management. Drug Saf 34(12):e1–e9. https://doi.org/10.2165/11593830-000000000-00000
117. Layton D, Shakir SA (2015) Specialist cohort event monitoring studies: a new study method for risk management in pharmacovigilance. Drug Saf 38(2):153–163. https://doi.org/10.1007/s40264-014-0260-x
118. New Zealand Pharmacovigilance Centre: About—IMMP. New Zealand Pharmacovigilance Centre. https://nzphvc.otago.ac.nz/about/. Accessed 16 Mar 2020
119. Ernst E, De Smet PAGM, Shaw D et al (1998) Traditional remedies and the "test of time". Eur J Clin Pharmacol 54(2):99–100. https://doi.org/10.1007/s002280050428
120. Layton D, Denham A, Whitelegg ME et al (2006) Methodology of a feasibility study to assess the application of prescription event monitoring (PEM) to monitor the safety of herbal medicines. In: Abstracts of the Pharmacovigilance of herbal medicines: current state and future directions, London, UK, 26–28 April 2006
121. Gauld NJ, Shaw JP, Emmerton LM et al (2000) Surveillance of a recently switched non-prescription medicine (Diclofenac) using a pharmacy-based approach. Pharmacoepidemiol Drug Saf 9(3):207–214. https://doi.org/10.1002/1099-1557(200005/06)9:3<207::AID-PDS483>3.0.CO;2-S
122. Sinclair HK, Bond CM, Hannaford PC (2000) Over-the-counter ibuprofen: how and why is it used? Int J Pharm Pract 8(2):121–127. https://doi.org/10.1111/j.2042-7174.2000.tb00996.x
123. Layton D, Sinclair HK, Bond CM et al (2002) Pharmacovigilance of over-the-counter products based in community pharmacy: methodological issues from pilot work conducted in Hampshire and Grampian. UK Pharmacoepidemiol Drug Saf 11(6):503–513. https://doi.org/10.1002/pds.734
124. Borden EK, Lee JG (1982) A methodologic study of post-marketing drug evaluation using a pharmacy-based approach. J Chron Dis 35(10):803–816. https://doi.org/10.1016/0021-9681(82)90092-3
125. Bond C, Hannaford P (2003) Issues related to monitoring the safety of over-the-counter (OTC) medicines. Drug Saf 26(15):1065–1074. https://doi.org/10.2165/00002018-200326150-00001
126. Aggarwal AM, Barnes J (2006) A pilot study of community-pharmacy-based pharmacovigilance of an over-the-counter herbal medicine ginkgo (Ginkgo biloba): methodological issues from work in progress. Drug Saf 29(4):358
127. Baybutt J, Sheridan J, Barnes J (2015) New Zealand community pharmacists' views on pharmacovigilance for natural health products. In: Abstracts of the Australasian Society of Clinical and Experimental Pharmacologists and Toxicologists (ASCEPT) and Australasian Pharmaceutical Science Association (APSA) Joint Scientific Meeting, Tasmania, Australia, 29 Nov–2 Dec 2015
128. Brown P, Bahri P (2019) 'Engagement' of patients and healthcare professionals in regulatory pharmacovigilance: establishing a conceptual and methodological framework. Eur J Clin Pharmacol 75(9):1181–1192. https://doi.org/10.1007/s00228-019-02705-1
129. Jonker CJ, van den Berg HM, Kwa MSG et al (2017) Registries supporting new drug applications. Pharmacoepidemiol Drug Saf 26(12):1451–1457. https://doi.org/10.1002/pds.4332
130. Schmitt-Egenolf M (2018) Patient registries for safetyness: clinical and quantitative aspects. In: Bate A (ed) Evidence-based pharmacovigilance. Humana Press, New York, pp 149–164

131. McGettigan P, Alonso Olmo C, Plueschke K et al (2019) Patient registries: an underused resource for medicines evaluation: operational proposals for increasing the use of patient registries in regulatory assessments. Drug Saf 42(11):1343–1351. https://doi.org/10.1007/s40264-019-00848-9
132. Olmo CA, McGettigan P, Kurz X (2019) Barriers and opportunities for use of patient registries in medicines regulation. Clin Pharmacol Ther 106(1):39–42. https://doi.org/10.1002/cpt.1414
133. Zhang L (2018) Pharmacovigilance of herbal and traditional medicines: clinical and quantitative aspects. In: Bate A (ed) Methods on pharmacology and toxicology. Humana Press, New York, pp 37–65
134. Liao X, Xie Y-m, Robinson N et al (2017) To establish a body of evidence on safety for postmarketing Chinese medicine: a new research paradigm. Chin J Integr Med 23(3):226–232. https://doi.org/10.1007/s11655-016-2534-x
135. Liang Z, Lai Y, Li M et al (2021) Applying regulatory science in traditional chinese medicines for improving public safety and facilitating innovation in China: a scoping review and regulatory implications. Chin Med 16:23
136. Guidance for Industry and Food and Drug Administration Staff. (2017). Use of real-world evidence to support regulatory decision-making for medical devices. U.S. Food and Drug Administration. https://www.fda.gov/media/99447/download. Accessed 4 Jun 2021
137. Dal Pan GJ (2019) Real-World data, advanced analytics, and the evolution of postmarket drug safety surveillance. Clin Pharmacol Ther 106(1):28–30. https://doi.org/10.1002/cpt.1415
138. Raghupathi W, Raghupathi V (2014) Big data analytics in healthcare: promise and potential. Health Inf Sci Syst 2:3–3. https://doi.org/10.1186/2047-2501-2-3
139. Bate A, Reynolds RF, Caubel P (2018) The hope, hype and reality of Big Data for pharmacovigilance. Ther Adv Drug Saf 9(1):5–11. https://doi.org/10.1177/2042098617736422
140. Liu F, Jagannatha A, Yu H (2019) Towards drug safety surveillance and pharmacovigilance: current progress in detecting medication and adverse drug events from electronic health records. Drug Saf 42(1):95–97. https://doi.org/10.1007/s40264-018-0766-8
141. Boyce RD, Ryan PB, Norén GN et al (2014) Bridging islands of information to establish an integrated knowledge base of drugs and health outcomes of interest. Drug Saf 37(8):557–567. https://doi.org/10.1007/s40264-014-0189-0
142. Hripcsak G, Duke JD, Shah NH et al (2015) Observational health data sciences and informatics (OHDSI): opportunities for observational researchers. Stud Health Technol Inform 216:574–578
143. Tabali M, Ostermann T, Jeschke E et al (2012) Adverse drug reactions for CAM and conventional drugs detected in a network of physicians certified to prescribe CAM drugs. J Manag Care Pharm 18(6):427–438. https://doi.org/10.18553/jmcp.2012.18.6.427
144. Milne BJ, Atkinson J, Blakely T et al (2019) Data resource profile: the New Zealand Integrated Data Infrastructure (IDI). Int J Epidemiol 48(3):677–677e. https://doi.org/10.1093/ije/dyz014
145. Patient-focused drug development glossary. U.S Food and Drug Administration. https://www.fda.gov/drugs/development-approval-process-drugs/patient-focused-drug-development-glossary. Accessed 4 Jun 2021
146. Nowell WB (2019) Information patients can provide will strengthen the real-world evidence that matters to them. Clin Pharmacol Ther 106(1):49–51. https://doi.org/10.1002/cpt.1460
147. PatientsLikeMe. Business Milestones 2015. https://news.patientslikeme.com/milestones. Accessed 4 Jun 2021
148. Balık CY, Goshgarian M, Donahue K et al (2015) Increasing patient engagement in pharmacovigilance through online community outreach and mobile reporting applications: an analysis of adverse event reporting for the Essure device in the US. Pharmaceut Med 29(6):331–340. https://doi.org/10.1007/s40290-015-0106-6
149. Pierce CE, de Vries ST, Bodin Parssinen S et al (2019) Recommendations on the use of mobile applications for the collection and communication of pharmaceutical product safety information: lessons from IMI WEB-RADR. Drug Saf 42(4):477–489. https://doi.org/10.1007/s40264-019-00813-6

150. de Vries ST, Wong L, Sutcliffe A et al (2017) Factors influencing the use of a mobile app for reporting adverse drug reactions and receiving safety information: a qualitative study. Drug Saf 40(5):443–455. https://doi.org/10.1007/s40264-016-0494-x
151. Golder S, Norman G, Loke Y (2015) Systematic review on the prevalence, frequency and comparative value of adverse events data in social media. Br J Clin Pharmacol 80(4):878–888. https://doi.org/10.1111/bcp.12746
152. Caster O, Dietrich J, Kürzinger ML et al (2018) Assessment of the utility of social media for broad-ranging statistical signal detection in pharmacovigilance: results from the WEB-RADR project. Drug Saf 41(12):1355–1369. https://doi.org/10.1007/s40264-018-0699-2
153. van Stekelenborg J, Ellenius J, Maskell S et al (2019) Recommendations for the use of social media in pharmacovigilance: lessons from IMI WEB-RADR. Drug Saf 42(12):1393–1407. https://doi.org/10.1007/s40264-019-00858-7
154. World Health Organization ICD-11 for mortality and morbidity statistics, 26 supplementary chapter traditional medicine conditions—module I. https://icd.who.int/browse11/l-m/en#/http%3a%2f%2fid.who.int%2ficd%2fentity%2f718687701. Accessed 9 Jun 2021
155. Choi SH (2020) A proposed revision of the International Classification of Diseases, 11th revision, chapter 26. Integr Cancer Ther 19:1534735420908334. https://doi.org/10.1177/1534735420908334
156. Choi SH, Chang IM (2010) A milestone in codifying the wisdom of traditional Oriental medicine: TCM, Kampo, TKM, TVM-WHO international standard terminologies on traditional medicine in the Western Pacific Region. Evid Based Complement Alternat Med 7(3):303–305. https://doi.org/10.1093/ecam/nen083
157. Brosch S, de Ferran AM, Newbould V et al (2019) Establishing a framework for the use of social media in pharmacovigilance in Europe. Drug Saf 42(8):921–930. https://doi.org/10.1007/s40264-019-00811-8
158. Cockayne NL, Duguid M, Shenfield GM (2005) Health professionals rarely record history of complementary and alternative medicines. Br J Clin Pharmacol 59(2):254–258. https://doi.org/10.1111/j.1365-2125.2004.02328.x
159. Wisniewski AF, Bate A, Bousquet C et al (2016) Good signal detection practices: evidence from IMI PROTECT. Drug Saf 39(6):469–490. https://doi.org/10.1007/s40264-016-0405-1
160. Schulz KF, Altman DG, Moher D (2010) CONSORT 2010 Statement: updated guidelines for reporting parallel group randomised trials. Trials 11:32. https://doi.org/10.1186/1745-6215-11-32
161. Ioannidis JP, Evans SJ, Gøtzsche PC et al (2004) Better reporting of harms in randomized trials: an extension of the CONSORT statement. Ann Intern Med 141(10):781–788. https://doi.org/10.7326/0003-4819-141-10-200411160-00009
162. Linde K, Berner MM, Kriston L (2008) St John's wort for major depression. Cochrane Database Syst Rev 2008(4):Cd000448. https://doi.org/10.1002/14651858.CD000448.pub3
163. van Puijenbroek EP, Bate A, Leufkens HG et al (2002) A comparison of measures of disproportionality for signal detection in spontaneous reporting systems for adverse drug reactions. Pharmacoepidemiol Drug Saf 11(1):3–10. https://doi.org/10.1002/pds.668
164. Bate A, Lindquist M, Edwards IR et al (1998) A Bayesian neural network method for adverse drug reaction signal generation. Eur J Clin Pharmacol 54(4):315–321. https://doi.org/10.1007/s002280050466
165. Waller PC, Evans SJW (2003) A model for the future conduct of pharmacovigilance. Pharmacoepidemiol Drug Saf 12(1):17–29. https://doi.org/10.1002/pds.773
166. Betz J, Anderson L, Avigan M et al (2009) Black Cohosh: considerations of safety and benefit. Nutr Today 44:155–162. https://doi.org/10.1097/NT.0b013e3181af63f9
167. Black cohosh. UK public assessment report. Medicine and Healthcare products Regulatory Agency. https://assets.publishing.service.gov.uk/government/uploads/system/uploads/attachment_data/file/852384/Black_cohosh_and_liver_injury.pdf. Accessed 4 Jun 2021
168. Mazzanti G, Di Sotto A, Vitalone A (2015) Hepatotoxicity of green tea: an update. Arch Toxicol 89(8):1175–1191. https://doi.org/10.1007/s00204-015-1521-x

169. Savage RL, Hill GR, Barnes J et al (2019) Suspected hepatotoxicity with a supercritical carbon dioxide extract of Artemisia annua in grapeseed oil used in New Zealand. Front Pharmacol 10:1448. https://doi.org/10.3389/fphar.2019.01448
170. Wechwithan S, Suwankesawong W, Sornsrivichai V et al (2014) Signal detection for Thai traditional medicine: examination of national pharmacovigilance data using reporting odds ratio and reported population attributable risk. Regul Toxicol Pharmacol 70(1):407–412. https://doi.org/10.1016/j.yrtph.2014.06.007
171. Li H, Deng J, Yue Z et al (2015) Detecting drug-herbal interaction using a spontaneous reporting system database: an example with benzylpenicillin and qingkailing injection. Eur J Clin Pharmacol 71(9):1139–1145. https://doi.org/10.1007/s00228-015-1898-8
172. Caster O, Juhlin K, Watson S et al (2014) Improved statistical signal detection in pharmacovigilance by combining multiple strength-of-evidence aspects in vigiRank. Drug Saf 37(8):617–628. https://doi.org/10.1007/s40264-014-0204-5
173. Uppsala Monitoring Cenre Signal detection at UMC. https://www.who-umc.org/research--scientific-development/signal-detection/signal-detection-at-umc. Accessed 27 Apr 2021
174. Barnes J, van Hunsel F (2020) Ginkgo biloba L. and cardiac arrhythmias. WHO Pharmaceuticals Newslett 3:14–22
175. Gogolak VV (2003) The effect of backgrounds in safety analysis: the impact of comparison cases on what you see. Pharmacoepidemiol Drug Saf 12(3):249–252. https://doi.org/10.1002/pds.823
176. Bate A, Ericsson J, Farah M (2006) International data mining for signals of herbal ADRs. In: Abstracts of the Pharmacovigilance of herbal medicines: current state and future directions, London, UK, 26–28 April 2006
177. Klein K, Hazell L, Stolk P et al (2020) The UK BIO-TRAC study: a cross-sectional study of product and batch traceability for biologics in clinical practice and electronic adverse drug reaction reporting in the UK. Drug Saf 43(3):255–263. https://doi.org/10.1007/s40264-019-00891-6
178. Heinrich M, Barnes J, Gibbons S et al (2018) Fundamentals of pharmacognosy and phytotherapy, 3rd edn. Elsevier
179. Arnaud M, Bégaud B, Thurin N et al (2017) Methods for safety signal detection in healthcare databases: a literature review. Expert Opin Drug Saf 16(6):721–732. https://doi.org/10.1080/14740338.2017.1325463
180. Sun X, Tan J, Tang L et al (2018) Real world evidence: experience and lessons from China. BMJ 360:j5262. https://doi.org/10.1136/bmj.j5262
181. Stricker BH (2018) Adverse reaction signal detection methodology in pharmacoepidemiology. Eur J Epidemiol 33(6):507–508. https://doi.org/10.1007/s10654-018-0417-5
182. Pottegård A, Hallas J, Wang SV et al (2018) Identifying signals of interest when screening for drug-outcome associations in health care data. Br J Clin Pharmacol 84(9):1865–1867. https://doi.org/10.1111/bcp.13634
183. Caster O, Aoki Y, Gattepaille LM et al (2020) Disproportionality analysis for pharmacovigilance signal detection in small databases or subsets: recommendations for limiting false-positive associations. Drug Saf 43(5):479–487. https://doi.org/10.1007/s40264-020-00911-w

Chapter 8
Botanical Nomenclature for Herbal Medicines and Natural Products: Its Significance for Pharmacovigilance

Bob Allkin and Kristina Patmore

8.1 Introduction

8.1.1 Names and Identity

In Lewis Carroll's book 'Through the Looking Glass' [1], Alice meets the 'White Knight' who offers to sing her a song. He tells Alice that the name of the song is called 'Haddocks' Eyes', that the song's name is 'Aged, Aged Man', that it is called 'Ways and Means', and that the song is, actually, 'A-sitting On a Gate'. Carroll had fun confounding the names given to things and their identity.

Humans use names (specific nouns) to communicate with one another about people, objects, and concepts. These names serve as shorthand, a convenient and effective means to indicate a particular person, town, colour or flower, avoiding long detailed descriptions or comparisons. They serve to communicate with others (in speech or text) and to find information (e.g. via Google).

This chapter explores how confusions similar to those encountered by Alice impact research, regulation and pharmacovigilance, and how scientific names hold the key to communicating effectively about plants and herbal substances.

8.1.2 Plants, Ingredients, Drugs and Names

Consumers in the Global North (the richest, more industrialised countries, found mainly in the northern part of the world) increasingly employ 'natural' products, and the sale, distribution and use of plant-based supplements and herbal drugs have

B. Allkin (✉) · K. Patmore
Medicinal Plant Names Services, Royal Botanic Gardens, Kew, Richmond, UK
e-mail: b.allkin@kew.org; K.Patmore@kew.org

J. Barnes (ed.), *Pharmacovigilance for Herbal and Traditional Medicines*,
https://doi.org/10.1007/978-3-031-07275-8_8

expanded significantly in recent years [2–4]. Simultaneously, an increasing number of countries are promoting the use of plant-based traditional remedies within formal public-health programmes, alongside more conventional medicines. Such treatments are even more widely used to treat chronic conditions less-well catered for by conventional medicines [5, 6]. Meanwhile, millions of individuals in less economically developed countries, particularly those living in rural environments, rely solely on traditional plant-based remedies for their primary healthcare [7].

Despite the substantial and increasing use of herbal substances, the diversity of plant species cited in national pharmacopoeias, which formally define the substances used in pharmacy, has consistently fallen over the last 100 years as the evidential requirements for 'efficacy' and 'safety' have increased. The Brazilian Pharmacopoeia, for example, has seen a reduction in the number of monographs relating to native plant species from 713 in the first edition (published in 1926) to just 44 in the fourth edition (published in 1996), and now cites more European or Chinese plants than it does native Brazilian plants [8, 9]. Investment in understanding the efficacy and safety of plant-based remedies, or in monitoring their use through national pharmacovigilance programmes, lags behind that of conventional drugs, partly because of the complex mixtures of molecules involved and the limitations of our understanding of human physiology.

Pharmacovigilance involves detecting, assessing, reporting, and preventing adverse effects from pharmaceutical products, including those sourced from plants. To achieve this for herbal drugs, pharmacovigilance needs to handle the names of the plants employed, the derived substances (plant parts and their preparation), as well as the drug and market (proprietary) names used in pharmacy and trade.

Pharmacopoeias contain the necessary formal definitions (descriptions) of specific herbal substances and establish in considerable detail which plant parts are to be used for that substance, and how that material should be identified and prepared. Increasingly they offer guidance as to how marketed products may be authenticated. Some pharmacopeial monographs also include indications or dosages relating to herbal drugs. However, given this level of detail and caution, it is surprising that many pharmacopoeias can be ambiguous or vague when establishing exactly which plant species should be used. Further confusion arises because of remarkable inconsistency between pharmacopoeias as to how to refer to particular herbal substances. The boundary between 'herbal drugs' and 'food supplements' is another source of ambiguity, and it is not uncommon for one herbal substance (species + plant part + preparation) to be marketed both as a food/dietary supplement and as a drug under alternative names and regulatory frameworks. This introduces yet further alternative uses of individual names, exacerbating an already complex set of terminologies, and confusing consumers, health practitioners and medicines' regulators alike.

Adverse drug reaction (ADR) reports may include different types of names for herbal drugs: common or scientific names (for plants), Latinised pharmaceutical names, drug names (in many languages), transliterations of those names (e.g. Pinyin names), or proprietary product names. Many countries allow direct patient reporting of suspected ADRs, with patients often unfamiliar with how to describe herbal medicines precisely, or the importance of doing so. Alternative names can occur in any language and in multiple scripts (e.g. Arabic, Chinese, Roman). Pharmacovigilance professionals, therefore, need to deal both with the multiple (and frequently ambiguous) names used for plants, and with names of the substances derived from those plants: pharmacopoeial, trade and ingredient names.

8.2 The Need for Scientific Names

8.2.1 Common Names and Their Limitations

Common names, in all countries, form part of everyday language. No formal controls exist: their meanings vary from place to place (even between neighbouring villages) and evolve over time. The plant names learned in childhood will differ depending on in which part of—for example—England (or China) you were born, and people moving to new places will frequently repurpose familiar names (e.g. 'robin') to refer to a similar-looking species found in their new locality. Single species are known by multiple common names ('synonyms'), even in the same language, and any one of those names may be employed, by different people, to refer to different species ('homonyms'). Even scientists can be surprisingly insistent that the common name that they personally use is 'correct', but in reality there is no 'right' or 'wrong'. All common names are equally valid.

The imprecision and ambiguity of common names makes them inappropriate for scientific, regulatory or pharmacovigilance purposes. A Google search for 'fleabane' may return images of scores of different species, each known to *someone* as 'fleabane'. Neither would that search retrieve all images of the plant of interest, since many images will have been uploaded using alternative names.

8.2.2 Pharmacopoeia Names: A Poor Remedy

The names employed in pharmacopoeias have particular significance in pharmacovigilance and are the names primarily employed by medicines or food regulators and by some health professionals. They generally refer to substance descriptions of the most exacting precision.

These names take different forms in different pharmacopoeias. Some repurpose the common name of the plant being used to also refer to a particular recipe for how to prepare that substance, introducing further ambiguity [10]. The British Pharmacopoeia [11], for example, defines 'Wormwood' as a substance obtained from drying the 'basal leaves or slightly leafy, flowering tops' of *Artemisia absinthium* L. 'Wormwood' thus refers to both the species and a substance derived from it. Some pharmacopoeias avoid this confusion by creating new labels for a particular definition of an herbal substance, while other pharmacopoeias employ several types of name to aid clarity. Table 8.1 lists alternative names used in pharmacopoeias and regulatory literature for herbal substances derived from the dried root of *Stephania tetrandra* S.Moore.

Creating completely new names for particular herbal preparations avoids confusion between the substance and the plant from which it is obtained. The use of Latin (e.g. in pharmaceutical names such as 'Stephaniae tetrandrae radix') may enhance

Table 8.1 Example of diversity of names employed for one herbal substance

Non-scientific name	Medicinal plant reference
ban fang ji	U.S. FDA Substance Registration System (2016)
fang ji	Herbs of Commerce (American Herbal Products Association 2000) Taiwan Herbal Pharmacop. 3rd Chinese ed. (Ministry of Health and Welfare 2018) U.S. FDA Substance Registration System (2016)
Fangji	Hong Kong Chinese Materia Med. Standards (2014) Pharmacopoeia of China (2005) Pharmacopoeia of China (2010) Pharmacopoeia of China (2015)
fen fang ji	Herbs of Commerce (American Herbal Products Association 2000) Taiwan Herbal Pharmacop. 3rd Chinese ed. (Ministry of Health and Welfare 2018) U.S. FDA Substance Registration System (2016)
fourstamen stephania root	European Pharmacopoeia, 7th edn. (2012) Pharmacopoeia of China (2005) Pharmacopoeia of China (2010)
han fang ji	Herbs of Commerce (American Herbal Products Association 2000) Taiwan Herbal Pharmacop. 3rd Chinese ed. (Ministry of Health and Welfare 2018)
radix stephaniae tetrandrae	Pharmacopoeia of China (2005)
Stephania	Herbs of Commerce (American Herbal Products Association 2000) U.S. FDA Substance Registration System (2016)
stephania tetrandra root	British Pharmacopoeia 2012, Vol. 4 (2011) British Pharmacopoeia 2015, Vol. 4 (2014)
stephaniae tetrandrae	Taiwan Herbal Pharmacop. 3rd Chinese ed. (Ministry of Health and Welfare 2018)
stephaniae tetrandrae radix	European Pharmacopoeia, 7th edn. (2012) Hong Kong Chinese Materia Med. Standards (2014) Pharmacopoeia of China (2010) Pharmacopoeia of China (2015) Taiwan Herbal Pharmacop. 3rd Chinese ed. (Ministry of Health and Welfare 2018)
漢防己	Taiwan Herbal Pharmacop. 3rd Chinese ed. (Ministry of Health and Welfare 2018)
粉防己	Taiwan Herbal Pharmacop. 3rd Chinese ed. (Ministry of Health and Welfare 2018)
防己	Hong Kong Chinese Materia Med. Standards (2014) Taiwan Herbal Pharmacop. 3rd Chinese ed. (Ministry of Health and Welfare 2018)

The dried root of *Stephania tetrandra* S.Moore is referred to in multiple pharmacopoeias, regulations and authoritative references using a remarkable diversity of names (Medicinal Plant Names Services V10, 2021) [12]

global accessibility and offer some sense of scholarly endeavour, but these names are established by pharmacists and, in practice, are under no more formal control than are the common names described above. A single pharmaceutical name may refer to different herbal preparations in different pharmacopoeias. Indeed, the drug definitions associated with one name may change between consecutive editions of the same pharmacopoeia. The use of Latinised names does not in itself confer legitimacy or scientific precision. Box 8.1 illustrates how the pharmaceutical name ‘Cimicifugae rhizoma’ is employed by different pharmacopoeias to refer to substances derived from 5 different plant species, which have differing chemistries and uses.

Box 8.1 Alternative Meanings of ‘Cimicifugae rhizoma’

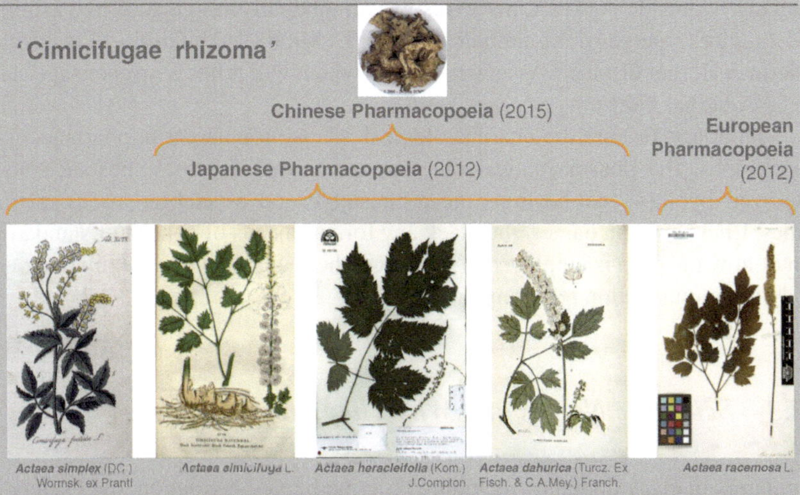

The pharmaceutical name ‘Cimicifugae rhizoma’ is widely cited. It is employed, for example, in the following Pharmacopoeias: EDQM Europe [13], Committee of the Chinese Pharmacopoeia [14] and Committee of the Japanese Pharmacopoeia (2012) [15]. Surprisingly, and confusingly, these three publications use the same term to refer to substances derived from different plant species with different properties and potential uses.

The growing popularity of Traditional Chinese Medicine (TCM) has resulted in herbal substances defined in the Chinese Pharmacopoeia [14] becoming increasingly well known outside China, with both the Chinese and English editions containing Pinyin names (Chinese language names transliterated into Roman script) as well as Chinese names. Adopting unfamiliar plants with names in another language can have consequences. One well-reported case involved a Belgian 'slimming clinic' confusing two herbal substances from the Chinese Pharmacopoeia which share the same Pinyin name ‘Fang Ji’ (‘防己’ in Chinese) [16]. These two substances derive from different plants: *Aristolochia fangchi* Y.C.Wu ex L.D.Chow & S.M.Hwang and *Stephania tetrandra* S.Moore. The substances have different purposes and are used at very different doses. These two substances were mistakenly substituted with tragic results: over 100 patients developed end-stage kidney failure following ingestion of *Aristolochia fangchi* at toxic doses[17]. Subsequent studies indicated that this confusion, and misuse of *Aristolochia* species was widespread [18], leading to bans on all substances derived from *Aristolochia* species.

The publication Herbs of Commerce [19] looks to standardise how pharmacopoeial and common names are to be employed in the USA for medicinal herbs and plant substances by mapping each to a scientific name (discussed below). It intends to establish good practice and improve communication across the 'herbal products' industry. The challenge, however, is that the meaning of the common and pharmacopoeial names indexed continues to vary geographically and evolve over time. The publication, while valuable, is in practice unable to fix the meaning of each term, particularly outside the USA, nor to resolve all inherent ambiguity. A new edition is due (authors' personal communication with Michael McGuffin, President of American Herbal Products Association [19]) which will reflect current usage across the USA herbal industry.

In summary, despite the pharmacological precision inherent in pharmacopoeial monographs, the pharmaceutical or other uncontrolled herbal substance names employed by these publications are as unstable and as ambiguous as common names. Like common names, they lack any formal means of control or standardisation of how the term is employed. This makes them unsuitable for regulation or any kind of communication requiring precision.

8.2.3 Scientific Names and Why They Should Be Used

Regulators in many domains (including drug/medicines' regulation, food safety and conservation), as well as scientists, publishers and professionals working with plants, rely on the use of scientific names. These, when employed appropriately, refer *unambiguously* to a *single* plant species. Scientific names enable us to be confident of being understood anywhere in the world and permit retrieval of information from publications, ADR report records or patents, secure in the knowledge that the information retrieved relates unambiguously to a single species.

Modern scientific nomenclature (using the 'binomial system', in which each biological species is assigned a two-part name) was established by Carl Linnaeus when he published *Species Plantarum*, a two-volume list of known plants [20]. Each name of a species comprises a genus epithet (equivalent to a surname shared by close relatives) and a species epithet (indicating a distinct subgroup within the genus), followed by the author of the name. This structure is discussed in more detail below. Scientific names are created, employed and monitored for plants, fungi, ferns, animals, viruses and bacteria according to specific 'nomenclatural codes', each relating to a biological kingdom [21–24]. These rather legalistic codes establish the precise naming protocols required in each discipline. Each has evolved independently of the others over the decades and is reviewed every few years.

The 'International Code of Nomenclature for algae, fungi, and plants' [21] establishes how scientific names of plants, fungi and algae are to be created, employed and managed. It is reviewed at a congress every 6 years and is intended primarily for use by practising taxonomists: botanists specialised in a single region or plant family, and directly involved in creating or revising the names used for those plants.

The 'Code' establishes how new scientific names for plants should be published and subsequently interpreted. It defines, for example, where and how a new scientific name should be published in order to be 'valid' and requires that this includes a 'diagnostic' description of how this new organism differs from those already known to science. Critically, the author(s) of each new scientific name must cite the physical plant specimen(s) they studied, which support their taxonomic conclusions, and upon which their description is based. The author of the new name will indicate the unique identifiers (the collector's name and collector's own collection number) for key specimens seen, and the herbaria where those specimens can be found and reviewed.

From the moment of publication, these key specimens become 'type' specimens and serve for all time as a physical reference point establishing beyond any dispute to what plant the associated scientific name refers (see Box 8.2). The distinct features (e.g. flower colour, type of leaf hair, DNA) of these type specimens establish which characteristics plants assigned this scientific name must have, such that the *meaning* of each individual scientific name is fixed for all time and cannot change. It is the use of 'type' specimens, serving as a constant physical reference, alongside their formal publication, which makes these names 'scientific'. The use of Latin is not, in itself, sufficient to make names 'scientific'. Latin is required by the 'Code', but offers no benefit over other languages other than as the global standard.

Box 8.2 'Type' Specimen as a Physical Reference Point for a Name

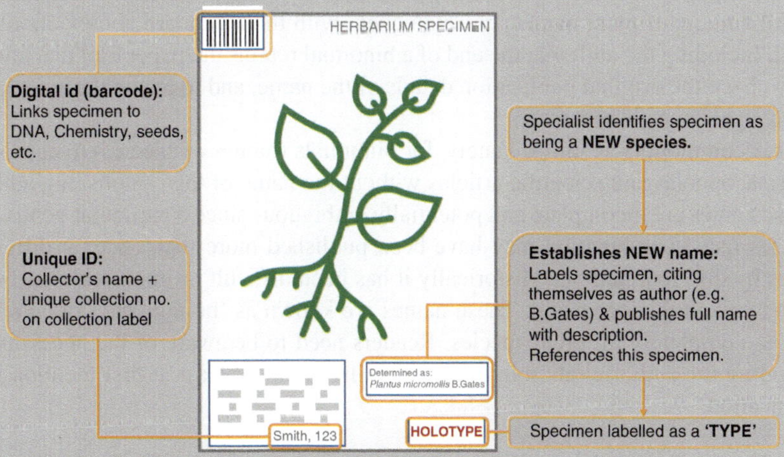

This illustrates how one particular herbarium specimen (identified by its collectors' name and unique collection number) is key to both the effective publication of a scientific name and to providing evidence of the meaning of that name for all time. Such specimens are categorised as 'type' specimens. Each scientific name will have one or more 'type' specimens. The author publishing the scientific name will also state in which herbaria the 'type' specimen(s) can be found. Duplicates of a 'type' specimen may exist in multiple herbaria.

Typically, new names are published by specialist botanists, but anyone can do so provided they follow the procedures and formats established in the 'Code', including citation of the 'type' specimens. In summary, scientific names are globally recognised and used for scientific purposes and in legislation because each is unique (unambiguous) and its meaning will not change over time. '*Hocus pocus* Bob' may appear to be in Latin and have the appropriate format but Bob never published this name. No description exists, no 'type' specimens are designated, and effectively this 'name' has no meaning.

8.2.4 Structure of Scientific Names

Mandragora officinarum L. is a valid scientific name. It was published by Linnaeus and is listed in the International Plant Names Index [25] with its place of publication and a persistent identifier for use in datasets. As the scientific name of a species, it consists of three components: the genus ('*Mandragora*'), the species within that genus ('*officinarum*'), and the standardised name of the author, Linnaeus, who published that scientific name ('L.'). Convention dictates that the genus and species epithets appear in italics, with the genus name having a capital letter. The author (or, in some cases, joint authors) who published that name will not be italicised. Linnaeus's name is conventionally abbreviated to 'L.'; other versions commonly seen include 'Linn.' and 'Linné'.

In some scientific names the 'author' component is compound, e.g. *Sorghum bicolor* (L.) Moench. The contributions of authors within and without parentheses will be explained below.

All authors of plant names are catalogued, with their standard abbreviations, in IPNI. Including the author at the end of a binomial retains the paper trail that allows us to check the original publication details of the name, and to locate the associated 'type' specimens if necessary.

It is commonplace, unfortunately, for binomials (names of species) to appear in pharmacopoeias and scientific articles without the name of their publishing author. These names are incomplete and potentially ambiguous since a particular genus and species name combination may have been published more than once, at different times by different authors. Historically it has been difficult or impossible for botanists to avoid this occurring. These names are known as 'homonyms', often referring to completely different species. Readers need to be aware of the necessity of including the author citation in every scientific name, at least at first mention in a publication.

8.2.4.1 Genera, Subspecies and Varieties

Further complications occur since scientific names may be used to refer to individual species, and to groups of closely related species at a higher taxonomic rank (e.g. using only a genus name), or to smaller groups at a lower rank, such as subspecies, varieties, or other subcategories, to differentiate between groups of individuals within a single species.

The names of genera have a simpler structure, consisting solely of the genus name followed by the name of the person who first described that genus. Just as for species names, genus names are defined using the 'type' concept. However, rather than citing one or more type specimens, authors of new genus names must establish in their publication which species within this group will serve as the 'type' (i.e. be the most typical). That species will, of course, itself be defined by the 'type' specimens cited when that species name was published. The same principles apply to taxonomic groupings above the genus level (e.g. Family); thus, the entire hierarchy is based conceptually on the collection of specimens cited as 'types' for individual species.

Plants belonging to 'infraspecific' groups (taxonomic ranks below the species level) will show small divergences (such as in their chemical composition) from plants identified as belonging to other infraspecific groups within the same species. The scientific names of plants belonging to these infraspecific groups will typically consist of five components: the names of the genus and species to which it belongs, the rank (usually 'subspecies' or 'variety'), the infraspecific epithet, and the author of that infraspecific name. For example, *Hypericum perforatum* subsp. *chinense* N.Robson is a subspecies within the species *Hypericum perforatum* L. The infraspecific name was published by Robson [26] to refer to a particular subset of plants identified as *H. perforatum*, but which share a unique set of characteristics, described by Robson, which are not found among other individuals of the species. Robson's publication established which specimens demonstrate these characteristics and serve as 'types' for this new name. 'N. Robson' is the standard abbreviation of the author's name [25]. Dauncey et al. [27] explore the significance of this and other names published by Robson, and the significance for herbal medicine of the underlying taxonomic changes implied.

8.2.4.2 Autonyms

Box 8.3 illustrates a concept which understandably causes confusion: why subspecies (or varietal) names sometimes appear with identical species and subspecies epithets and lacking a publishing author.

Box 8.3 Subspecies of *Hypericum perforatum* L.

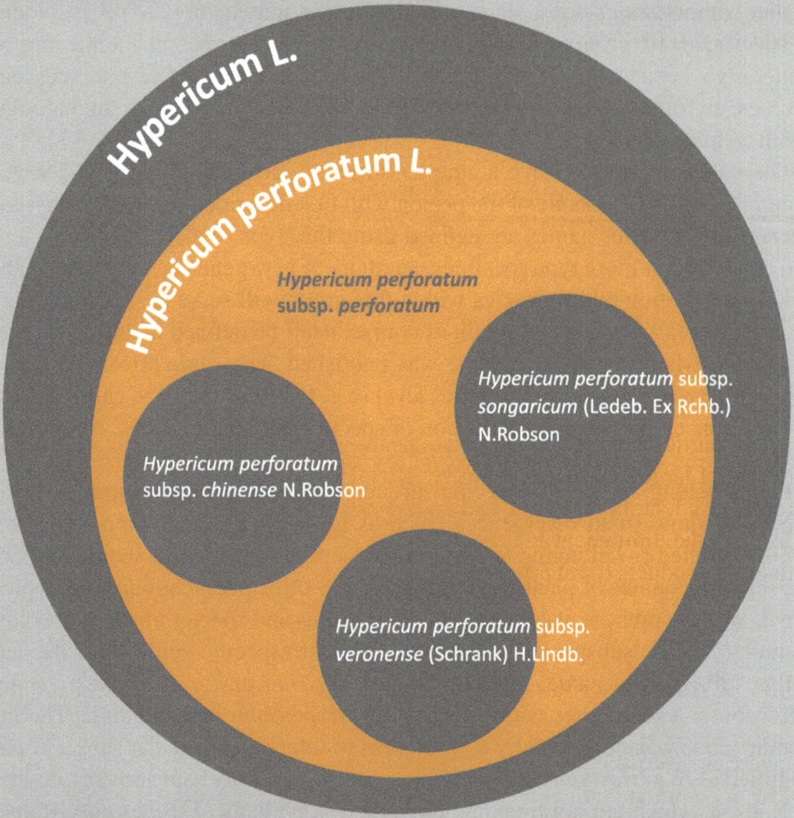

Lindberg [28] and Robson [26] published new scientific names for subspecies of *Hypericum perforatum* L. Their descriptions establish the particular characteristics to be expected of individual plants belonging to each of these subspecies, differentiating them from individuals belonging to other subspecies. No individual can belong to more than one subspecies.

Individual plants belonging to *Hypericum perforatum* L., but **not** matching the description of any published subspecies (in the orange area of the graphic), belong to a fourth subspecies: *Hypericum perforatum* subsp. *perforatum*—referred to as the 'type' subspecies. The name of this fourth 'type' subspecies has no author: it was created automatically as the 'default' subspecies at the point when the first subspecies description, deviating from the 'type' concept, was published.

Hypericum perforatum subsp. *perforatum*, for example, appears to lack the author citation required of valid scientific plant names. This name refers to all individual plants belonging to the species whose characteristics match the species description perfectly: they are 'typical', and shown none of the variations described in particular subspecies (or varieties). These names (technically called 'autonyms') lack an author because the 'Code' requires that they be created automatically whenever a subspecies or varietal name is first published for that species. Autonyms become a necessary tag for referring to that otherwise nameless subset of the species *not* found within any of the named subspecies (or varieties).

8.2.5 Nomenclature Versus Taxonomy

So far, only the nomenclatural aspects of using scientific names have been discussed. Another fundamental dimension is the use of names to reflect how plants relate genetically, and to establish where that plant belongs within the taxonomic tree. It is a feature of botanical nomenclature that a plant's scientific name places it within a genus, which in turn belongs to one (plant) family.

Plant systematics generates ever-increasing volumes of data enabling us to better appreciate which plants are most closely related. Linnaeus used primarily morphological observations to classify plants, but more recent physiological, chemical, and, increasingly, DNA data have led to ever more reliable conclusions. Where evidence suggests that a species is better placed in another genus, then, by definition, it must be given a name which correctly places it in the taxonomic hierarchy. Both old and new names may be sound from a nomenclatural perspective and each refer unambiguously to that plant, i.e. either can be used. However, only the newer, 'accepted' or preferred name indicates our current understanding of that plant's evolutionary past. This is covered in more detail below.

'Nomenclature' deals purely with the validity of structure and publication of scientific names. 'Taxonomy' nominates an 'accepted name', placing a species in a genus and thereby within a hierarchy which groups genera into families. Opinions as to the correct taxonomic position of a plant may differ among experts and may change over time as new evidence becomes available.

8.2.6 Taxonomic Reference Sources of Relevance to Pharmacovigilance

Numerous resources have been created by botanists to catalogue names, organise taxonomies, share floristic information or create checklists. It can be difficult to navigate the many options available, or to recognise the strengths and limitations of each. A few key resources, however, should provide the information needed by most pharmacovigilance professionals.

The Royal Botanic Gardens, Kew, UK ('Kew'), created and continues to maintain the *International Plant Names Index* (IPNI) [25], *The World Checklist of Vascular Plants* (WCVP) [29] and *Plants of the World Online* (POWO) [30]. Kew also hosts *Medicinal Plant Names Services* (MPNS) [12]. Each of these resources is actively curated: erroneous data are corrected, new names are added, and novel taxonomic arrangements are adopted.

IPNI catalogues all scientific plant names ever published; it is purely nomenclatural. WCVP catalogues vascular plant species: providing an accepted name with its synonyms. POWO aims to be more encyclopaedic, and slowly adds descriptive, visual and geographical data that Kew, and its collaborators, wish to make available and subsequently share through initiatives such as the *World Flora Online* [31]. MPNS specifically targets the health and regulatory communities and those working with natural products. It tracks citations of plants being used medicinally, the parts of the plants used, and the diverse common and pharmacopoeial names used for those plants and the drugs derived from them.

Kew's four resources are linked and offer different views onto the same plant taxonomy, which is constantly being updated as curators add, edit and change the relationships between plants, reflecting the many taxonomic publications appearing each year. MPNS enriches this with names and parts derived from an increasingly comprehensive medical literature: those working with vascular plants are relatively well-served.

Finding equivalent resources for organisms other than plants is more challenging. Algaebase [32] offers similar coverage for algae. Index Fungorum [33] and Species Fungorum [34] have similar functionality to IPNI and WCVP, respectively, for fungi, without currently being blessed with equivalent levels of curatorial support. These major fungal resources were adopted by Kew, but have yet to be fully integrated [35]. Nomenclatural resources for animals naturally reflect our variable level of understanding and knowledge of different groups: birds, mammals and fish being well catered for, but many insect groups being poorly researched. Catalogue of Life [36] is an 'aggregator', periodically taking copies of data subsets from multiple more specialist taxonomic sources, merging these, and aiming to provide a single point of reference for all biological organisms. This serves some purposes but, as might be expected, offers datasets of variable reliability and currency (including contributions which no longer reflect their owners' views).

It is beyond the scope of this chapter to discuss details of the nomenclature and taxonomy for organisms other than plants. The principles and practices in zoology and virology, for example, are largely the same as for plants, although there do exist small variations in how these are implemented in practice. These issues are, of course, also relevant for pharmacovigilance since many traditional medicine systems utilise preparations which also include ingredients sourced from other organisms.

8.3 Challenges Inherent in Using Scientific Names

Scientific names provide the only means of precise, unambiguous communication about plants. Scientific names do not change in meaning overtime, refer uniquely to single plant species and will be recognised globally in all countries and disciplines. Despite our dependence on scientific names, there are practical challenges to employing them effectively, which complicates their use by non-taxonomists when interpreting published research, patient records or regulations, and when publishing. These challenges are discussed below, with examples and statistics relevant to vascular plants. The issues and situations described are, however, equally relevant to fungi, fish, algae or insects.

8.3.1 Too Many Scientific Names

Approximately two thousand new plant species are discovered every year, mostly in the tropics where the plants have been less intensively studied [37]. Evidence for this comes from the two thousand scientific plant names for new species consistently published annually in the botanical literature. A further 8000 names change as species are moved between genera, or are otherwise reclassified following molecular study. Increasingly, scientific journals automatically record newly published plant names in IPNI, but typically there is some delay between publication and incorporation into IPNI.

IPNI has about 1.6 million records of which more than one million are binomials (names for species). In comparison, the World Checklist of Vascular Plants [29] recognises approximately 345,000 vascular plant species [38], placing many IPNI names in 'synonymy'. On average, each plant has three alternative scientific names [39]. Many medicinal plants, however, with long histories of economic importance, have been more intensively studied, resulting in many more synonyms. The Medicinal Plant Names Services (MPNS) records about ten scientific synonyms for each of the 33,000 medicinal plants registered.

8.3.1.1 Why Do Synonyms Occur?

Synonyms occur for various reasons. One reason, mentioned earlier, will typically follow revision of a genus. Increasingly such studies employ molecular (DNA) or chemical evidence of shared evolutionary pathways. Such revision may move one or more species into other genera, and require that different scientific names be used to reflect that.

For example, Linnaeus [20] published *Cassia occidentalis* L. A subsequent study [40] moved the species into the genus Senna, creating the name '*Senna occidentalis*

(L.) Link'. This is known as a 'new combination': the species epithet ('occidentalis') remains unchanged in recognition of Linnaeus's original description, and Linnaeus's own name (in abbreviated form 'L.') appears in parentheses prior to that of Link, marking him as author of the original description, while Link is cited as moving the plant into the genus '*Senna*'. The plant now has two synonyms. This creation of a new scientific name to recognise a change in our taxonomic understanding is an inherent consequence of the binomial system. In this situation, both names are based on the *same* 'type' specimens: there can be no dispute as to whether there are one or two species involved. Synonyms sharing the same 'types' are termed 'homotypic synonyms'. In these cases, synonymy is absolute and fact, not subject to debate nor requiring further evidence.

Other causes of synonymy exist. Imagine botanists studying plants in Bolivia and discovering a species unknown to them. They successfully publish a new scientific name citing the specimens they collected as 'types'. Unfortunately, these botanists were unaware of an earlier study of Peruvian plants which had already described that species, publishing a different name and citing different 'type' specimens. A subsequent review of all South American material would detect this duplication and the authors of this review would publish an article placing these names into synonymy and presenting morphological, chemical and molecular evidence indicating the Peruvian and Bolivian plants belong to the same species. The two scientific names are 'heterotypic synonyms': based on *different* 'types'. However, it remains possible that new evidence may subsequently appear requiring the decision to merge the species to be reversed.

8.3.1.2 Accepted Names: Establishing Taxonomic Position

The presence of multiple names for the same plant raises the obvious question of *which* name should be used. MPNS, drawing on WCVP, advises users of all possible scientific synonyms for each plant (all of which can still be used unambiguously if necessary), and which of those alternative names is the 'accepted name' within Kew's current classification. Each genus, species, subspecies or botanical variety has a single 'accepted' scientific name which places that plant unambiguously within the taxonomic hierarchy: *Senna occidentalis* (L.) Link is a binomial (the name of a species) explicitly recognised as sharing more in common with other species in the genus 'Senna' than with species placed in other genera. The significance and benefits of this are explored below.

8.3.1.3 Issues with Multiple Synonyms

The existence of synonyms hinders effective and reliable research, ADR reporting and signal detection, and herbal medicinal product regulation. Individuals or agencies using different scientific names for the same plant (and unaware that other names exist) will fail to communicate effectively. Regulators using one scientific name will fail to detect ADR reports relating to an herbal recorded elsewhere using alternative synonyms. Multiple synonyms similarly prevent comprehensive data retrieval from databases or online resources. MPNS studies demonstrate that searching PubMed [41] for medicinal plants using a single scientific name, on average, retrieves only 15% of all publications indexed by PubMed referring to that plant. MPNS users can search PubMed employing all scientific synonyms simultaneously, thereby guaranteeing that they retrieve 100% of the relevant articles.

MPNS has long collaborated with the USA Food and Drug Administration (FDA). In 2014, MPNS analysed all of the scientific plant names employed in FDA's regulatory datasets (unpublished report for FDA). In total, 58% of these names were valid and 'currently accepted' scientific names, unambiguously referring to a single plant and taxonomically up to date. This figure is significantly higher than for datasets analysed from other regulators, since FDA is aware of the complexity inherent in botanical nomenclature and employs staff to address this. A further 25% of FDA's entries were valid scientific names which were no longer the 'currently accepted' names for those plants, but were older synonyms. For most purposes, this might not be an issue. Each name uniquely relates to a single plant species: there could be no confusion. However, issues did arise for FDA because it was unaware of this synonymy. As a consequence, 24 species of plant were listed more than once, using alternative synonyms, and plants were therefore regulated inconsistently, depending on which synonym was used. Further, despite FDA's careful curation, 22% of the plant names in its database were invalidly published, misspelt, ambiguous, or referred to organisms other than plants. The challenges are clear; some solutions are offered below.

8.3.2 Homonyms

Homonyms occur where two or more botanists independently publish different scientific names using exactly the same genus and species epithets. Approximately 4% of scientific names (between 40,000 and 50,000 scientific names in total) have 'homonyms'. Use of the binomial alone (genus and species without the publishing author) risks being ambiguous. Two homonyms will frequently refer to completely different plants. Much confusion derives from the lack of appreciation that homonyms exist, erroneous assumptions being common in both research and regulation. Box 8.4 illustrates a well-known case. The EU Commission [42] sought to ban the import into Europe of a poisonous Japanese plant (*Illicium anisatum* L.). It looked to reduce the mistaken import of this toxic species, driven at the time by high demand for 'staranise' (*Illicium verum* Hook.f.), required for pharmaceutical drug production during the 'bird-flu' pandemic [43]. Unknown to the Commission, the existence of homonyms would lead to confusion: one synonym of *Illicium verum* Hook.f. is *Illicium anisatum* Lour. The EU regulation simply banned '*Illicium anisatum*', without including an author, and so was ambiguous. This failure to specify could feasibly be interpreted as banning the import of an economically important spice. The regulation was hastily replaced to avoid confusion in the trade. The lesson is simple: citing a scientific name without including the name of the publishing author risks ambiguity.

Box 8.4 Homonyms: An Example of How the Failure to Cite the Publishing Author of a Scientific Plant Name in Regulation Led to Ambiguity

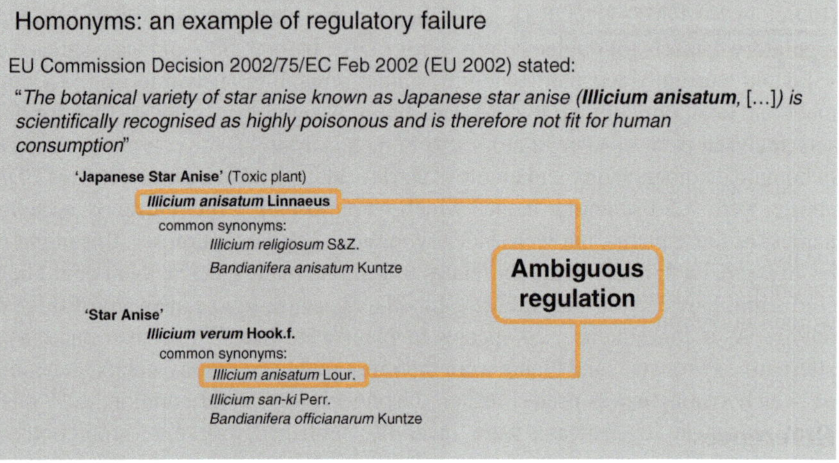

8.3.3 *Names Keep Changing*

Systematists study the diversity of plants and the relationships between them. Linnaeus published species descriptions based on his morphological and ecological observations. Later generations added phytochemistry as a source of evidence, and DNA studies today provide fresh insights into how closely plants are related. Sophisticated methodologies and data analyses add to our ability to understand how plants evolved over time. A reliance on the binomial system means that rearrangement of taxonomic hierarchies will frequently require changes to the accepted scientific names used since these indicate the position of that plant in the hierarchy. Previously employed accepted names will be demoted to become synonyms.

Individual scientific names retain their meaning for all time, always referring to the same plant. Which names are 'preferred' by taxonomists when referring to a particular plant, however, does change, and surprisingly quickly. As mentioned, approximately 10,000 changes to the scientific names of higher plants are published each year. Table 8.2 summarises the major causes of change. At least 1500 changes annually are estimated to relate to medicinal plants.

Table 8.2 Rates of change of scientific plant names

Rates of change for scientific species names[a]		
Reason for name change	Approximate number published/year[b]	Approximate number/year for medicinal plants[c]
Newly described species (new names)	2200–2700	380
Species moved from one Genus into another (new name combinations)	2000–6000	300–900
Changing species delimitation (scope and description) (a) One species split into many or (b) two or more species merged into one	4000[d]	600

[a]Statistics are for species (binomials alone) and do not include counts for genera, subspecies or varieties

[b]Numbers vary from year to year (IPNI 2020) [25]. Molecular studies increasingly drive these taxonomic changes

[c]Assumes 15% of all vascular plants have a medicinal use (c.50,000 species)

[d]Estimates for changing delimitations more difficult to obtain

The following example demonstrates the practical significance of this. Kew chemists had extracted a novel molecule, 'castanospermine', from an Australian tree (*Castanospermum australe* A.Cunn. ex Mudie) [44], which proved to play a role in inhibiting the human immunodeficiency virus (HIV) [45, 46]. Given this important activity, Kew biochemists sought to isolate similar compounds from other plants to further understand and treat HIV. Initially, they thought to look among the most closely related plants, i.e., among other species of the genus *Castanospermum* A.Cunn ex Mudie. However, there are no other species belonging to that genus. Conversations with Kew taxonomists familiar with this group of plants, however, alerted the biochemists to the existence of a South American genus *Alexa* Moq. The taxonomists considered *Alexa* to be 'congeneric' with *Castanospermum* [47], all evidence indicating that these two genera should be merged into one. With this information, the biochemists successfully isolated similar, but slightly different, molecules from several species of *Alexa* [48]. The biochemists were fortunate to work at Kew and to know who to ask. All biochemists would have had access to this information if taxonomists had found the time and resources to publish this research, merge the genera, and create the new scientific names necessary to reflect this.

Gardeners and herbalists may feel aggrieved when plants long-known by one name are given new 'accepted' names. But these changes are not random and reflect our better appreciation of that plant's taxonomic position and evolutionary past, permitting us to anticipate its chemistry or other properties: evidently important in drug discovery, for example. A modern realistic taxonomic hierarchy can similarly benefit pharmacovigilance, aiding our ability to anticipate characteristics and to analyse large volumes of ADR records more meaningfully. Box 8.5 illustrates how a recent study reclassified the popular culinary herb 'rosemary' to reflect its very close genetic resemblance to 'sage'.

Box 8.5 Why 'Rosemary' Came to be Renamed

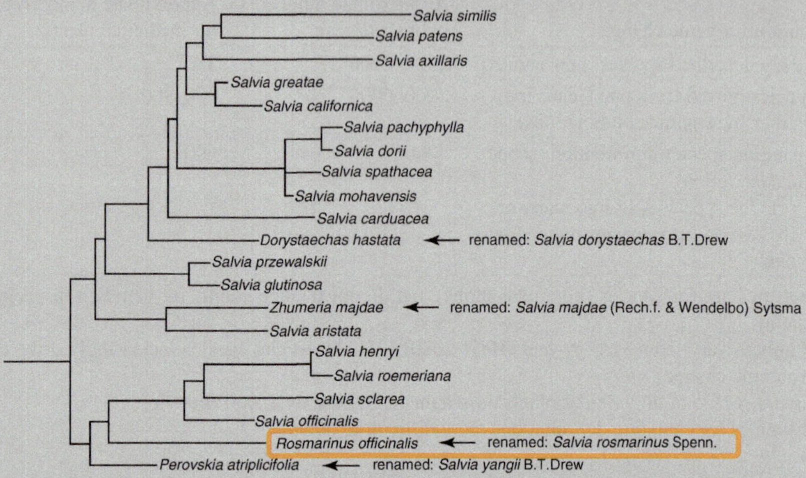

This figure illustrates a portion of a recent phylogenetic tree [49] showing the genetic similarities found between some closely related species in the Lamiaceae (mint) family. Most of these species already belonged to the genus *Salvia*, but four near relatives had previously been placed in other genera. Based on this evidence, Drew reclassified these four species, renaming them as species of *Salvia* to reflect their genetic similarity. One consequence was that the accepted name of the culinary herb *Rosmarinus officinalis* L. ('rosemary') became *Salvia rosmarinus* Spenn.

8.3.4 Conflicting and Out-of-Date Reference Sources

Botanists have made multiple attempts to publish taxonomies but, in doing so, have not always served non-botanists well. An array of websites offers lists of plants, often with synonyms and occasionally their geographic distribution. Unfortunately, these resources frequently overlap, or present out-of-date taxonomies and contradictory opinions, causing further confusion for non-specialists.

These sites have different purposes, rarely made sufficiently clear to users. As discussed, IPNI sets out to catalogue scientific names and makes no pretence to record data about plants or establish accepted names, yet users often fail to appreciate this. Other sites differ in their geographic scope (studies of plant material from one country may be expected to arrive at different conclusions to those based on material from others). The most important cause of botanical resources offering conflicting answers, however, is the quality and currency of their data. The larger the resource, the more difficult data maintenance becomes.

Catalogue of Life (COL) [36], an 'aggregator', pools data from other specialist sites. The data presented are of variable reliability depending on both the quality of the data shared and how long ago those data were retrieved from the curator. Resource limitations prevent data from being replaced, resulting in failure to keep abreast of taxonomic changes. Data subsets for some plant families supplied to COL by Kew are more than ten years old despite significant enhancements having been made to data currently visible via Kew's website. Other sites take other classifications and attempt to synthesise taxonomies through automated comparisons [50].

A challenge facing 'aggregator' sites, and those automatically synthesising taxonomies, is to detect and resolve all conflicting taxonomic opinions contained within the datasets supplied. In 2012, this chapter's first author managed the build and publication of The Plant List (TPL) [51], which remains a popular source of plant nomenclature. When TPL was published, it contained errors and gaps deriving from the various datasets used, and the taxonomic views expressed within these datasets often conflicted, making it necessary to find and then resolve these differences. Time did not allow for manual expert review. Instead, scores of logical 'rules' (implementing taxonomic procedures) were created and applied sequentially, thereby introducing further errors. It transpired that there was no perfect order in which to apply these rules. Each sequence introduced some form of bias: removing some types of error, but introducing others, and 22% of species names remained 'unresolved' since evidence was lacking as to how they should be interpreted. There has been no curation or correction of TPL data since 2012; known errors and gaps remain. The datasets that contributed to TPL, in contrast, have been enhanced by their curators, but sadly none of these improvements are reflected in TPL; thus, it is static and out of date[52].

8.3.5 Cultivars, Landraces and Chemical Variants

All the taxonomic ranks described above (genus, species, subspecies and variety) sit within the realm of botanical science and employ scientific names. The 'Code' controls their form and use, and these names can be assigned to populations of individual plants sharing a common biology and chemistry. How species should be defined is widely debated [53], however, and for both biological and practical reasons, varies between different classes of organism. Botanists studying global diversity are rarely well-equipped to detect or classify minor variations among plants of the same species. Differences between individuals may have an environmental cause (temperature, soil pH or altitude), or result from small, but consistent, genetic variations, possibly as a product of selection by humans.

Humans seek wines from specific grape varieties grown in particular soils or conditions. The individual properties of *materia medica* may vary similarly. 'Fennel' may be an example familiar to readers. The swollen leaf-base of 'Florence fennel' [54] is used as a vegetable, while two other forms without swollen leaf-bases ('sweet fennel' and 'bitter fennel', containing differing proportions of key constituents) are grown primarily for seed, and used medicinally. These differences are important to growers and users, yet are not particularly stable or genetically distinct: the characteristics of each may vary depending on conditions, and individuals of different groups may interbreed. All three forms belong to a single species: *Foeniculum vulgare* Mill. To refer to these plant groups individually, it becomes necessary to use names for specific cultivars or 'selections', as recognised in trade, in addition to the scientific species names, but this comes with its own challenges.

The International Code of Nomenclature for Cultivated Plants [55] establishes standards and procedures for the use of plant names in horticulture, agriculture and forestry. This plays a similar role to the botanical 'Code', but focusing on cultivated varieties and trade names [56].There are, however, differences in how effectively such names are controlled and employed between continents and domains. Major commercial agricultural crops, for example, have different regional reference resources (e.g. EU Commission [57] or USDA [58]). Resources used by traders in timber or wine overlap, or contradict, one another. The horticultural and agricultural reference resources that do exist fail to list all varieties, certainly of less commercial species or locally recognised landraces (populations of domesticated plants which have, informally and over time, developed characteristics distinct from other populations of that species as an adaptation to localised growing conditions and cultivation practices), and many horticultural growers prefer using commercially beneficial 'trade names' and region-specific name registration schemes. A single core index, equivalent to IPNI, does not exist for cultivated varieties and selections, nor is such a resource ever likely to be practical.

8.3.6 Misapplied Names

You cannot always believe what you read. The scientific literature includes many publications reporting poorly executed research, or well-executed research that was nonetheless carried out using mistakenly identified plants. Such publications wrongly assign properties of one plant to another. Strictly, this is not a nomenclatural issue, but misapplication of a valid scientific name clearly leads to published research or ADR reports being misleading. Such reports may be omitted from (or included in error in) analyses focusing on a particular medicinal plant or ingredient. Ultimately, only access to a voucher specimen [59], or sophisticated chemical or molecular authentication of the plant material used, can enable 100% certainty of the identity of the plant involved.

8.4 Consequences of Misusing or Not Using Scientific Names

The considerable challenges to appropriate use of scientific nomenclature outlined above lead, inevitably, to misuse by regulators, authors of pharmacopoeias, those submitting and analysing ADR reports, coders analysing clinical trials data, traders, health practitioners and scientists publishing or using published research. The following sections outline some consequences.

8.4.1 Imprecision in the Scientific Literature

A lack of appreciation that scientific names are necessary at all is surprisingly common with, for example, research publications reporting laboratory analysis of 'ginseng', offering no further indication of which species was tested. 'Ginseng' is used by different pharmacopoeias to refer to herbal products derived from at least 16 different plant species (including *Panax quinquefolius* L., *Eleutherococcus senticosus* (Rupr. & Maxim.) Maxim., *Panax notoginseng* (Burkill) F.H.Chen, *Hebanthe erianthos* (Poir.) Pedersen, *Lepidium meyenii* Walp.) [12]. There is no guarantee, therefore, that the authors of research into ginseng actually analysed material belonging to the species *Panax ginseng* C.A.Mey., unless they explicitly say so. Even if they are precise with their terminology, there is no guarantee that they have correctly authenticated the material under study. Given the complexity and potential confusion surrounding common substitute or adulterant species [60] effective authentication is beyond the scope of this chapter. Further confusion arises because some infraspecific names published under *Panax ginseng* are now recognised to be synonyms of other species. *Panax ginseng var. repens* (Maxim.) Makino, for example, is a synonym of *Panax japonicus* (T.Nees) C.A.Mey., which has quite different properties and uses in the Japanese and Chinese Pharmacopoeias [12]. Failure to cite a scientific name in full, or to establish where voucher specimens of the plants studied are deposited, undermines the scientific merit of a publication [61, 62].

Nesbitt et al.[63] reviewed the names employed in fifty peer-reviewed scientific journal papers relating to food composition data from 502 plant species in Ethiopia. The articles were published between 1991 and 2003 and indexed in major abstracting resources, e.g. Medline [41]. Only 37 (about 7%) of the 502 plant citations followed best practice for plant nomenclature (and only 36 followed best practice as to plant identification). Overall, 27% of plants were listed using names not in current use, incorrectly spelt, or both. A search of citation or abstract databases with accepted scientific names found only 159 of the 502 plant citations, despite all having been indexed.

Rivera et al. [61] studied 428 articles relevant to herbal pharmacovigilance from two well-regarded journals (Journal of Ethnopharmacology, Phytomedicine) whose editors are aware of the challenges of using scientific plant names. These articles were published between 2012 and 2013. In total, 308 articles (72% of total number published) were found to have cited incorrect, incomplete, ambiguous or imprecise names. These articles employed 9178 scientific plant names in total, of which 3445 (37%) were incorrect, incomplete, ambiguous, or imprecise. The situation is likely to be worse among those journals paying less attention to scientific nomenclature than do the editors of the above journals.

Nesbitt et al. [63], Rivera et al. [61] and Bennett and Balick [64] all suggest ways to avoid imprecision. The editors of the journals *Journal of Ethnopharmacology* and *Phytomedicine* introduced guidelines for authors, although given the volume of articles submitted, there remain challenges in detecting the use of incorrect or outdated names, particularly where journals suggest use of out-of-date taxonomic resources (e.g. TPL). Heinrich et al. [65] offer guidelines in the context of ethnopharmacological fieldwork.

8.4.2 A Failure to Protect Traditional Knowledge

Precise, unambiguous names are required in many other contexts, including for establishing patents and protecting intellectual property. Simmonds et al. [66] evidenced how poorly this is currently achieved with 35% of the plant names cited in the patents studied being found to be misspelt, ambiguous or meaningless (i.e. they were not published scientific names). The Indian Government looks to protect the intellectual property inherent in traditional practices employing plants for human health. The Traditional Knowledge Digital Library (TKDL) [67] catalogues traditional use, aiming to prevent unscrupulous third parties from patenting long-known uses. The scientific nomenclature within TKDL, however, are no better than the scientific literature and employ older and misspelled names: in some cases, plants are listed twice under alternative synonyms. Most significantly, however, TKDL lacks most scientific synonyms for the plants listed. Someone wishing to register a patent for a plant covered by TKDL can do so simply by citing one of its older synonyms. Patents Officers are unfamiliar with botanical nomenclature and may authorise the patent unaware of that plant's previously documented use.

8.4.3 Ambiguity in Monographs and Pharmacopoeias

Previous sections outlined how common names and pharmacopoeial names are inherently ambiguous. Using scientific names to refer to plants removes one cause of ambiguity, but fails to resolve others inherent in the descriptions of some herbal substances in monographs and pharmacopoeias. Some herbal substances are explicitly defined citing the possible use of two or more species. Scientifically, these alternative species may be closely related and, therefore, assumed to have similar properties. However, as separate species, it must be assumed that they will show some differences in the chemical constituents present, which could potentially lead to differences in the herbal substance(s) created. A second cause of imprecision occurs when monographs are not explicit as to whether only one of the cited species can be used, or whether mixtures of those species are permissible.

Further ambiguity derives from the practice in some monographs of citing genera: implying that use of any species from that genus would be permissible. This raises several issues. For wide-ranging genera (with scores of species found across multiple continents), it is reasonable to doubt whether the monograph's authors have carried out laboratory analysis of all species. Even if genera contain relatively few species, as explained previously, the exact group of species considered to form that genus will vary between different taxonomies. Species will move in and out of a genus over time. To our knowledge, the authors of herbal monographs fail to indicate which taxonomic delimitation of the genus they follow, leaving the reader uncertain as to which species can be used.

Some genera (in some taxonomies) may contain only one species ('monotypic'). In such cases, the author of the monograph may feel that it is 'obvious' which species is intended. Nevertheless, some specialists may group more species within that genus, and ambiguity persists for the reader. There is, therefore, no alternative other than to specify the scientific name of each species which is intended for use (see best practice below).

8.4.4 Difficulties Recording and Utilising Adverse Reaction Reports

The World Health Organisation's (WHO) global database of individual case safety reports, VigiBase, is developed and maintained by the Uppsala Monitoring Centre (UMC) [68]. It holds over 25 million (in May 2021) centralised ADR records gathered by National Pharmacovigilance centres globally. UMC searches these combined records for 'signals' of safety concerns regarding specific drugs (refer to Chap. 9 in this volume). MPNS has reviewed, corrected and enriched the scientific botanical nomenclature underlying UMC's VigiBase and WHODrug Global systems. UMC and MPNS are working together to implement annual revision and updates to ensure UMC data resources remain current.

More widely, the data quality of ADR reports presents challenges to national pharmacovigilance centres, to UMC who centralise these records, and, ultimately, to those undertaking signal analyses of these records. Individual ADR reports may use differing names for a herbal or plant; scientific names may be incomplete, incorrect or misspelt. As described above, pharmacopoeial names of herbal substances are inherently and frequently ambiguous, being imprecise as to the exact formulation or species employed. Records may, surprisingly frequently, lack any specificity and cite, for example, only the plant's genus name without specifying which particular species was involved, or simply record use of 'traditional medicines'. Such imprecision clearly limits the ability of pharmacovigilance professionals to analyse these datasets adequately, quantitatively, or to safely draw conclusions regarding harms associated with use of plants or herbal medicines.

The following section offers some guidance for improving the quality of ADR reports. Given the existing body of ADR reports, however, it is imperative that analyses of these records consider all possible interpretations of the 'names' (of plants or herbal drugs) employed, consider all possible synonyms, and avoid false assumptions as to the significance of records employing potentially ambiguous terms.

Clinical trials of conventional pharmaceutical medicines face similar issues. 'Coders' of clinical trial data need an understanding of the medical history of patients participating in the trial. Increasingly, this means ascertaining if trial participants are using any herbal medicines or other 'natural health products', traditional medicines or dietary supplements. Understandably, coders struggle to establish precisely the plant substances being taken concomitantly by study participants, given the product, substance and plant names passed to them by participants. Such imprecision will again confound attempts to detect signals of safety concerns.

8.5 Solutions

People from many walks of life (medicines' regulators, health practitioners and rural communities employing traditional remedies) need to access information about medicinal plants and herbal substances reliably. The success and significant use of

The Plant List (TPL) demonstrated the demand for a comprehensive, simple to use, catalogue of medicinal plants [69]. However, TPL, like most botanical references, lacks any common herbal drug or pharmacopoeial names. TPL also fails to record or resolve the many misspelt or misused scientific names occurring within health regulation and research literature. Medicinal Plant Names Services (MPNS) was conceived to overcome these limitations: it serves as an intermediary between authoritative, constantly curated botanical references and health professionals (Box 8.6).

Box 8.6 The Intermediary Role of MPNS Between Botanical Systematics and Health Professionals

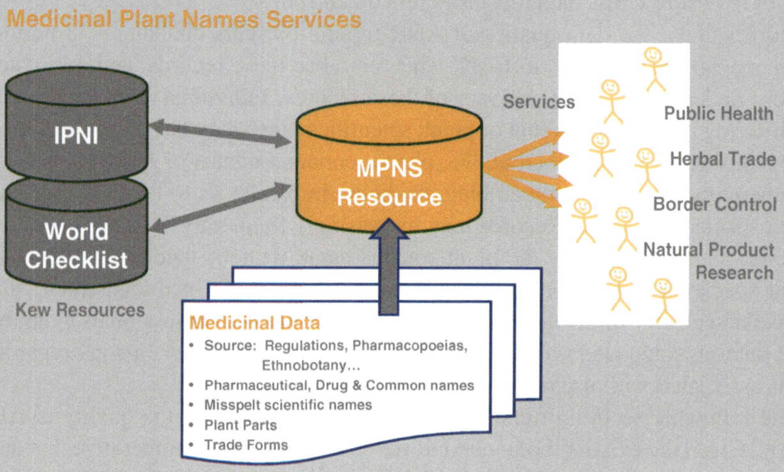

8.5.1 The MPNS Data Resource

MPNS was first published online in 2014 and has grown over time. Version 10 [12] contains more than 500,000 name records relating to about 33,000 plants. MPNS catalogues the common, drug, pharmacopoeial and scientific names (even where misspelt) employed in 190 major medical, regulatory, natural product or ethnobotanical references. It comprehensively covers the major pharmacopoeias from all continents and tracks the changing use of pharmacopoeial names and herbal substance definitions between editions of these pharmacopoeias. MPNS maps these herbal substances to Kew's plant taxonomy from which it draws the currently accepted scientific names, taxonomy and scientific synonyms for these plants.

Each new version of MPNS catalogues the herbal substances cited in an increasing number of medical, chemical and ethnobotanical publications, prioritising literature from countries less well documented to date. The botanical nomenclature remains static for each version and is then refreshed at the next release to reflect enhancements made to WCVP [29] and IPNI [25]. Citing the version of MPNS used to obtain information about a plant helps retain a paper trail should the taxonomic position of the plant change.

8.5.2 *Using the MPNS Portal*

MPNS's freely available search portal (www.kew.org/mpns) is the easiest means of accessing MPNS data. Users can search using common, pharmacopoeial, or scientific names and retrieve all plants, plant parts, other names and medicinal plant references associated with their search term. Searches can include names in Chinese, Arabic and an increasing number of other scripts. Boxes 8.7 and 8.8 illustrate an example search. MPNS detects and exposes the inherent ambiguity in the Pinyin name 'Mu Xiang', revealing its use in 14 drugs and medicines references to refer to herbal substances derived from seven different plants. These will each have differing chemistries and uses. MPNS lists the accepted scientific names of each plant involved. Additional information is available, such as the names of references citing the search term, and all the various types and permutations of associated scientific and non-scientific names.

Box 8.7 Searching the MPNS Portal [12]

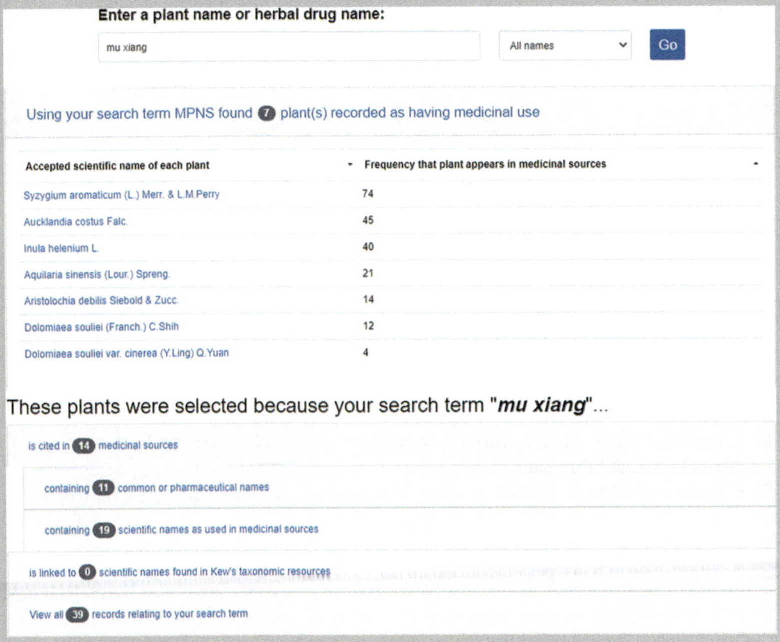

Box 8.8 Exploring MPNS Data for Each Plant

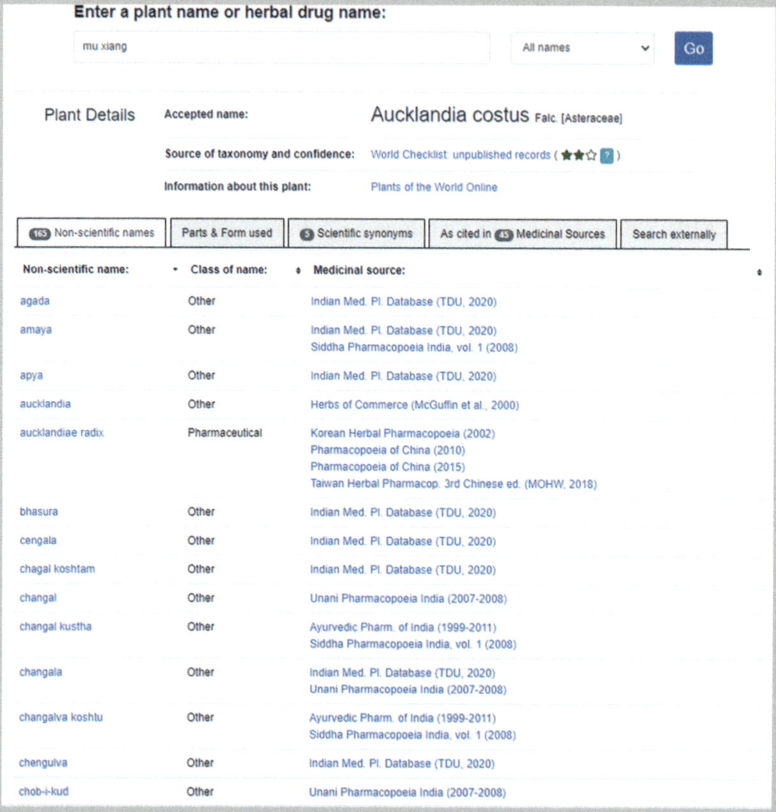

Enter a plant name or herbal drug name:

mu xiang | All names | Go

Plant Details

Accepted name: Aucklandia costus Falc. [Asteraceae]

Source of taxonomy and confidence: World Checklist: unpublished records (★★☆ ?)

Information about this plant: Plants of the World Online

165 Non-scientific names | Parts & Form used | 5 Scientific synonyms | As cited in 45 Medicinal Sources | Search externally

Non-scientific name:	Class of name:	Medicinal source:
agada	Other	Indian Med. Pl. Database (TDU, 2020)
amaya	Other	Indian Med. Pl. Database (TDU, 2020) Siddha Pharmacopoeia India, vol. 1 (2008)
apya	Other	Indian Med. Pl. Database (TDU, 2020)
aucklandia	Other	Herbs of Commerce (McGuffin et al., 2000)
aucklandiae radix	Pharmaceutical	Korean Herbal Pharmacopoeia (2002) Pharmacopoeia of China (2010) Pharmacopoeia of China (2015) Taiwan Herbal Pharmacop. 3rd Chinese ed. (MOHW, 2018)
bhasura	Other	Indian Med. Pl. Database (TDU, 2020)
cengala	Other	Indian Med. Pl. Database (TDU, 2020)
chagal koshtam	Other	Indian Med. Pl. Database (TDU, 2020)
changal	Other	Unani Pharmacopoeia India (2007-2008)
changal kustha	Other	Ayurvedic Pharm. of India (1999-2011) Siddha Pharmacopoeia India, vol. 1 (2008)
changala	Other	Indian Med. Pl. Database (TDU, 2020) Unani Pharmacopoeia India (2007-2008)
changalva koshtu	Other	Ayurvedic Pharm. of India (1999-2011) Siddha Pharmacopoeia India, vol. 1 (2008)
chengulva	Other	Indian Med. Pl. Database (TDU, 2020)
chob-i-kud	Other	Unani Pharmacopoeia India (2007-2008)

Users of the MPNS search portal [12] clicking on the accepted scientific name '*Aucklandia costus* Falc.' (see Box 8.7) can explore all data held by MPNS for this species. The screen is divided into five 'information tabs', each displaying a subset of the available information. From Left to Right, these are:

- Tab 1 listing all (165) non-scientific names for this plant or derived herbal substances.
- Tab 2 detailing the parts of the plant with medicinal use and the forms employed.
- Tab 3 listing all (5) known scientific synonyms for this plant.
- Tab 4 providing details of the (45) plant-derived medicines publications from which the above information is taken, along with the scientific name employed for this plant in each publication.
- Tab 5 allowing users to search external third-party references for further information about this plant.

Typically, users would now wish to explore or compare one or more of the seven plants involved. Box 8.8 contains a screen from the MPNS portal if a user clicked on the first plant listed in Box 8.7: *Aucklandia costus* Falc. Selecting an accepted name makes all information held by MPNS about this plant accessible across a number of 'tabs', including which publications refer to this species, and the various common, drug and scientific names, plant parts and trade forms cited in each.

The 'Search externally' tab enables users to explore third-party online resources for further information about the plant. This illustrates how important it can be to know all possible synonyms when looking for data. For example, one source which MPNS links to is PubMed [41]. PubMed is a significant digital research library, hosted by the USA National Institutes of Health, and comprises more than 30 million citations from biomedical literature, life science journals, and online books. Searching PubMed manually (directly) using the name *Aucklandia costus* Falc. retrieves two published articles relating to that plant. In contrast, using MPNS, PubMed can be searched indirectly using the accepted name and all five known scientific synonyms simultaneously. This search now retrieves 627 publications from PubMed that refer to this plant using at least one of those synonyms. Thus, comprehensive retrieval requires knowing all possible synonyms.

8.5.3 MPNS Network and Services

MPNS deploys its data and expertise in other ways to help organisations and individuals manage their data about medicinal plants and herbal products. Long-standing collaborators include the USA Food and Drug Administration (USFDA), the World Health Organisation's Uppsala Monitoring Centre (WHO-UMC) [70], Medicinal Plant Resources of the World (MAPROW) [71] with the International Union for Conservation of Nature: Medicinal Plant Specialist Group [72]. MPNS now collaborates with medicinal plant and natural products specialists in many countries that are themselves building catalogues of local use of medicinal plants. MPNS helps enhance the integrity and nomenclatural richness of their databases while facilitating their use of and links to other resources. Typically, MPNS will 'validate' the scientific names compiled by collaborators (e.g. are they valid, meaningful and unique?), check integrity (e.g. are any plants listed twice under alternative synonyms?), update taxonomy (supplying accepted scientific names and families) and enrich a dataset with all possible synonyms. As a result, MPNS' network of partners now shares a common taxonomy enabling them to reliably link or exchange data. MPNS benefits, in return, by adding the local and pharmacopoeial names and plant parts employed, as recorded by collaborators. MPNS also offers consultancy (e.g. advising organisations on how to structure databases containing plant records, or on managing dataflows) and training as to botanical nomenclature and best practice [73].

8.5.4 IDMP: A New Drug Standard

Identification of Medicinal Products (IDMP) [74–76] is a recent global data standard for all medicinal drugs, developed and adopted by many regulators including USFDA [77], the European Medicines Agency [78] and WHO. It covers all drugs, including herbal substances. IDMP facilitates the exchange of information and consistent regulation of individual drugs despite their being known by different names in different countries and disciplines. Working together with International Organization for Standardization (ISO) and FDA (who implemented IDMP software used by multiple regulators) to ensure adequate data structures for herbal products [79], MPNS provides 'controlled vocabularies' for plant names and plant parts.

8.6 What Does Best Practice Look Like?

8.6.1 Documenting Identity

Successfully citing a scientific name clearly has no value if the plants studied were misidentified. Although not the primary topic of this chapter, it is clear that citing voucher specimens, where these are deposited, and how specimens were identified adds credibility to published research, enabling readers to confirm the identity of the plants studied were this to be questioned. It ensures the reproducibility of research [59, 61, 62, 65].

8.6.2 Using Scientific Names to Communicate and Publish Effectively

Authors of publications, reports, pharmacopoeias, regulations and adverse reaction records have a responsibility (and self-interest) in being as precise and unambiguous as possible regarding the plant(s) to which they refer. Since common names and pharmacopoeial names are inadequate and open to misinterpretation, complete scientific plant names are obligatory for achieving precision and avoiding ambiguity. The questions these authors must address therefore become 'What is the identity of this plant?', 'Am I using a valid scientific name?', 'Is this the current scientific name for this species?' and 'Which synonyms of this plant may be employed by my intended audience?'.

Dauncey et al (2016) [62] summarise the types of mistake commonly made when using plant names and offer simple recommendations for avoiding them.

Choice of the 'accepted' scientific name for each plant will ideally follow that of a recognised authoritative taxonomic reference which authors can cite. Incomplete or older (not regularly updated) reference sources should be avoided. MPNS [12,

52] serves pharmacovigilance professionals well in providing a modern authoritative and comprehensive nomenclature for all medicinal plants, and linking these to the pharmacopoeial names and plant parts from medicinal literature. By being versioned, it also provides a static, citable snapshot of taxonomic information for use by, for example, professionals in health/medicines' regulation. Kew's wider resources (e.g. WCVP [29]) serves for all other plants. MPNS is considering inclusion of all plants, flagging those for which no medicinal use has been found to date. Authors might also cite widely used synonyms in their publications since it avoids any doubt for readers more accustomed to other names.

Papers including data tables with long lists of plants face particular challenges and risk including plants several times under alternative synonyms. To avoid such illogicality, authors again need to check all names against a comprehensive source such as MPNS [12]. MPNS seeks to automate such a service for journals and editors.

8.6.3 Locating Publications and Data

As explained previously, to be certain of locating all research publications, or all ADR reports, associated with a plant requires knowing all of a plant's synonyms and searching using each in turn. MPNS [12] provides convenient access to a comprehensive and current list of synonyms of medicinal plants, and the ability to automatically search online medical resources, such as PubMed [41], using all synonyms.

Searching databases using a genus name alone will retrieve all records citing that genus name. However, this wider trawl will also find species names now considered to be synonyms of species in other genera. A search of MPNS using 'Hypericum', for example, retrieves records relating to well over 100 species. Most of these do belong to the genus '*Hypericum* Tourn. ex L.', but a significant percentage of these plants is now considered to be genetically closer to entirely different genera. Thus, '*Hypericum loureiroi* K.Koch' is now a synonym of the species '*Cratoxylum cochinchinense* (Lour.) Blume'. As our understanding advances, the list of species considered to belong to a genus will change. Unlike most databases which are unaware of synonymy or employ incomplete synonymy, MPNS enables users to detect and resolve such apparent contradictions.

8.6.4 Interpretation of the Literature

Articles employing a pharmaceutical name (e.g. 'Cimicifugae rhizoma') and no scientific plant name will leave the reader uncertain as to which herbal substance (and plant) is involved (see Box 8.1). Pharmaceutical terms are used differently between pharmacopoeias. Such ambiguity risks readers drawing false conclusions by making assumptions as to which plant/herbal substance was involved. MPNS [12] catalogues ambiguous use of common and pharmacopoeial names.

As described previously (see Box 8.4) scientific binomials lacking a publishing author may also be ambiguous. Caution is necessary in determining to which plant species the authors referred. It may not be possible to be certain of their intention, but readers can explore all possible interpretations using MPNS [12].

8.6.5 *Creating and Interpreting Adverse Reaction Reports Involving Herbal Substances*

Adverse reaction reports relating to herbal substances or plant material are often imprecise, limiting their value and preventing their effective use in signal detection. Circumstances, of course, may make imprecision unavoidable, but what advice might be given to practitioners or others when preparing such reports that could improve their precision or reliability?

Establishing the precise identity of the substance or plant material should be the objective. Scientific plant names are necessary for recording an unambiguous identity but cannot themselves ensure that the plant material was correctly identified. Wherever possible, therefore, samples of the plant itself (or the herbal substance(s) involved) should be captured, stored as a voucher, and identified either through traditional morphological analysis or using chemical and molecular validation. The name of a marketed herbal product (including its batch identifier) may indicate its content. Where manufactured herbal products are involved, the original container showing how the product was labelled should be retained. Again, although a necessary precaution, this may be insufficient since labels may not cite a scientific name and, unauthorised products at least, may not contain the correct ingredients. Even regulated products suffer from supply chain issues putting into question the validity of the content of the commercial product. There is no substitute for an authenticated physical specimen of the substance consumed.

Where identification to species is impossible, then a genus name may provide some information. Any indication of how the plant material was identified, the diagnostic features observed and images may provide supplementary evidence. Photographs of plants, unfortunately, cannot be assumed to provide sufficient evidence for botanists to identify a plant remotely. Details necessary to distinguish between similar species (e.g. root, fruit or flower) may not be evident when photographed or only be visible through a microscope.

Common names for plants (or herbal drugs) must be avoided as a primary identifier. They have no formal definition and provide no certainty as to which plant was involved. Pharmacopoeial names are similarly ambiguous, being associated with multiple herbal substances, and are only useful if accompanied by the precise definition of the substance or its source (monograph or pharmacopoeia) to differentiate between alternative meanings. Ultimately, to be most useful, an ADR report would cite the scientific name of the plant (or plants), the plant parts used and, if appropriate, the prepared herbal drug(s) involved. Physical voucher specimens are essential if any subsequent investigation is to be possible.

8.7 Conclusion

Which names are selected to refer to plants or herbal substances, and how precisely these names are used, has enormous significance for pharmacovigilance. Each individual's inherent preferences and interpretation of common names and pharmacopoeial names are informed by their context and cultural heritage. Even when such names are written in Latin or appear in prestigious scientific publications, they may be interpreted and employed differently by people in different countries or in different disciplines.

Scientific and regulatory integrity requires that each herbal substance be defined as precisely and unambiguously as possible, which requires absolute clarity as to the plant (and parts of that plant) to be used. Use of formal scientific plant names, including the publishing author, and citing the physical 'voucher' specimens studied or analysed, can unequivocally establish what plant material is involved.

Understanding how plants have evolved enables us to predict with greater confidence the chemical allegiances among plants, including where molecules with particular patterns are most likely to be found. Scientific nomenclature serves not only as a global means to specify a plant, but also indicates its position in the taxonomic hierarchy. As new evidence accrues, particularly from molecular biology, our understanding of a plant's evolutionary past continues to improve. As a result, the most appropriate ('accepted') scientific name may change to reflect this new understanding of where that plant belongs taxonomically.

Though challenging, effective and responsible pharmacovigilance requires that scientific names, and changes to the taxonomic status of those names, are given attention and respect.

Acknowledgements The authors are indebted to Christine Leon MBE both for inspiring creation of MPNS and for advice concerning particular TCM examples cited here. We thank the editor for the opportunity to publish this chapter and for invaluable guidance as to scope and language.

References

1. Carroll L (1871) Through the looking glass. Macmillan & Co., London, UK
2. Lange D (2006) International trade in medicinal and aromatic plants. In: Bodgers RJ, Craker LE, Lange D (eds) Medicinal and aromatic plants. Springer, Netherlands, pp 155–170
3. Jenkins M, Timoshyna A, Cornthwaite M (2018) Wild at home: exploring the global harvest, use and trade of wild plant ingredients, TRAFFIC, Cambridge, UK. https://www.traffic.org/publications/reports/wild-at-home-an-overview-of-the-harvest-and-trade-in-wild-plant-ingredients/. Accessed 1 Feb 2021
4. Howes MJ, Allkin B, Simmonds MSJ et al (2020) Molecules from nature: reconciling biodiversity conservation and global healthcare imperatives for sustainable use of medicinal plants and fungi. Plants People Planet 2(5):463–481. https://doi.org/10.1002/ppp3.10138
5. Payyappallina U, Subramanian SM (2015) Traditional medicine. In: Connecting global priorities: biodiversity and human health, a state of knowledge review. World Health Organisation

and Secretariat of the Convention on Biological Diversity. https://www.who.int/publications/i/item/connecting-global-priorities-biodiversity-and-human-health. Accessed 1 Feb 2021
6. Dauncey EA, Howes MJ (2020) Plants that cure: plants as a source of medicines, from pharmaceuticals to herbal remedies. Princeton University Press, US. and Kew Publishing, UK
7. World Health Organisation (2019) WHO Global Report on Traditional and Complementary Medicine. https://www.who.int/traditional-complementary-integrative-medicine/WhoGlobalReportOnTraditionalAndComplementaryMedicine2019.pdf?ua=1. Accessed 1 Feb 2021
8. Brandão MG, Cosenza G, Moreira RA, Monte-Mor RLM (2006) Medicinal plants and other botanical products from the Brazilian official pharmacopoeia. Rev Bras de Farmacognosia 16(3). https://doi.org/10.1590/s0102-695x2006000300020
9. Allkin B,Patmore K, Black N et al (2017) Chapter 4: Medicinal plants: current resource and future potential. In: Willis K (ed) State of the world's plants. Royal Botanic Gardens, Kew, London, UK.
10. Allkin B (2014) Communicating safely and effectively using plant names. In: Mukerjee PK (ed) Traditional medicines and globalisation: the future of ancient systems of medicine. Maven Publishers, Kolkata, India
11. British Pharmacopoeia Commission (ed) (2013) British Pharmacopoeia 2014 (Vol. IV): herbal drugs, herbal drug preparations and herbal medicinal products. TSO, London, UK
12. Royal Botanic Gardens, Kew (2021) Medicinal plant names services, version 10. www.kew.org/mpns. Accessed 1 Mar 2021
13. European Directorate for the Quality of Medicines & Health Care (EDQM) (2012) European Pharmacopoeia (7.8 edn). Council of Europe, Strasbourg, France
14. Chinese Pharmacopoeia Commission (2015) Pharmacopoeia of the People's Republic of China 2015, Chinese edition. China Medical Science and Technology Press, Beijing, China.
15. Committee of the Japanese Pharmacopoeia (2006) The Japanese Pharmacopoeia, 15th ed. English edition (JP XV). Ministry of Health Labour and Welfare, Tokyo, Japan.
16. Vanherweghem JL, Depierreux M, Tielemans C et al (1993) Rapidly progressive interstitial renal fibrosis in young women: association with slimming regimen including Chinese herbs. Lancet 341(8842):387–391. https://doi.org/10.1016/0140-6736(93)92984-2
17. Debelle FD, Vanherweghem JL, Nortier JL (2008) Aristolochic acid nephropathy: a worldwide problem. Kidney Int 74(2):158–169. https://doi.org/10.1038/ki.2008.129
18. Jadot I, Declèves AE, Nortier J, Caron N (2017) An integrated view of aristolochic acid nephropathy: update of the literature. Int J Mol Sci 18(2):297. https://doi.org/10.3390/ijms18020297
19. American Herbal Products Association (2000) Herbs of commerce, 2nd edn. AHPA, Silver Spring, MD
20. Linnaeus C (1753) Species plantarum (2 vols). Salvius, Stockholm, Sweden.
21. Turland NJ, Wiersema JH, Barrie FR et al (2018) International Code of Nomenclature for algae, fungi, and plants (Shenzhen Code) adopted by the Nineteenth International Botanical Congress Shenzhen, China, July 2017. Koeltz Botanical Books, Glashütten
22. International Committee on Systematics of Prokaryotes (ICSP) (2019) International Code of Nomenclature of Prokaryotes 2019. https://www.the-icsp.org/bacterial-code. Accessed 1 Feb 2021
23. International Committee on Taxonomy of Viruses (ICTV) (2017). International Committee on Taxonomy of Viruses. https://talk.ictvonline.org/. Accessed 1 Feb 2021
24. International Commission on Zoological nomenclature (ICZN) (2012) International Code of Zoological Nomenclature (4th ed.). https://www.iczn.org/the-code/the-international-code-of-zoological-nomenclature/the-code-online/. Accessed 1 Feb 2021
25. The Royal Botanic Gardens, Kew, The Harvard University Herbaria, and The Australian National Herbarium (2021) International Plant Names Index (IPNI). https://www.ipni.org/. Accessed 1 Feb 2021

26. Robson NKB (2002) Studies in the Genus *Hypericum* L. (Guttiferae) 4(2). Section 9. *Hypericum* sensu lato (part 2): subsection 1. Hypericum series 1. Hypericum. Bull Nat Hist Mus Bot 32(2):61–123. https://doi.org/10.1017/S096804460200004X
27. Dauncey EA, Irving JTW, Allkin R (2019) A review of issues of nomenclature and taxonomy of Hypericum perforatum L. and Kew's Medicinal Plant Names Services. J Pharm Pharmacol 71(1):4–14. https://doi.org/10.1111/jphp.12831
28. Lindberg H (1906) Öfversigt af Finska Vetenskaps-Societetens Förhandlingar. Helsinki, Finland.
29. Royal Botanic Gardens, Kew (2021) World Checklist of Vascular Plants (WCVP), version 2.0. http://wcvp.science.kew.org. Accessed 1 Feb 2021
30. Royal Botanic Gardens, Kew (2021) Plants of the world online. http://www.plantsoftheworldonline.org/. Accessed 1 Feb 2021
31. World Flora Online Consortium (2020) An online flora of all known plants. http://www.worldfloraonline.org/. Accessed 1 Feb 2021
32. Algaebase (2020) Global algal database of taxonomic, nomenclatural and distributional information. https://www.algaebase.org/. Accessed 1 Feb 2021
33. Index Fungorum (2021) Index of scientific names of fungi,. http://www.indexfungorum.org/Names/Names.asp. Accessed 01 Feb 2021
34. Species Fungorum (2021) Checklists for fungi. http://www.speciesfungorum.org/. Accessed 1 Feb 2021
35. Royal Botanic Gardens, Kew (2021). Name and synonymy resources. https://www.kew.org/science/collections-and-resources/data-and-digital/names-and-taxonomy. Accessed 1 Feb 2021
36. Catalogue of Life (2021) Catalogue of life. http://www.catalogueoflife.org/. Accessed 1 Feb 2021
37. Cheek M, Nic Lughadha E (2016) New plant species discovered in 2015. In: Willis K (ed) State of the world's plants. Royal Botanic Gardens, Kew, London, UK
38. Nic Lughadha E, Govaerts R, Belyaeva I et al (2016) Counting counts: revised estimates of numbers of accepted species of flowering plants, seed plants, vascular plants and land plants with a review of other recent estimates. Phytotaxa 272(1):82. https://doi.org/10.11646/phytotaxa.272.1.5
39. Barker A, Belyaeva I, Black N, et al (2016) Naming and counting the world's plants. In: Willis K (ed) State of the world's plants. Royal Botanic Gardens, Kew, London, UK
40. Link JHF (1831). Handbuch zur Erkennung der Nutzbarsten und am Haufigsten Vorkommenden Gewächse. Berlin, Germany
41. National Center for Biotechnology Information, National Institute for Health (2021) PubMed: 30 million citations for biomedical literature from MEDLINE, life science journals, and online books. https://pubmed.ncbi.nlm.nih.gov/. Accessed 21 Nov 2020
42. EU Commission (2002) Decision of 1 February 2002 laying down special conditions on the import from third countries of star anise (notified under document number C(2002) 379) 2002/75/EC. https://www.legislation.gov.uk/eudn/2002/75/adopted. Accessed 1 Feb 2021
43. Awang DVC, Blumenthal M (2006) Tamiflu and star Anise: securing adequate supplies of the oral antiviral for Avian flu treatment. HerbalGram 70:58–60
44. Nash RJ, Bell EA, Fleet GJW, Jones RJ, Williams JM (1985) The identification of a hydroxylated pyrrolidine derivative from *Castanospermum australe*. J Chem Soc Chem Commun 11:738–740
45. Walker BD, Kowalski M, Goh WC et al (1987) Inhibition of human immunodeficiency virus syncytium formation and virus replication by castanospermine. Proc Natl Acad Sci U S A 84(22):8120–8124. https://doi.org/10.1073/pnas.84.22.8120
46. Fellows LE, Kite GC, Nash RJ, Simmonds MSJ, Scofield AM (1989) Castanospermine, Swainsonine and related polyhydroxy alkaloids: structure, distribution and biological activity. In: Poulton JE, Romeo JT (eds) Plant nitrogen metabolism. Recent advances in phytochemistry, vol 23. Springer, Boston, MA, pp 395–427

47. Lewis GP, Schrire B, Mackinder B, Lock M (eds) (2005) Legumes of the world. Royal Botanic Gardens, Kew, London, UK
48. Nash RJ, Fellows LE, Dring JV et al (1988) Castanospermine in *Alexa* species. Phytochemistry 27(5):1403–1404. https://doi.org/10.1016/0031-9422(88)80203-6
49. Drew BT, González-Gallegos JG, Xiang C-L et al (2017) Salvia United: the greatest good for the greatest number. Taxon 66(1):133–145. https://doi.org/10.12705/661.7
50. Global Biodiversity Information Facility (GBIF) Secretariat (2019) GBIF Backbone taxonomy. Checklist dataset. https://doi.org/10.15468/39omei. Accessed 1 Feb 2021
51. Royal Botanic Gardens, Kew and Missouri Botanical Garden (2013) The plant list, Version 1.1. http://www.theplantlist.org/. Accessed 1 Feb 2021
52. Medicinal Plant Names Services (2018) Why use MPNS rather than The Plant List? https://us11.campaign-archive.com/?e=&u=c82fe6feb1542c251a3b2b97f&id=fdbb815bfe. Accessed 1 Feb 2021
53. Mayr E (1982) Chapter 6: Microtaxonomy, the science of species. In: The growth of biological thought: diversity, evolution, and inheritance. Belknap Press of Harvard University Press, Massachusetts, US
54. Royal Horticultural Society (2021) The RHS Plant Finder. https://www.rhs.org.uk/about-the-rhs/publications/plant-finder. Accessed 1 Feb 2021
55. International Code for the Nomenclature for Cultivated Plants 9th edn. (2016). https://www.ishs.org/scripta-horticulturae/international-code-nomenclature-cultivated-plants-ninth-edition. Accessed 1 Feb 2021
56. Andrews S, Leslie A, Alexander C (eds) (1999) Taxonomy of cultivated plants. Royal Botanic Gardens Kew, London, UK
57. EU Commission (2020) Plant variety catalogues, databases & information systems. https://ec.europa.eu/food/plant/plant_propagation_material/plant_variety_catalogues_databases_en
58. United States Department of Agriculture (2020). US National Plant Germplasm System. https://npgsweb.ars-grin.gov/gringlobal/taxon/taxonomysearch. Accessed 1 Feb 2021
59. Culley TM (2013) Why vouchers matter in botanical research. Appl Plant Sci 1(11):130076. https://doi.org/10.3732/apps.1300076
60. Leon C, Yu-Lin L (2017) Chinese medicinal plants, herbal drugs and substitutes: an identification guide. Kew Publishing, London, UK
61. Rivera D, Allkin R, Obón C, Alcaraz F, Verpoorte R, Heinrich M (2014) What is in a name? The need for accurate scientific nomenclature for plants. J Ethnopharmacol 152(3):393–402. https://doi.org/10.1016/j.jep.2013.12.022
62. Dauncey EA, Irving J, Allkin R, Robinson N (2016) Common mistakes when using plant names and how to avoid them. Eur J Integr Med 8(5):597–601. https://doi.org/10.1016/j.eujim.2016.09.005
63. Nesbitt M, McBurney R, Broin M, Beentje HJ (2010) Linking biodiversity, food and nutrition: the importance of plant identification and nomenclature. J Food Compos Anal 23(6):486–498. https://doi.org/10.1016/j.jfca.2009.03.001
64. Bennett BC, Balick MJ (2014) Does the name really matter? The importance of botanical nomenclature and plant taxonomy in biomedical research. J Ethnopharmacol 152(3):387–392. https://doi.org/10.1016/j.jep.2013.11.042
65. Heinrich M, Lardos A, Leonti M et al (2018) Best practice in research: consensus statement on ethnopharmacological field studies—ConSEFS. J Ethnopharmacol 211:329–339. https://doi.org/10.1016/j.jep.2017.08.015
66. Simmonds MJ, Fang R, Wyatt L et al (2020) Biodiversity and patents: overview of plants and fungi covered by patents. Plants People Planet 2(5):546–556. https://doi.org/10.1002/ppp3.10144
67. Council of Industrial and Scientific Research India (CSIR) (2020) Traditional knowledge digital library: representative database of Ayurvedic, Unani, Siddha and Sowarigpa formulations. http://www.tkdl.res.in/tkdl/langdefault/common/Home.asp?GL=Eng. Accessed 1 Feb 2021

68. Uppsala Monitoring Centre, WHO-UMC (2020) VigiBase: the unique global resource at the heart of the drive for safer use of medicines. https://www.who-umc.org/VigiBase/VigiBase/. Accessed 1 Feb 2021
69. Paton A, Allkin R, Belyaeva I et al (2016) Plant name resources: building bridges with users. In: Rakotoarisoa NR, Blackmore S, Riera B (eds) Botanists of the twenty-first century: roles, challenges and opportunities. United Nations Educational, Scientific and Cultural Organisation (UNESCO), Paris, France
70. Uppsala Monitoring Centre, WHO-UMC (2020). WHODrug Global: the most comprehensive and actively used drug reference dictionary in the world. https://www.who-umc.org/whodrug/whodrug-portfolio/whodrug-global/. Accessed 1 Feb 2021
71. Schippmann U (2020) Medicinal Plant Resources of the World (MAPROW). Local MS-Access database
72. International Union for Conservation of Nature (IUCN) (2017). Medicinal plant specialist group report. https://www.iucn.org/sites/dev/files/2016-2017_medicinal_plant_sg_report.pdf. Accessed 1 Feb 2021
73. Royal Botanic Gardens, Kew (2021) Use of scientific plant names for herbal products: best practice, safety and labelling. professional development course: https://www.kew.org/science/training-and-education/continuing-professional-development/herbal-plant-scientific-names. Accessed 1 Feb 2021
74. International Organization for Standardization (2017) Revised IDMP standards to improve description of medicinal products worldwide. https://www.iso.org/news/ref2234.html. Accessed 1 Feb 2021
75. International Organization for Standardization (2018) ISO 11238:2018. Health informatics. Identification of medicinal products. Data elements and structures for the unique identification and exchange of regulated information on substances. https://www.iso.org/standard/69697.html. Accessed 1 Feb 2021
76. International Organization for Standardization (2018) ISO/TS 19844:2018. Health informatics. Identification of medicinal products (IDMP)—Implementation guidelines for ISO 11238 for data elements and structures for the unique identification and exchange of regulated information on substances. https://www.iso.org/standard/71965.html. Accessed 1 Feb 2021
77. United States Food and Drug Administration (2019) Identification of Medicinal Products (IDMP). https://www.fda.gov/industry/fda-resources-data-standards/identification-medicinal-products-idmp. Accessed 1 Feb 2021
78. European Medicines Agency (2020) Data on Medicines (ISO IDMP standards): research and development. https://www.ema.europa.eu/en/human-regulatory/overview/data-medicines-iso-idmp-standards-overview. Accessed 1 Feb 2021
79. Dauncey EA, Black N, Edwards SE, Irving JTW, Patmore K, Allkin B (2016) ISO Identification of Medicinal Products (IDMP) and its implication for herbals. Planta Med 82(S01):S1–S381. https://doi.org/10.1055/s-0036-1597061

Chapter 9
Coding Reports Involving Herbal Medicines in a Pharmacovigilance Database

Florence van Hunsel and Souad Skalli

9.1 Background

Many pharmacovigilance centres worldwide receive spontaneous reports for herbal medicines (including unapproved/unlicensed products) in addition to reports for conventional medicines. Herbal medicines include *herbs*, *herbal materials*, *herbal preparations* and *finished herbal products*. In some countries, the term 'herbal' medicines may also be used to describe products that contain, by tradition, natural organic or inorganic active ingredients that are not of plant origin (e.g. animal and mineral materials) [1]. Spontaneous reports involving herbal medicines have to be stored in pharmacovigilance databases that are mainly developed for storing, coding, assessing, analysing and transferring data to other databases in the context of conventional medicines. The major aim of pharmacovigilance is the early detection of signals of previously unrecognized adverse (drug) reactions (ADRs). Early signals may be strengthened by combining the experiences reported in various countries. Coding herbal medicines so that they may be identified and assessed on different levels (e.g. medicinal plant species, part of the plant used, type of extract, specific manufacturer's product, particular chemical constituents, mode of action and indication) is needed both for national and global pharmacovigilance activities.

F. van Hunsel (✉)
Netherlands Pharmacovigilance Centre Lareb, 's-Hertogenbosch, The Netherlands
e-mail: f.vanhunsel@lareb.nl

S. Skalli
Faculty of Science, University Mohammed V in Rabat, Rabat, Morocco

J. Barnes (ed.), *Pharmacovigilance for Herbal and Traditional Medicines*,
https://doi.org/10.1007/978-3-031-07275-8_9

9.2 Challenges in Coding and Classifying Herbal Medicinal Products

It can be challenging to store reports on herbal products in a database without the fixed structure that is in place for regular drugs, while still needing to code and be able to analyse reports at all the different hierarchical levels, such as plant species, specific plant part used, type of extract and specific manufacturer's product (proprietary) name. For licensed medicines, the WHO Collaborating Centre for Drug Statistics Methodology in Oslo has devised, and maintains, the Anatomical Therapeutic Chemical (ATC) classification system. In the ATC system, the active substances in a medicine are classified into different groups according to the organ or body system on which they act, as well as their therapeutic, pharmacological and chemical properties [2]. Also, for registered medicines, other important information, such as batch/lot number, is often available.

While nomenclature for conventional medicines is (usually) unambiguous, nomenclature for herbal products is particularly complex. This can be particularly challenging in pharmacovigilance, especially when trying to code and analyse reported ADRs associated with a particular plant species, specific herbal substance, and/or specific manufacturer's product [3]. Herbal product nomenclature lacks uniformity and several types of names are currently in use: botanical or scientific names (and synonyms) of medicinal plants; common or vernacular names; Latinised pharmaceutical names or plant pharmacopoeial names (where they exist); for some ingredients and products, the pinyin name, which is often used in traditional Chinese medicine (TCM) [4]. Further, product labels may be written in a language that is different to the one usually used by the pharmacovigilance centre handling the report, particularly for some products that users may obtain through internet purchases.

Herbal 'prescriptions' (such as those compounded and supplied by a traditional medicine practitioner), and product packaging and/or labels may list one or more of these names depending on the source and regulatory status of the product. Some 'prescriptions' or other crude preparations of plant material may have no label at all. Where present, names must be interpreted with care as even the scientific names may be incorrect, or synonyms may be used, sometimes incorrectly. The common or vernacular name is the least precise, as the same name may be used for plants from completely different genera or species. A single vernacular name may refer to several different plants; these plants may have other vernacular names in different geographical areas, and a single plant may have several vernacular names. Moreover, a single plant may be called by different names in different communities. Thus, botanical identification by a specialist is necessary in these cases. Plant common names may be misleading or confusing if used for raw plant material or on labels of unlicensed herbal products, and so should be avoided. To avoid ambiguity, it is desirable that the genus, species, botanical authority and part of the plant are described on the herbal product label, or on packaging in the context of raw material [5]. In a pharmacovigilance setting, this information should also be added to the ADR report data in a pharmacovigilance database.

Ambiguous vernacular and pharmaceutical names, scientific synonyms and the incorrect use of scientific names can cause confusion and could result in attributing reported ADRs to the 'wrong' plant. Unravelling the diversity of nomenclature used in suspected ADR reports is time-consuming, and it is not always possible to be certain regarding precisely which plant species, plant parts, types of extracts and so forth might be involved in an ADR [3]. An example illustrating how plant nomenclature can be difficult comes from the plant family Aristolochiaceae, used in TCM, where at least three systems of nomenclature can be identified [6].

In addition to the difficulties with nomenclature, there may be many 'unknowns' about herbal products, such as the lack of reported exact ingredients. Herbal practitioners, such as herbalists, often prescribe compound and/or dispense preparations containing herbal mixtures, and preparations or products in processed or powdered forms, which may make identification of a product difficult in cases where ADRs occur. These issues are common to all forms of traditional medicines used in traditional medicine systems globally. This raises many questions about how to code and classify these kinds of products and preparations.

The product information and package leaflet (where present, i.e. usually only for approved/authorized/registered herbal medicinal products) constitute an important source of information for health practitioners and patients/consumers, as a guide for rational use and correct administration of herbal medicines. Where regulations exist, licensed products are required to carry information on ingredients, dosage, indications and cautions, contraindications and potential interactions on their labels [7]. In the absence of regulations, this information may be absent from the product label, or may be substantially incomplete. In some African countries, although herbal medicines are often the main form of primary healthcare, there is an absence of herbal medicines on National Essential Medicines lists. A lack of standard treatment guidelines or a national herbal medicines pharmacopoeia is a major challenge for the implementation of coding and classification for these products.

Another challenge is that even if the plant used is declared on the product's label, the plant part may not be specified. Different plant parts contain different chemical constituents; for optimal pharmacovigilance causality assessment and coding, it is important to know which part(s) of the plant was/were used in the manufacturing process. In addition, for some traditional medicines, information on how the herbal material is prepared, e.g. with honey, or 'cooked', is also often lacking. The use of oil, vinegar and honey for their biological activities, or to aid the processing of traditional herbal medicines, is well documented [8].

Finished herbal products are herbal preparations comprising one or more herbal ingredients. Finished herbal products, particularly those formulated as tablets, capsules, and liquid dose forms, may contain excipients in addition to the active (herbal) ingredients. However, finished herbal products to which chemically defined active compounds have been added, including synthetic compounds and/or isolated chemical constituents from herbal materials, are not usually considered to be herbal (or, at least, do not meet regulatory definitions for herbal products, depending on the country) [1]. The use of non-herbal 'natural' substances in herbal products, such as animal parts (for example, powdered animal horns, animal thyroid hormone), vitamins/minerals, or the addition of conventional medicines to a herbal product, present substantial difficulties for coding these products. In many cases, particularly with the latter category, these non-herbal substances are undeclared on the product label and are, therefore, adulterations. There are numerous examples in the literature where such adulterations have led to serious harm in patients [9]. It is very important to be able to code these products in a pharmacovigilance database, so that they can be easily recognized and flagged as adulterated products.

9.3 Current Approaches in Coding and Classification

9.3.1 Botanical Nomenclature

The Uppsala Monitoring Centre (UMC), Uppsala, Sweden, the Royal Botanic Gardens at Kew, UK, and the Department of Systematic Botany, Uppsala University, Uppsala, Sweden, have collaborated on botanical nomenclature in pharmacovigilance and have specified the following criteria [3]:

- A plant name should indicate only one species of plant.
- The source for this name must be authoritative.
- The name should indicate which part of the plant is used. The collaboration has proposed that using the binomial name for each plant species that is included as an ingredient of an herbal medicinal product is best practice for avoiding confusion about the precise plant(s) used. In botanical nomenclature, there is only one 'accepted' scientific name for each plant species in a given taxonomy, with only one accepted spelling: these names are unique and refer to only one species [3]. A scientific botanical name has three parts: a genus (generic) component, a species epithet (specific name) and an author's name (usually abbreviated). The genus name and the species epithet are in Latin and are italicized, for example, *Hypericum perforatum* L.

The binomial name alone gives no information on which part of the plant is used. As mentioned above, having information about the plant part used and its chemical composition (as far as possible) is important for a comprehensive causality assessment to be undertaken. However, the packaging of an herbal medicine may lack this important information. Also, the extraction and processing methods used in the preparation of herbal products are not captured by using the binomial name for a plant. This information, where provided, allows the type of extract (and, therefore, its chemical composition) to be considered, along with similar information from other reports, and thus should, ideally, be included in the report [1].

As discussed above, for spontaneous reports for herbal products, best practice for naming the ingredient(s) of herbal products associated with suspected ADRs is to use the binomial name for the herbal substances, i.e. including both the genus and species epithets, together with the botanical author (e.g. *Hypericum perforatum* L.).

However, for many marketed products or traditionally prepared products, it is not always possible to determine to which species (within the genus) the stated ingredient corresponds. In these instances, the term 'spp.' substances (e.g. *Aloe* spp.) is used. 'Spp.' indicates that the genus is known, but not the precise species, where more than one species exists for the genus in question [10]. This occurs with several other popular herbal substances, such products containing 'echinacea'; where this is not further described, and in the context of ADR reports, 'Echinacea spp.' must be used.

9.3.2 The Herbal Anatomical Therapeutic Chemical (ATC) Classification System

For conventional medicines, the ATC coding system, where drugs are divided into different groups in accordance with the organ or system on which they act, and their chemical, pharmacological, and therapeutic properties, has been in use since 1976; the system was initially developed as a tool for drug utilization research with the aim of improving the quality of drug use [11]. In this system, conventional medicines are divided into fourteen main groups ('first' level), with pharmacological/therapeutic subgroups ('second' level). The third and fourth levels are chemical/pharmacological/therapeutic subgroups and the fifth level is the chemical substance [2].

In 1991, de Smet proposed a new method for the classification of herbal medicines [12, 13], based on the ATC system. This Herbal-ATC (HATC) classification provides a scientific framework for a harmonized, global nomenclature and therapeutic classification of herbal substances and combinations of them for herbal medicines [14].

By placing an 'H' before the existing ATC classes at the "0-level" of the code, a system is produced that is compatible with the regular ATC classification and which can be used for classifying herbal medicines [12, 13]. The first level comprises 14 anatomical groups designated by the letters A–V. These are the same in both the ATC and the HATC systems. The Uppsala Monitoring Centre (UMC) has further developed this classification tool to permit the inclusion of individual herbal products in the global WHO database of ADR reports for pharmacovigilance purposes. The HATC classification, unlike the regular ATC system, is based on botanical science, pharmacognosy, phytochemistry, literature search, and documented traditional use, rather than chemistry and evidence-based medicine (as for conventional medicines). It is linked to botanical synonyms and vernacular names via the substance register of the WHODrug Global® dictionary, which contains all ingredients, herbal and chemical, of medicinal products mentioned on spontaneous reports of suspected ADRs in the global WHO database [15]. The UMC always aims to assign codes on the fourth level of the ATC-code (chemical subgroup). Often a herbal product may have several indications/uses and, therefore, will appear in several places in the HATC classification. For instance, *Aesculus hippocastanum* (horse chestnut) fruit is used to treat haemorrhoids [16], the leaves are used in arthritis and rheumatism [17], and seeds are used as an anti-varicose therapy [17].

In data-analyses, the hierarchical HATC structure supports both the broader overview and in-depth analysis, by allowing grouping and aggregation of data on different levels of specificity [18]. The structure of the HATC coding system is shown below (Table 9.1), using as an example the complete classification of preparations of *Aloe ferox* Mill. dry leaf juice, which is used as a laxative [14, 18].

For pharmacovigilance centres using the WHODrug Global® dictionary, maintained by the UMC, the HATC code can be used as it is included in this dictionary. Users of this dictionary can look up herbal products based on proprietary name or on product ingredients. The system identifies the 'preferred names' of ingredients of products listed on ADR reports in the global WHO database. The logic for identifying 'preferred names' for herbal substances follows, as far as possible, that for identifying preferred chemical substance names in the WHODrug Global® dictionary. There is a system checklist for cross-referencing of botanical and vernacular names used as names of ingredients. In cases where only the product name is known in a report, the UMC searches its global ADR database to see if there are existing reports for the same product and where the ingredients are already coded. If the product is not already in the global WHO ADR database, it will be added, together with the available information [1].

Although the HATC coding system represents a valuable attempt at coding herbal medicines, it may not be perfect for covering all types of herbal medicinal products [1]. For instance, traditional Chinese medicines often have indications/uses that are not listed in the ATC classification, such as 'Yin Deficiency' or 'Qi-deficiency' [19], and are, therefore, difficult to capture with the HATC.

The HATC can be considered a valid approach if a herbal medicinal product implicated in a spontaneous report consists of a single medicinal plant ingredient, where the part used is known (i.e. provided), the active principle/constituent(s) is/are known, the dose taken is known, and the product always meets the same quality criteria. This is often the case in certain countries where herbal medicines fall under legislation that governs and that guarantees all these requirements. In practice, there are many marketed or traditionally prepared herbal products available that contain multiple herbal (and, sometimes, non-herbal) ingredients and it is not always

Table 9.1 Herbal-Anatomical Therapeutic Chemical (HATC) classification system structure, using *Aloe ferox* Mill. dry leaf juice as an example [14]

Level 0	Herbal Remedy designated by letter H
Level 1	A—Alimentary tract and metabolism (first level, anatomical main group)
Level 2	A06—Drugs for constipation (second level group, therapeutic main group)
Level 3	A06A—Drugs for constipation (third level group, therapeutic/pharmacological subgroup)
Level 4	A06AB—Contact laxatives (fourth level group, therapeutic/pharmacological/chemical subgroup)
Level 5	A06AB5001—Aloe ferox Mill., dry leaf juice (fifth level group, individual crude drug)

possible to identify them all. In many countries, herbal medicines are used in the form of raw or crude herbal substances, also making it difficult to apply the HATC system to coding their ingredients. Also, the quality of these products and preparations varies, and it is not possible to capture this with the HATC system.

In the UMC database, the ATC code V90—'unspecified herbal and traditional medicine'—is assigned to every product given an herbal ATC code, or if no ATC (herbal or chemical) is suitable for products with at least one herbal ingredient. The ATC V90 is not included in the official ATC codes. The UMC aims to avoid assigning 'V90' only. However, there are situations, such as multi-ingredient products with ingredients that cannot all be identified, where it is impossible to identify an appropriate HATC code, and so the ATC V90 code is used. The WHODrug Global® dictionary also contains substances used in traditional methods of processing herbal products, such as 'honey', 'oil' and 'vinegar', that can be used as coded ingredients.

9.3.3 *Other Coding Methods*

WHO HATC codes are not universally used by all pharmacovigilance centres. For pharmacovigilance centres that do not use the WHODrug Global® dictionary, it is difficult to use the HATC because, without the use of a WHO drug dictionary, the herbal ATC codes are not automatically linked to substance and product-level.

There are also other reasons for using different coding systems. For instance, in the pharmacovigilance centre in Morocco (WHO Collaborating Centre for Strengthening Pharmacovigilance Practices) the HATC classification is not used because of some of the limitations mentioned above. For recording and coding the identity of herbals, mainly consisting of raw materials, the Moroccan Pharmacovigilance Centre uses the binomial nomenclature as described above. In addition, after a comma, the part of the plant used is added, if this is specified in the spontaneous report (e.g. *Aesculus hippocastanum*, seeds). For unlicensed finished herbal products found in pharmaceutical dosage forms, the Moroccan Pharmacovigilance Centre classifies them by name (with binomial nomenclature of the herbal medicine if specified) with the country of origin of the product and ingredients listed on the product label. The Moroccan Pharmacovigilance Centre has developed an MS Excel® herbal medicines database according to VigiFlow® (a management system for recording, processing and sharing reports of adverse effects) which meets its needs and its own specifications to have all information available to analyse data. To assist with coding accuracy, the Moroccan centre has a single reporting form covering all medicinal products, with an adaptation for herbal medicines raw material to specify the part of the plant used, the type of extract, and the dose used. All Moroccan herbal medicine reports, now around 3000, are in the VigiFlow® database with all needed information. Each report can be flagged if there is a suspicion that it involves an adulterated product containing undeclared conventional medicines. Products suspected to be adulterated, along with a sample where available, are analysed by the Centre Anti Poison et de Pharmacovigilance laboratory.

The Netherlands Pharmacovigilance Centre (Lareb) has developed a reporting database, PV Report, where reports relating to conventional medicines and non-registered/unapproved products can be stored, coded and analysed. As Lareb currently uses a specific Dutch drug dictionary, the HATC is not available. Reports involving registered herbal medicinal products are coded with the regular ATC-coding system (for instance, a registered *Valeriana officinalis* product has the ATC code N05CM09).

All reports involving non-registered herbal medicinal products are flagged in the Lareb database as 'herbal'. If known, the product name and the manufacturer are stored. All active ingredients are coded with their binomial plant names using an in-house built dictionary that allows for additions if the herbal medicine involved has not been reported to Lareb previously. If known, the part of the plant used, extract type and dosage are also stored with the report. The system also allows for the addition of names of vitamins and minerals, or other non-herbal ingredients, as ingredients for an herbal product itself. In the summary of the report, additional information can be added. In cases where information is not available, for instance, for reports where a product is simply described as 'valerian', the 'spp.' substance is coded. Each report can be flagged if there is a suspicion it is an adulterated product with undeclared registered medicinal ingredients. Products with these suspected adulterations, for which a sample is available, are sent for laboratory testing to the National Institute for Public Health and Environment (RIVM) in the Netherlands. If a product indeed contains an undeclared ingredient, this is described in the narrative and summary of the report. Test-results are stored with the report.

9.4 Solutions Towards Better Coding

It is important for pharmacovigilance centres to identify the specific herbal product(s) involved in a spontaneous report, including label and manufacturer information, specific ingredients and dose used. Also, assessment of reports would benefit substantially from having results of analysis of the suspect product(s) used, for contamination and adulteration, or species identification, where possible [7, 20]. If possible, national pharmacovigilance centres could collaborate with pharmacognosy departments of universities, and with botanists or botanical garden staff, regarding taxonomic (botanical and chemical) identification and botanical and vernacular nomenclature [20]. Of course, the reporter of the information plays an important role here as they are the primary source of information for a pharmacovigilance centre for obtaining precise information about products involved for a particular report. Having a reporting form in place with additional questions that prompt the reporter for specific information for herbal drugs, such as that designed by the WHO, or used by the Moroccan Pharmacovigilance Centre, can ensure that reports are more complete to start with. Also, asking specific follow-up questions to reporters can help to make the report as complete as possible, which enables more precise coding. If the finished herbal product(s) concerned, or its raw materials, were imported from other countries, the drug regulatory authority of the exporting country may be able to provide helpful information [20]. This, of course, requires comprehensive, reliable traceability throughout the supply chain for herbal medicinal products from field to finished product.

9.4.1 Coding Options

The ultimate goal of coding spontaneous reports involving herbal products as precisely as possible is to be able to search for and aggregate reports in order to detect signals of safety concerns. This approach can be undertaken either on a national level and/or in a global database, such as that maintained by the WHO-UMC. Despite the limitations mentioned, the HATC coding system as developed by the UMC is a valuable option for centres already using the WHODrug Global® dictionary. It should be noted that not all countries who are members of the WHO Programme for International Drug Monitoring use the HATC [21].

For countries not using the HATC, coding on multiple levels for herbals in a pharmacovigilance database is needed in order to perform searches and signal detection: the herbal product name; the herbal ingredients described using binominal nomenclature; for each ingredient, the part of the plant used, and the preparation method; the other ingredients in the product should all be coded. If a product contains non-herbal ingredients, such as vitamins or animal parts, these should also be coded. Storing a photograph of the product or the original packaging can be useful to be able to differentiate between products and also helpful if a manufacturer changes the ingredients of a product over time [20].

In addition to coding systems for medicinal products, there are other systems for coding of adverse drug reactions, such as MedDRA®, the Medical Dictionary for Regulatory Activities [22]. The MedDRA® coding system also includes terms for 'herbal interaction', 'herbal supplement' and 'herbal toxicity'.

9.5 Final Considerations

Herbal medicines have their own specificities and characteristics that are different from those for conventional drugs. This creates challenges in coding the precise details and ingredients of implicated herbal products for spontaneous reports of suspected ADRs in pharmacovigilance databases. Pharmacovigilance systems were developed according to the principles and conditions for conventional medicines and require modifications to address the specific features of herbal medicines. This is even more the case for medicinal plants, or parts of plants, used in the form of fresh or dried plant material (crude or raw material), in the form of herbal teas, for plants that are sold in bulk quantities, and for other plants without essential information on their label or for herbal products without national legislation regarding quality, efficacy and safety [23]. Inaccurate coding at a national pharmacovigilance centre level makes it difficult to conduct further analysis later at both the national and global levels [24]. Independent of the method used, the coding of the product in a pharmacovigilance database should be as accurate as possible, without losing information and, equally important, without implying more about the precise herbal ingredients of a product than is actually known.

References

1. World Health Organization (2004) WHO guidelines on safety monitoring of herbal medicines
2. Norwegian Institute of Public Health—WHO Collaborating Centre for Drug Statistics Metodology (2011) ATC structure and principles. https://www.whocc.no/atc/structure_and_principles/. Accessed 1 Jun 2020
3. Farah MH, Olsson S, Bate J et al (2006) Botanical nomenclature in pharmacovigilance and a recommendation for standardisation. Drug Saf 29(11):1023–1029. https://doi.org/10.2165/00002018-200629110-00002
4. Zhao Z, Liang Z, Chan K et al (2010) A unique issue in the standardization of Chinese materia medica: processing. Planta Med 76(17):1975–1986. https://doi.org/10.1055/s-0030-1250522
5. Skalli S, Jordan SA (2017) Herbal and traditional medicines, now and future. In: Edwards R, Lindquist M (eds) . Springer International Publishing, Pharmacovigilance critique and ways forward, pp 145–159
6. Wu KM, Farrelly JG, Upton R, Chen J (2007) Complexities of the herbal nomenclature system in traditional Chinese medicine (TCM): lessons learned from the misuse of Aristolochia-related species and the importance of the pharmaceutical name during botanical drug product development. Phytomedicine 14(4):273–279. https://doi.org/10.1016/j.phymed.2006.05.009
7. Jordan SA, Cunningham DG, Marles RJ (2010) Assessment of herbal medicinal products: challenges, and opportunities to increase the knowledge base for safety assessment. Toxicol Appl Pharmacol 243(2):198–216. https://doi.org/10.1016/j.taap.2009.12.005
8. Wu X, Wang S, Lu J et al (2008) Seeing the unseen of Chinese herbal medicine processing (Paozhi): advances in new perspectives. Chin Med 13:4. https://doi.org/10.1186/s13020-018-0163-3
9. Reeuwijk NM, Venhuis BJ, de Kaste D, Hoogenboom RL, Rietjens IM, Martena MJ (2014) Active pharmaceutical ingredients detected in herbal food supplements for weight loss sampled on the Dutch market. Food Addit Contam Part A Chem Anal Control Expo Risk Assess 31(11):1783–1793. https://doi.org/10.1080/19440049.2014.958574
10. Centre for Australian National Biodiversity Research and Australian National Herbarium. https://www.anbg.gov.au/cpbr/databases/names.html. Accessed 16 Jun 2017
11. Bergman U (2006) The history of the drug utilization research group in Europe. Pharmacoepidemiol Drug Saf 15(2):95–98. https://doi.org/10.1002/pds.1171
12. Farah HM, Edwards R, Linquist M, Leon C, Shaw D (2000) International monitoring of adverse health effects associated with herbal medicines. Pharmacoepidemiol Drug Saf 9(2):105–112. https://doi.org/10.1002/(SICI)1099-1557(200003/04)9:2<105::AID-PDS486>3.0.CO;2-2
13. De Smet PA (1993) An introduction to herbal pharmacoepidemiology. J Ethnopharmacol 38(2–3):197–208. https://doi.org/10.1016/0378-8741(93)90016-x
14. The Uppsala Monitoring Centre (2019) The herbal anatomical therapeutic chemical classification system. https://www.who-umc.org/whodrug/whodrug-portfolio/whodrug-global/herbal-atc/. Accessed 11 Nov 2019
15. Dighe NS, Nirmal SA, Musmade SA, Dhasade VV (2010) Herbal database management. Syst Rev Pharm 2(1):152–157

16. Gurel E, Ustunova S, Ergin B et al (2013) Herbal haemorrhoidal cream for haemorrhoids. Chin J Physiol 56(5):253–262. https://doi.org/10.4103/0975-8453.75067
17. Stepantschitz G, Kresbach E (1954) Treatment of rheumatism with horse chestnut extracts (venostasin). Z Rheumaforsch 13(9–10):266–275
18. the Uppsala Monitoring Centre (2004) Guidelines for herbal ATC classification. The Uppsala Monitoring Centre, Sweden
19. Yi YD, Chang IM (2004) An overview of traditional Chinese herbal formulae and a proposal of a new code system for expressing the formula titles. Evid Based Complement Alternat Med 1(2):125–132. https://doi.org/10.1093/ecam/neh019
20. Shetti S, Kumar CD, Sriwastava NK, Sharma IP (2011) Pharmacovigilance of herbal medicines: current state and future directions. Pharmacogn Mag 7(25):69–73. https://doi.org/10.4103/0973-1296.75905
21. Skalli S, Bencheikh RS (2015) Pharmacovigilance of herbal medicines in Africa: questionnaire study. J Ethnopharmacol 171:99–108. https://doi.org/10.1016/j.jep.2015.05.033
22. The International Conference on Harmonisation of Technical Requirements for Registration of Pharmaceuticals for Human Use (ICH) (2019) Welcome to MedDRA. https://www.meddra.org/. Accessed 16 Jun 2017
23. Barnes J (2003) Pharmacovigilance of herbal medicines: a UK perspective. Drug Saf 26(12):829–851. https://doi.org/10.2165/00002018-200326120-00001
24. van Hunsel F, van de Koppel S, Skalli S et al (2019) Analysis of hepatobiliary disorder reports associated with the use of herbal medicines in the global suspected ADR database Vigibase. Front Pharmacol 10:1326. https://doi.org/10.3389/fphar.2019.01326

Chapter 10
Causality Assessment in Pharmacovigilance for Herbal Medicines

Rolf Teschke and Gaby Danan

10.1 Introduction

Herbal medicines, including traditional herbal medicines (THMs), are widely used throughout the world in developed and developing countries [1–7]. THMs include traditional Chinese medicines (TCMs) or, more broadly, traditional Oriental Medicines [3–8], Indian Ayurvedic medicines [8], traditional Arab medicines [9], and traditional South African medicines [10], to name a few of many worldwide examples [4]. Herbal medicinal products are used as treatment options for minor ailments, prophylaxis of diseases, or improvement of general health conditions [1]. The use of these products continues to increase despite discussions on unclear health benefits when compared with conventional drugs, including a limited evidence-base for efficacy [11], questionable herbal product quality [12–15], insufficient regulatory surveillance [15–17], and adverse reactions [17–24], including herb-drug interactions [25, 26]. A comprehensive overview of herbal ingredients most commonly used in traditional or modern herbal medicinal products [1] highlights and critically analyses the limited data on clinical efficacy and safety aspects.

The focus of this chapter is on pharmacovigilance approaches, namely on how to detect suspected adverse reactions and how to validly establish, or refute, a causal association. Assessing such adverse reactions is a particular regulatory challenge due to the issues around the precise content and quality of the suspected herbal medicinal product(s).

R. Teschke (✉)
Division of Gastroenterology and Hepatology, Department of Internal Medicine II, Klinikum Hanau, Hanau, Teaching Hospital of the Medical Faculty, Goethe University, Frankfurt/Main, Germany

G. Danan
Pharmacovigilance Consultancy, Paris, France

J. Barnes (ed.), *Pharmacovigilance for Herbal and Traditional Medicines*,
https://doi.org/10.1007/978-3-031-07275-8_10

10.2 Definitions

10.2.1 Herbal Medicines

Herbal medicines are widely used, although their efficacy and safety typically have not been evaluated systematically in randomized clinical trials, whereas risks of certain adverse reactions are well documented [15, 18, 22]. Herbal products often contain a single herbal ingredient containing a limited number of phytochemicals; some contain several herbal ingredients, with multiple phytochemicals [15, 23]. According to the World Health Organization definition, herbal medicines include herbs, herbal materials, herbal preparations, and finished herbal products [3]. Crude plant material includes leaves, flowers, fruit, seed, stems, wood, bark, roots, rhizomes, or other plant parts, which may be entire, fragmented, or powdered. However, finished herbal products or mixture herbal products, to which chemically defined active substances have been added, including synthetic compounds and/or isolated constituents from herbal materials, are not considered to be herbal in a strict sense. To simplify the discussion, however, the present analysis also includes under the broad sector of herbal medicines marketed herbal dietary supplements, which contain various herbal ingredients among other ones [27–31].

10.2.2 Traditional Herbal Medicines

Traditional herbal medicines refer to the long historical use of these herbal medicines [3], which often comprise mixtures of many crude herbal substances containing abundant plant chemicals [23, 24]. According to the WHO, their use is widely considered by users and practitioners to be safe and effective [3]. However, this official statement downgrades reports of severe and serious adverse reactions, some of which are fatal [24] and overlooks published clinical trials showing limited efficacy [11].

10.3 Principles of Causality Assessment

Case reports of suspected adverse reactions in connection with the use of herbal products are usually submitted to the manufacturer [15] and/or regulatory agencies [15–17] on a voluntary basis. Reporters are physicians, other healthcare professionals, including pharmacists and nurses, and patients and other non-healthcare professionals. Problematic are these spontaneous cases reported to national spontaneous reporting databases, such as MedWatch in the USA [31], or in each European country, because many are of limited quality [31]. Indeed, these principles include (1) the identification, quality and usage modalities of the suspected herbal product, (2) the temporal association between herbal use and development of the adverse effect, (3) the exclusion of the alternative causes of the adverse events, and finally (4) the systematic causality assessment based on algorithms combining the previous data elements.

10.3.1 Herbal Product Identification, Quality and Usage Modalities

Whenever a patient is reporting a suspected adverse reaction in assumed connection with the use of a herbal medicine, the suspected herbal product must be clearly identified and be available for possible product quality analysis, particularly if the suspected adverse reaction is serious (Table 10.1) [3–6, 12–24, 32–35]. Treatment details, including indication and past medical history, must be documented to exclude that the signs, the symptoms or the syndrome were present prior to herbal use and were not, in fact, the indication of the use of the herbal medicine, a common issue known as protopathic bias.

Table 10.1 Minimum quality requirements, suggested for some herbal medicines, regulatory approved herbal drugs, and herbal supplements, including traditional herbal medicines, for assessing causality of adverse liver effects

Quality specifications for herbal products
• Herbal product declaration of the manufacturer with address, phone, fax number, and e-mail • Expiration date of the herbal product • Batch number • Detailed recommendation for indication and contraindication • Advice for daily dose and maximum use duration • Correct labelling of all ingredients • Definition of plant family, subfamily, species, subspecies, and variety. • Definition of plant part • Definition of used solvents and solubilizers • Exclusion of impurities, adulterants, and misidentifications • Minimum or lack of batch-to-batch variability • Minimum or lack of product-to-product variability • Lack of variety-to-variety variability • Good Agricultural Practices (GAPs) • Good Manufacturing Practices (GMPs) • Regulatory surveillance

Respective products refer to a wide range of herbal products. Minimum requirements do not apply to all herbal products and depend on the individual herbal products as well as regulatory guidelines, which may vary among different countries. Clearly, regulatory requirements are stronger for regulatory approved herbal drugs as compared to other herbal products. Details are adapted from previous reports [12–20, 23, 24, 32–35]

10.3.2 Temporal Association

Verification of a clear temporal association between herbal use and adverse reaction is an essential aspect of causality assessment. This should include examination of dates of the first intake and those of discontinuation of the suspected herbal product [15–17]; these conditions are the same as those assessed in the reporting of adverse reactions associated with conventional medicines [15]. Acute adverse reactions must be recognized during herbal use, or shortly thereafter, since a long time interval between last use of the herbal product and the occurrence of the adverse reaction argues against a causal association.

10.3.3 Confounding Variables

Reports of herbal use and associated adverse reactions may be confounded by several variables. These include product quality issues [12–17, 32–35], incorrect use of products, such as overdosing [34], concurrent use with conventional medicines, other herbal products or 'dietary supplements' [34], pre-existing diseases [30–33], and incorrect reporting [29–31]. Confounding variables impede a clear causality imputation and must be examined and specified through sophisticated causality assessments.

10.3.4 Systematic Causality Evaluation

Expectations are that healthcare professionals reporting suspected adverse reactions, or, at least, regulators receiving such reports provide systematic causality assessment, but this is not always done [28–31]. In addition, spontaneous reports can lack key information [31].

10.4 Causality Assessment Methods

Regardless of the severity and seriousness of suspected adverse reactions, which may include fatal outcomes [15–20, 23, 24], causality should be assessed despite variabilities in clinical features, complex clinical diagnoses, and confounding variables [15–20]. Prerequisites for case evaluation include a documented past medical history and a clinical physical examination of the patient, although this is rarely done.

In approaching formal causality assessment of adverse events, three options are to be considered (Table 10.2) [36–45]. First, the assessor builds up an opinion, which is based on global introspection taking into consideration personal

Table 10.2 Causality assessment methods used for suspected herbal-induced adverse reactions

Causality assessment method	Details	Additional information	References
Ad hoc approach	Subjective global introspection approach Lacking structure Lacking individual element specification Lacking individual element scoring Lacking organ specificity Lacking disease specificity Lacking method validation Lacking use of RUCAM	Unstructured approach, not recommended for assessing suspected HILI	[36]
WHO method	Subjective global introspection approach Limited structure Lacking individual element specification Lacking individual element scoring Lacking organ specification Lacking disease specification Lacking appropriate method validation	Method not specified for injured organ or disease	[43]
DILIN method	Subjective global introspection approach Complex, cumbersome structure Limited element specification Lacking individual element scoring Assumed liver injury specification Lacking method validation	Global introspection-based, non-transparent method lacking validation, no clearly published diagnostic elements, and no published scorings, not for prospective use.	[42]
Naranjo scale	Objective standardized algorithm approach Limited element specification Individual element scoring Lacking organ specification Lacking disease specification Lacking appropriate method validation	Method not specified for disease or injured organ	[38]
RUCAM scale	Objective standardized algorithm approach Individual element specification Individual element scoring Final causality scoring Final causality gradings Liver injury specification Method validation	Worldwide first structured, validated, liver and hepatotoxicity specific method with clear element criteria and scoring Preferential prospective use to ensure complete data	[37]

(continued)

Table 10.2 (continued)

Causality assessment method	Details	Additional information	References
MV scale	Objective standardized algorithm approach Individual element specification Individual element scoring Final causality scoring Final causality grading Lacking appropriate method validation	Heavily modified from RUCAM	[44]
TKK scale	Objective standardized algorithm approach Individual element specification Individual element scoring Final causality scoring Final causality grading Lacking appropriate method validation	Heavily modified from RUCAM	[45]
Probabilistic approach	Complicated and complex approach Derivative of Bayes theorem Requires previous causality probability data Calculated from available knowledge with background data, final causation	Problematic acquisition of previous incidence data of specific adverse effects caused by the herbal product under consideration	[39–41]

Several causality assessment methods have been used in suspected herb-induced adverse effects [36–45], with focus on the ad hoc approach [36], WHO method [43], DILIN method [42], Naranjo scale [38], RUCAM scale [37], MV scale [44], TKK scale [45], and the probabilistic approach [39–41]. Abbreviations: DILIN, Drug-Induced Liver Injury Network; MV scale, Maria and Victorino scale; RUCAM, Roussel Uclaf Causality Assessment Method; TTK scale, Takikawa, Takamori, and Kumagi scale

experience, but without predefined key elements and quantitative scorings leading to vague conclusions that are hardly re-assessable; for sake of clarity, such approaches are summarized as subjective global introspective approaches [36]. Second, the assessor follows an objective standardized, transparent approach, which may, or may not, be specific for the injured organ and the disease, and uses clearly predefined items with corresponding scores, enabling re-assessment by others and comparison of the results with those of other assessors to reach finally an objective causality level; such approaches are summarized as objective standardized algorithm approaches [37, 38]. Third, the assessor relies on a probabilistic method derived from Bayes´ theorem [39–41].

Causality assessment methods show substantial variabilities in their characteristics (Table 10.3) [36–45]. This applies to the ad hoc approach [36], the WHO method for standardized case causality assessment [43], the US DILIN (Drug-Induced Liver Injury Network) method [42], the Naranjo scale [38], the RUCAM

Table 10.3 Causality assessment methods for adverse reactions associated with herbal medicines, including traditional herbal medicines

Causality assessment method	Suitability for herbal medicines	Organ specificity	Disease specificity	Individual item scoring	Evaluation	Transparency
Ad hoc approach	–	–	–	–	+ Prospective + Not quantitative	–
WHO method	–	–	–	–	+ Retrospective + Not quantitative	–
DILIN method	–	+ Liver	+ Hepatotoxicity	–	+ Retrospective + Not quantitative	–
Naranjo scale	+	–	–	+	+ Prospective + Quantitative	+
RUCAM scale	+	+ Liver	+ Hepatotoxicity	+	+ Prospective + Quantitative	+
MV scale	–	+ Liver	+ Hepatotoxicity	+	+ Prospective + Quantitative	+
TTK scale	–	+ Liver	+ Hepatotoxicity	+	+ Prospective + Quantitative	+
Probabilistic approach	–	–	–	–	+ Retrospective + Not quantitative	–

Compilation of several causality assessment methods (CAMs) to assess causality of adverse effects caused by herbal medicines and traditional herbal medicines: ods7: Ad hoc approach [36], WHO method [43], DILIN method [42], Naranjo scale [38], RUCAM scale [37], MV scale [44], TTK scale [45], probabilistic approach [39–41]

scale (Roussel Uclaf Causality Assessment Method) [37], the 'MV' (Maria and Victorino) scale [44], the 'TTK' (Takikawa, Takamori, and Kumagi) scale [45], and the probabilistic method [39–41]. Approaches occasionally lack specificity for the injured organ, whereas others are organ-specific, such as the RUCAM scale for liver injury; some methods use a system of individual item scoring, present a final quantitative causality grading, and provide data transparency; some are prepared for prospective use, whereas others are limited to retrospective analysis (Table 10.3). For assessing causality of suspected adverse reactions possibly caused by herbal products and/or traditional herbal medicines, two approaches are recommended (Table 10.3): the Naranjo scale is suitable for general effects without specification of the involved organ or body system of the established injurious disease, but it is not recommended for liver injury cases [38]; instead, the RUCAM scale validly assesses causality in liver injury cases, but cannot be applied to causality assessment for suspected adverse reactions affecting other organs/systems [37].

10.4.1 Global Introspection Approaches

By definition, global introspection provides exclusively subjective results, which reflects the personal experience the assessor may have had in patients with adverse reactions unrelated to the liver or with DILI (drug-induced liver injury) or HILI (herb-induced liver injury). Such assessments are problematic on various grounds, which include the undefined assessor's expertise and, particularly, the uncertainty as to how to handle cases with incomplete data, a common limitation in cases with suspected adverse reactions [32, 33, 46]. As no formal algorithms combining clearly defined elements together with their specific scores is used that may guide the assessor, the obtained results remain vague and then require expert rounds with the aim of reaching consensus among the various opinions, conditions conflicted by substantial difficulties as outlined previously [43]. By definition, none of the global introspection approaches is based on a gold standard, which rules out any possible method validation and comparative studies.

10.4.1.1 Ad hoc Approach

As an example, at first presentation of a patient with acute liver injury and a possible temporal association with the use of a herbal product, the physician will consider a quick ad hoc approach regarding the question whether HILI may be a diagnostic option, based on previous experience and subjective opinion. With a tentative preliminary HILI diagnosis, the further diagnostic workup can then be initiated. As it is based on global introspection and, therefore, not possible to validate, the ad hoc approach is imprecise due to many shortcomings resulting from missing data such as core elements (Table 10.4) [36]; it cannot be recommended for causality assessment for general adverse reactions, nor for liver injury [37]. It is a rapid method and, therefore, possibly the kind most commonly used in cases of suspected adverse reactions presented as spontaneous reports to regulatory agencies or pharmaceutical manufacturers, or published as short case reports in the scientific literature. This ad hoc approach is sometimes also referred to as 'guilt-by-association' [36]. As a

Table 10.4 Ad hoc causality approach mostly lacking essential elements

Missing items
1. Robust causality assessment method such as the updated RUCAM for liver injury cases
2. Signature of clinical manifestation
3. Latency period
4. Dechallenge features
5. Definitive exclusion of alternative causes
6. Risk factors
7. Track record of the herb

Mostly missing items of an ad hoc approach using global introspection in suspected cases with adverse effects caused by herbs, modified from a published report [36]

cautionary note, whenever adverse reactions are assumed in temporal connection with herbal medicine use, not all herbal products or ingredients are guilty just by this association.

10.4.1.2 WHO Method for Standardized Case Causality Assessment

With the WHO global introspection method, or WHO standardized case causality assessment method [43], attempts have been made to assess causality of any and unspecified adverse reactions, using a range of items as listed that are not necessarily precisely defined. There is, for instance, a lack of time frame for the challenge and dechallenge period, exclusion criteria for other causes, or how to handle missing data elements. This is reflected to some extent in that reports with important missing data are classed as 'unassessable'.

Not surprisingly, such an approach does not allow for a specific scoring of individual items or for a final scoring with a valid causality grading. Despite these shortcomings, causality gradings range from 'certain/definite' to 'unlikely', 'unclassified', or 'unassessable' causality. These lower causality classifications reflect retrospective causality assessment and associated incomplete data. Shortcomings of the WHO method are partially compensated by the quantitative Naranjo scale that uses some defined items and an individual item scoring system [38].

Although structured, the WHO method is not a quantitative system, lacks transparent results, and represents an invalid approach without possible validation. Of note, the WHO itself came under scientific pressure when the WHO method was applied to assess causality in cases of liver injury caused by drugs [47] or herbal products [48–53], dismissing thereby the use of robust liver and hepatotoxicity specific quantitative causality assessment methods, such as RUCAM [37]. The WHO method is inappropriate for use in liver injury cases associated with use of herbal products or conventional medicines [47–53] and is also not recommended (by this author) for suspected adverse reactions in general due to the shortcomings discussed above and summarized (Tables 10.2 and 10.3).

10.4.1.3 DILIN Method

Based on global introspection and not validated against a gold standard [42], the US DILI Network (DILIN) established its own method to assess causality in cases of liver injury associated with use of conventional medicines [42]. This was in addition to the existing RUCAM, which was published in 1993 [54, 55], and which, since then, has become the most commonly used causality assessment method for liver injury cases associated with drugs and herbal products worldwide and is considered the gold standard [37]. The DILIN method was used in cases of suspected liver injury by herbs and dietary supplements, but it performed poorly [56]. In particular, there was little transparency of case data, case evaluation, and causality assessment;

of most concern, the DILIN method did not allow for individual product causality attribution in cases where patients had used up to six different 'dietary supplements', and comedication with conventional dugs also remained unconsidered [56].

The DILIN method was also applied in a case series concerning suspected liver injury associated with use of the product OxyELITE Pro (a multi-ingredient product promoted for weight loss and 'body building'). In this context, DILIN again performed poorly, neglecting overt comedication and alternative diagnoses, such as chronic hepatitis B infection with cirrhosis, and dismissing hepatitis E virus infection, thereby overreporting HILI cases, as discussed in detail [57]. Assessments using the DILIN method are restricted to the USA and require expert opinion rounds with delayed results that are not available in real time for clinical purposes; data are not transparently presented, are without clearly defined items, and missing individual item scoring; only approximate percentage ranges of vague causalities are published [37, 42]. Clearly, the DILIN method cannot be used for causality assessment for suspected adverse reactions in general and, for liver-related reactions, there is no recommendation to use it as a substitute for the preferred RUCAM [37, 58].

10.4.2 Standardized Algorithm Approaches

10.4.2.1 Naranjo Scale

The Naranjo scale is the preferred algorithm to be used for causality assessment for suspected adverse reactions associated with herbal product use where there is no specification of injury disease or injured organ (Table 10.2) [38]. However, its use in liver injury cases is obsolete [32, 33, 37, 48, 58–60]. Originally established for assessing causality for suspected adverse reactions associated with conventional medicines, the Naranjo scale is standardized and quantitative with individual element scoring (Tables 10.2 and 10.3) [38]. However, clear key items of the scale, such as challenge, dechallenge, and rechallenge criteria, as well as suggestions on how to exclude alternative causes are only marginally described, or missing. Despite these and many other shortcomings, preference should be given to the Naranjo scale that can be used for general non-liver adverse reactions associated with herbal product use and, due to its item scoring [38].

10.4.2.2 RUCAM

RUCAM represents a standardized, structured, quantitative, transparent, and validated causality assessment method for liver injury associated with use of herbal products and conventional medicines in real time when the patient is under medical care. Defined key elements with specific scores are the cornerstones of RUCAM and provide objective results [37, 54, 55]. An example of how to use RUCAM is provided for a patient with liver injury, who was treated with four different Indian Ayurvedic herbs; RUCAM allows adjudication of an individual causality classification for each herbal used (Table 10.5) [61]. RUCAM was also used for assessing HILI causality for green tea and its extracts, an approach that could be considered with respect to previous conclusions regarding a cautionary on green tea extracts due to a negative benefit: risk constellation with discouraging now the use of the outdated Naranjo method [62]. With special reference to China, other recent HILI case analyses also benefited from the use of RUCAM [63–70].

In support of the original RUCAM of 1993 and its updated version of 2016, various aspects were summarized at the occasion of its use for the last 25 years [71]. RUCAM is viewed as a user-friendly method with a simple work sheet and clear recommendations to users. Case management with RUCAM is quick, effective and cost saving, as no network or rounds are needed. RUCAM cannot compensate for poor quality in medical records, but to meet these issues HILI cases should be assessed for causality using a prospective rather than a retrospective study protocol [71]. Recent attempts to build electronic RUCAM or automatic applications of this method were successful, although some weaknesses need to be corrected [72]. It is also expected to improve RUCAM with biomarkers provided they were validated by robust standards like RUCAM. Overall, RUCAM was applied successfully without any overt problems in 46,266 DILI cases published from 2014 until early 2019, in line with the assumption that no other method can outperform RUCAM [73].

In brief, the scoring system of RUCAM attributes scores to key items and ranges from −3 to +3 points. The sum of the individual scores provides the final score for each suspected herb. Final scores range from +14 to −9 points and allow for grading causality: ≤ 0, relationship excluded; 1–2, unlikely; 3–5, possible; 6–8, probable; ≥ 9, highly probable [37]. Therefore, causality grading is transparent and objective, does not require expert rounds, and differs substantially from expert-based opinions, which, by definition, are subjective and not transparent [37]. RUCAM-based causality gradings and individual item scores should be included in any report of liver injury associated with herbal products, listing also alternative causes, and providing case narratives.

Most importantly, RUCAM criteria and element scoring were developed using data from a cohort consisting of real DILI cases with positive rechallenge, recognized as gold standard to confirm the diagnosis [54, 55]. RUCAM was also validated using cases with positive rechallenge to determine the performance indicators (sensitivity, specificity, positive and negative predictive values) and by external independent assessors using injury cases to determine the reproducibility of the method [55]. As the preferred tool for assessing liver injury cases associated with medicines' use, RUCAM is the most commonly used causality assessment method worldwide, and its updated version should be used in future HILI cases [37]. However, RUCAM was not developed for general use in causality assessment for adverse reactions associated with herbal products.

Table 10.5 RUCAM scale as an example with items required for causality assessment in a patient with HILI by four different Indian Ayurvedic herbs

Items for hepatocellular injury	Possible score	Psoralea corylifolia	Acacia catechu	Eclipta alba	Vetivexia zizaniodis
1. Time to onset from the beginning of the herb					
• 5–90 days (rechallenge: 1–15 days)	+2				
• <5 or > 90 days (rechallenge: >15 days)	+1	+1	+1	+1	+1
Alternative: Time to onset from cessation of the herb					
• ≤15 days (except for slowly metabolized herbal chemicals: >15 days)	+1				
2. Course of ALT after cessation of the herb					
Percentage difference between ALT peak and ULN					
• Decrease ≥50% within 8 days	+3	+3	+3	+3	+3
• Decrease ≥50% within 30 days	+2				
• No information or continued herbal use	0				
• Decrease ≥50% after the 30th day	0				
• Decrease <50% after the 30th day or recurrent increase	−2				
3. Risk factors					
• Alcohol use (drinks/day: >2 for women, >3 for men)	+1				
• Alcohol use (drinks/day: ≤2 for women, ≤3 for men)	0	0	0	0	0
• Age ≥ 55 years	+1	+1	+1	+1	+1
• Age < 55 years	0				
4. Concomitant herbs(s) and drug(s)					
• None or no information	0				
• Concomitant herb or drug with incompatible time to onset	0				
• Concomitant herb or drug with compatible or suggestive time to onset	−1	−1			
• Concomitant herb or drug known as hepatotoxin and with compatible or suggestive time to onset	−2		−2	−2	−2
• Concomitant herb or drug with evidence for its role in this case (positive rechallenge or validated test)	−3				
5. Search for non-herb causes					
Group I (7 causes)					
• Anti-HAV-IgM		–	–	–	–

Table 10.5 (continued)

Items for hepatocellular injury	Possible score	Psoralea corylifolia	Acacia catechu	Eclipta alba	Vetivexia zizaniodis
• HBsAg, anti-HBc-IgM, HBV-DNA		–	–	–	–
• Anti-HCV, HCV-RNA		–	–	–	–
• HEV (anti-HEV-IgM, anti-HEV-IgG)		–	–	–	–
• Hepatobiliary sonography/colour Doppler sonography of liver vessels/ endosonography/CT/MRC		–	–	–	–
• Alcoholism (AST/ALT ≥2)		–	–	–	–
• Acute recent hypotension history (particularly if underlying heart disease)		–	–	–	–
Group II (5 causes)					
• Complications of underlying disease(s) such as sepsis, autoimmune hepatitis, chronic hepatitis B or C, primary biliary cirrhosis or sclerosing cholangitis, genetic liver diseases		–	–	–	–
• Infection suggested by PCR and titre change for					
• CMV (anti-CMV-IgM, anti-CMV-IgG)		–	–	–	–
• EBV (anti-EBV-IgM, anti-EBV-IgG)		–	–	–	–
• HSV (anti-HSV-IgM, anti-HSV-IgG)		–	–	–	–
• VZV (anti-VZV-IgM, anti-VZV-IgG)		–	–	–	–
Evaluation of group I and II					
• All causes—groups I and II—reasonably ruled out	+2	+2	+2	+2	+2
• The 6 causes of group I ruled out	+1				
• 5 or 4 causes of group I ruled out	0				
• Less than 4 causes of group I ruled out	−2				
• Non-herb cause highly probable	−3				
6. Previous information on hepatotoxicity of the herb					
• Reaction labelled in the product characteristics	+2				
• Reaction published but unlabelled	+1	+1			
• Reaction unknown	0		0	0	0
7. Response to rechallenge					
• Doubling of ALT with the herb alone, provided ALT below 5 × ULN before rechallenge	+3				
• Doubling of ALT with the herb(s) and drug(s) already given at the time of first reaction	+1				

(continued)

Table 10.5 (continued)

Items for hepatocellular injury	Possible score	Psoralea corylifolia	Acacia catechu	Eclipta alba	Vetivexia zizaniodis
• Increase of ALT but less than 1 × ULN in the same conditions as for the first administration	−2				
• Other situations	0				
Total score for each individual herb used by the patient		+7	+5	+5	+5

The data of the patient with severe hepatotoxicity by four different Indian Ayurvedic herbs are modified from a published report [61], using the RUCAM subscale for the hepatocellular type of liver injury of the updated RUCAM [37]. The '–' symbol signifies that this particular item has been evaluated and no abnormality was found. For the four herbs, the total score was either +7 (probable causality) or + 5 (possible causality). Abbreviations: ALT, Alanine aminotransferase; AST, Aspartate aminotransferase; CMV, Cytomegalovirus; CT, Computer tomography; DILI, Drug-induced liver injury; EBV, Epstein Barr virus; HAV, Hepatitis A virus; HBc, Hepatitis B core; HBsAg, Hepatitis B antigen; HBV, Hepatitis B virus; HCV, Hepatitis C virus; HEV, Hepatitis E virus; HILI, Herb-induced liver injury; HSV, Herpes simplex virus; MRC, Magnetic resonance cholangiography; RUCAM, Roussel Uclaf Causality Assessment Method; ULN, upper limit of the normal range; VZV, Varicella zoster virus. Total RUCAM score and resulting RUCAM causality grading [37]: ≤0, excluded; 1–2, unlikely; 3–5, possible; 6–8, probable; ≥ 9, highly probable

10.4.2.3 MV Scale

In an attempt to improve the original RUCAM [54, 55], the MV scale was developed by deleting laboratory items and adding clinical elements, along with simplifying and changing the relative weight of elements in the algorithm [44], as discussed in detail [37]. Therefore, the MV scale has fewer specific criteria; evaluates dechallenge as the time necessary for ALT (alanine aminotransferase) or ALP (alkaline phosphatase) to fall below 2 × ULN (upper limit of normal); and considers a shorter latency period [44]. It also asks for less accurate exclusion criteria for alternative causes, ignores concomitant medicines or other herbal product/'dietary supplement' use, overemphasizes drugs with more than 5 years marketing without published hepatotoxicity, and overestimates extrahepatic manifestations. Consequently, the MV scale is not recommended for assessing causality in suspected liver injury associated with use of herbal products [37] and is certainly not a substitute for the original RUCAM [54, 55]. The MV scale cannot be used for general use in causality assessment for adverse reactions.

10.4.2.4 TTK Scale

The TTK scale was established for DILI cases specifically in Japan [45] and is another attempt to modify the original RUCAM [54], with different evaluations of the chronology, exclusion of comedication, inclusion of the drug lymphocyte stimulation test (DLST) and eosinophilia in its assessment [45]. Limited access and lack of standardization have prevented general clinical use of the DLST and, consequently, TTK scale applications outside Japan; this may be due to methodological difficulties with false positive and false negative DLST results, which prevents using this parameter as a valid diagnostic criterion [37]. The TTK scale cannot replace RUCAM for liver injury cases and is not applicable for general use in causality assessment for adverse reactions associated with herbal products.

10.4.3 Probabilistic Approach Derived from Bayes' Theorem

This tool is cumbersome and has rarely been applied in assessing causality in adverse reactions associated with herbal product use [39–41]. Its principle lies in the use of specific findings in a case to transform a prior probability into a posterior probability for product causation [41]. In most cases, however, valid prior causality is hardly assessable.

10.5 Diagnostic Biomarkers

For cases of adverse reactions associated with herbal product use, diagnostic biomarkers to establish a specific herbal ingredient, or even specific chemical constituent, as causative are rare and not usually available for clinic use. Exceptions exist for intrinsic liver injury caused by a small number of herbal ingredients [24, 74] such as germander (*Teucrium chamaedrys* L.) [75, 76], and pyrrolizidine alkaloid (PA) containing TCM herbs, causing clinical hepatic sinusoidal obstruction syndrome (HSOS) [24, 75, 77–79]. For liver injury associated with germander, serum anti-microsomal epoxide hydrolase autoantibodies are the specific diagnostic, mechanism-based biomarkers [24, 75, 76]. For HSOS caused by PA containing herbs, specific diagnostic biomarkers of pyrrole protein adducts are available [24, 75, 77–79]. The clinical value of microRNAs as diagnostic biomarkers is still under clinical evaluation [24].

10.6 Herb-Herb and Herb-Drug Interactions

Some suspected adverse reactions have been interpreted on the basis of herb-herb interactions or herb-drug interactions [41, 80–82]. However, such interpretation is based on vague clinical impressions [82–84] rather than a valid quantitative diagnostic approach, such as the Drug Interaction Probability Scale (DIPS) developed by Horn et al. [83]. Similar to the Naranjo scale, DIPS uses 10 questions, and answers receive scores that provide after addition an estimated likelihood of drug interaction [41, 83]. DIPS was conceptualized for interactions related primarily to conventional medicines, rather than for herbal products. Of note, a recent study applied DIPS to assess suspected adverse reactions following concomitant use of natural health products and prescription drugs, providing several cases as examples [84]. Currently, DIPS is not widely used for herb-herb interactions, which limits a thorough clinical appreciation of this tool. Major problems may emerge if products with multiple herbal ingredients are to be evaluated regarding possible herb-herb interactions.

10.7 Herbal Pharmacovigilance Challenges and Future Perspectives

Not specified for liver injury, general guidelines for submitting adverse event reports for publication are available [85]. Nevertheless, pharmacovigilance for herbal medicines remains challenging, and robust causality assessment depends primarily on the quality of HILI reports [37]. Problematic are cases with incomplete data, because they are provided by non-healthcare professionals. Future efforts should focus on analyses of cases with good data quality, rather than on case quantity based on poorly documented cases. For suspected general adverse reactions unrelated to the liver as the target organ, causality assessment for herbal medicines should be evaluated using the Naranjo scale, or the RUCAM scale specifically for cases describing suspected liver injury.

10.8 Conclusions

Suspected adverse reactions associated with the use of herbal medicines require a careful causality assessment. For general causality assessment of suspected adverse reactions not related to a specific disease or an injured organ, such as the liver, the quantitative Naranjo scale is the most appropriate objective tool with its defined key elements and their individual scores; this author considers that the global introspection-based WHO method has unclear key elements and, without a scoring system, is subjective. For causality assessment of cases of liver injury associated with the use of herbal products, the quantitative updated RUCAM as the most commonly used causality assessment method worldwide is recommended, since it provides objective transparent results.

References

1. Barnes J, Anderson LA, Phillipson JD (2007) Herbal Medicines, 3rd edn. Pharmaceutical Press, London
2. Pelkonen O, Xu O, Fan TP (2014) Why is research on herbal medicinal products important and how can we improve quality? J Tradit Complement Med 4(1):1–7. https://doi.org/10.4103/2225-4110.124323
3. WHO (2019) WHO global report on traditional and complementary medicine 2019. http://www.who.int/medicines/areas/traditional/definitions/en/. Accessed 2 Sept 2019
4. WHO (2013) WHO traditional medicines strategy 2014–2023, Geneva. http://www.who.int/medicines/publications/traditional/trm_strategy14_23/en/. Accessed 2 Sept 2019
5. Leonti M, Casu L (2013) Traditional medicines and globalization: current and future perspectives in ethnopharmacology. Front Pharmacol 4:92. https://doi.org/10.3389/fphar.2013.00092
6. National Institutes of Health (2019) Traditional Chinese Medicine: what you need to know. https://nccih.nih.gov/health/whatiscam/chinesemed.htm. Accessed 2 Sept 2019
7. Park HL, Lee HS, Shin BC et al (2012) Traditional medicine in China, Korea, and Japan: a brief introduction and comparison. Evid Based Complement Alternat Med. https://doi.org/10.1155/2012/429103
8. Patwardhan B, Warude D, Pushpangadan BN (2005) Ayurveda and traditional Chinese medicine: a comparative overview. Evid Based Complement Alternat Med 2(4):465–473. https://doi.org/10.1093/ecam/neh140
9. Saad B, Azaizeh H, Abu-Hijleh G, Said O (2006) Safety of traditional Arab medicine. Evid Based Complement Alternat Med 3(4):433–439. https://doi.org/10.1093/ecam/nel058
10. Popat A, Shear NH, Malkiewicz I et al (2001) The toxicity of *Callilepis laureola*, a South African traditional herbal medicine. Clin Biochem 34(3):229–236. https://doi.org/10.1016/s0009-9120(01)00219-3
11. Teschke R, Wolff A, Frenzel C, Eickhoff A, Schulze J (2015) Herbal traditional Chinese medicine and its evidence base in gastrointestinal disorders. World J Gastroenterol 21(15):4466–4490. https://doi.org/10.3748/wjg.v21.i15.4466
12. Shaw D (2010) Toxicological risks of Chinese herbs. Planta Med 76(17):2012–2018. https://doi.org/10.1055/s-0030-1250533
13. Efferth T, Kaina B (2011) Toxicities by herbal medicines with emphasis to traditional Chinese medicine. Curr Drug Metab 12(10):989–996. https://doi.org/10.2174/138920011798062328
14. Ernst E (2002) Heavy metals in traditional Indian remedies. Eur J Clin Pharmacol 57(12):891–896. https://doi.org/10.1007/s00228-001-0400-y
15. Barnes J (2003) Pharmacovigilance of herbal medicines: a UK perspective. Drug Saf 26(12):829–851. https://doi.org/10.2165/00002018-200326120-00001
16. Shaw D, Ladds G, Duez P, Williamson E, Chan K (2012) Pharmacovigilance of herbal medicine. J Ethnopharmacol 140(3):513–518. https://doi.org/10.1016/j.jep.2012.01.051
17. Zhang L, Yan J, Liu X et al (2012) Pharmacovigilance practice and risk control of traditional Chinese medicine drugs in China: current status and future perspective. J Ethnopharmacol 140(3):519–525. https://doi.org/10.1016/j.jep.2012.01.058
18. Ekor M (2014) The growing use of herbal medicines: issues relating to adverse reactions and challenges in monitoring safety. Front Pharmacol 4:177. https://doi.org/10.3389/fphar.2013.00177
19. Hung SK, Hillier S, Ernst E (2011) Case reports of adverse effects of herbal medicinal products (HMPs): a quality assessment. Phytomedicine 18(5):335–343. https://doi.org/10.1016/j.phymed.2010.07.007
20. Posadzki P, Watson LK, Ernst E (2013) Adverse effects of herbal medicines: an overview of systematic reviews. Clin Med 13(1):7–12. https://doi.org/10.7861/clinmedicine.13-1-7

21. Lai JN, Tang JL, Wang JD (2013) Observational studies on evaluating the safety and adverse effects of traditional Chinese medicine. Evid Based Complement Alternat Med. https://doi.org/10.1155/2013/697893
22. Calitz C, du Plessis L, Gouws C et al (2015) Herbal hepatotoxicity: current status, examples, and challenges. Expert Opin Drug Metab Toxicol 11(10):1551–1565. https://doi.org/10.1517/17425255.2015.1064110
23. Teschke R, Eickhoff A (2015) Herbal hepatotoxicity in traditional and modern medicine: actual key issues and new encouraging steps. Front Pharmacol 6:72. https://doi.org/10.3389/fphar.2015.00072
24. Teschke R, Larrey D, Melchart D, Danan G (2016) Traditional Chinese medicine (TCM) and herbal hepatotoxicity: RUCAM and the role of novel diagnostic biomarkers such as microRNAs. Medicines 3(3):18. https://doi.org/10.3390/medicines3030018
25. Izzo AA, Ernst E (2009) Interactions between herbal medicines and prescribed drugs: an updated systematic review. Drugs 69(13):1777–1798. https://doi.org/10.2165/11317010-000000000-00000
26. Hu Z, Yang X, Ho PCL et al (2005) Herb-drug interactions: a literature review. Drugs 65(9):1239–1282. https://doi.org/10.2165/00003495-200565090-00005
27. Teschke R, Frenzel C, Schulze J, Schwarzenboeck A, Eickhoff A (2013) Herbalife hepatotoxicity: evaluation of cases with positive reexposure tests. World J Hepatol 5(7):353–363. https://doi.org/10.4254/wjh.v5.i7.353
28. Zambrone FAD, Corrêa CL, Sampaio do Amaral LM (2015) A critical analysis of the hepatotoxicity cases described in the literature related to Herbalife products. Braz J Pharm Sci 51(4):785–796. https://doi.org/10.1590/S1984-82502015000400004
29. Teschke R, Schulze J, Eickhoff A, Wolff A, Frenzel C (2015) Mysterious Hawaii liver disease case—naproxen overdose as cause rather than OxyELITE Pro? J Liver Clin Res 2(2):1013
30. Teschke R, Schwarzenboeck A, Frenzel C, Schulze J, Eickhoff A, Wolff A (2016) The mystery of the Hawaii liver disease cluster in summer 2013: a pragmatic and clinical approach to solve the problem. Ann Hepatol 15(1):91–109. https://doi.org/10.5604/16652681.1184237
31. Teschke R, Eickhoff A (2016) The Honolulu liver disease cluster at the medical center: its mysteries and challenges. Int J Mol Sci 17(4):476. https://doi.org/10.3390/ijms17040476
32. Teschke R, Bahre R, Genthner A, Fuchs J, Schmidt-Taenzer W, Wolff A (2009) Suspected black cohosh hepatotoxicity-challenges and pitfalls of causality assessment. Maturitas 63(4):302–314. https://doi.org/10.1016/j.maturitas.2009.05.006
33. Teschke R (2010) Black cohosh and suspected hepatotoxicity: inconsistencies, confounding variables, and prospective use of a diagnostic causality algorithm. A critical review. Menopause 17(2):426–440. https://doi.org/10.1097/gme.0b013e3181c5159c
34. Teschke R, Lebot V (2011) Proposal for a kava quality standardization code. Food Chem Toxicol 49(10):2503–2516. https://doi.org/10.1016/j.fct.2011.06.075
35. Teschke R, Schwarzenboeck A, Eickhoff A, Frenzel C, Wolff A, Schulze J (2013) Clinical and causality assessment in herbal hepatotoxicity. Expert Opin Drug Saf 12(3):339–366. https://doi.org/10.1517/14740338.2013.774371
36. Kaplowitz N (2001) Causality assessment versus guilt-by-association in drug hepatotoxicity. Hepatology 33(1):308–310. https://doi.org/10.1053/jhep.2001.21083
37. Danan G, Teschke R (2015) RUCAM in drug and herb induced liver injury: the update. Int J Mol Sci 17(1):14. https://doi.org/10.3390/ijms17010014
38. Naranjo CA, Busto U, Sellers EM et al (1981) A method for estimating the probability of adverse drug reactions. Clin Pharmacol Ther 30(2):239–245. https://doi.org/10.1038/clpt.1981.154
39. Théophile H, André M, Miremont-Salamé G, Arimone Y, Bégaud B (2013) Comparison of three methods (an updated logistic probabilistic method, the Naranjo and Liverpool algorithms) for the evaluation of routine pharmacovigilance case reports using a consensual expert judgment as reference. Drug Saf 36(10):1033–1044. https://doi.org/10.1007/s40264-013-0083-1

40. Szarfman A, Machado SG, O'Neill RT (2002) Use of screening algorithms and computer systems to efficiently signal higher-than-expected combinations of drugs and events in the US FDA's spontaneous report database. Drug Saf 25(6):381–392. https://doi.org/10.2165/00002018-200225060-00001
41. Agbabiaka TB, Savovic J, Ernst E (2008) Methods for causality assessment of adverse drug reactions: a systematic review. Drug Saf 31(1):21–37. https://doi.org/10.2165/00002018-200831010-00003
42. Hayashi PH (2016) Drug-induced liver injury network causality assessment: criteria and experience in the United States. Int J Mol Sci 17(2):201. https://doi.org/10.3390/ijms17020201
43. WHO (2000) The use of the WHO-UMC system for standardised case causality assessment. WHO Collaborating Centre for International Drug Monitoring (Uppsala Monitoring Centre, UMC), Database https://www.who.int/medicines/areas/quality_safety/safety_efficacy/WHOcausality_assessment.pdf. Accessed 2 Sept 2019
44. Maria VAJ, Victorino RMM (1997) Development and validation of a clinical scale for the diagnosis of drug-induced hepatitis. Hepatology 26(3):664–669. https://doi.org/10.1002/hep.510260319
45. Takikawa H, Takamori Y, Kumagi T et al (2003) Assessment of 287 Japanese cases of drug induced liver injury by the diagnostic scale of the International Consensus Meeting. Hepatol Res 27(3):192–195. https://doi.org/10.1016/s1386-6346(03)00232-8
46. Agarwal VK, McHutchison JG, Hoofnagle JH, Drug-Induced Liver Injury Network (DILIN) (2010) Important elements for the diagnosis of drug-induced liver injury. Clin Gastroenterol Hepatol 8(5):463–470. https://doi.org/10.1016/j.cgh.2010.02.008
47. Björnsson E, Olsson R (2006) Suspected drug-induced liver fatalities reported to the WHO database. Dig Liver Dis 38(1):33–38. https://doi.org/10.1016/j.dld.2005.06.004
48. Teschke R, Wolff A (2011) Regulatory causality evaluation methods applied in kava hepatotoxicity: are they appropriate? Regul Toxicol Pharmacol 59(1):1–7. https://doi.org/10.1016/j.yrtph.2010.09.006
49. WHO (2007) Assessments of the risk of hepatotoxicity with kava products. Switzerland, Geneva
50. Teschke R, Frenzel C, Schulze J, Eickhoff A (2012) Spontaneous reports of primarily suspected herbal hepatotoxicity by *Pelargonium sidoides*: was causality adequately ascertained? Regul Toxicol Pharmacol 63(1):1–9. https://doi.org/10.1016/j.yrtph.2012.02.009
51. Teschke R, Frenzel C, Schulze J, Eickhoff A (2012) Suspected herbal hepatotoxicity: the pharmacovigilance dilemma with disputed and obsolete evaluation methods. Regul Toxicol Pharmacol 64(2):343–344. https://doi.org/10.1016/j.yrtph.2012.06.010
52. Teschke R, Frenzel C, Wolff A et al (2012) Initially purported hepatotoxicity by *Pelargonium sidoides*: the dilemma of pharmacovigilance and proposals for improvements. Ann Hepatol 11(4):500–512
53. Teschke R, Eickhoff A, Wolff A, Frenzel C, Schulze J (2013) Herbal hepatotoxicity and WHO global introspection method. Ann Hepatol 12(1):11–21
54. Danan G, Bénichou C (1993) Causality assessment of adverse reactions to drugs—I. A novel method based on the conclusions of international consensus meetings: application to drug-induced liver injuries. J Clin Epidemiol 46(11):1323–1330. https://doi.org/10.1016/0895-4356(93)90101-6
55. Bénichou C, Danan G, Flahault A (1993) Causality assessment of adverse reactions to drugs—II. An original model for validation of drug causality assessment methods: case reports with positive rechallenge. J Clin Epidemiol 46(11):1331–1336. https://doi.org/10.1016/0895-4356(93)90102-7
56. Navarro VJ, Barnhart H, Bonkovsky HL et al (2014) Liver injury from herbals and dietary supplements in the U.S. Drug-Induced Liver Injury Network. Hepatology 60(4):1399–1408. https://doi.org/10.1002/hep.27317
57. Teschke R, Eickhoff A (2017) Suspected liver injury and the dilemma of causality. Dig Dis Sci 62(4):1095–1098. https://doi.org/10.1007/s10620-016-4442-5

58. Teschke R, Danan G (2018) Causality assessment methods in drug-induced liver injury. In: Chen M, Will Y (eds) Drug-induced liver toxicity. Springer Nature, Berlin, Germany, pp 555–594
59. Teschke R, Schulze J (2012) Suspected herbal hepatotoxicity: requirements for appropriate causality assessment by the US Pharmacopeia. Drug Saf 35(12):1091–1097. https://doi.org/10.2165/11631960-000000000-00000
60. García-Cortés M, Lucena MI, Pachkoria K et al (2008) Evaluation of naranjo adverse drug reactions probability scale in causality assessment of drug-induced liver injury. Aliment Pharmacol Ther 27(9):780–789. https://doi.org/10.1111/j.1365-2036.2008.03655.x
61. Teschke R, Bahre R (2009) Severe hepatotoxicity by Indian Ayurvedic herbal products: a structured causality assessment. Ann Hepatol 8(3):258–266
62. Teschke R, Xuan TD (2019) Suspected herb induced liver injury by green tea extracts: critical review and case analysis applying RUCAM for causality assessment. Jpn J Gastroenterol Hepatol V(6):1–16
63. Zhang P, Ye Y, Yang X, Jiao Y (2016) Systemic review on Chinese herbal medicine induced liver injury. Evid Based Complement Alternat Med. https://doi.org/10.1155/2016/3560812
64. Wang R, Qi X, Yoshida EM et al (2018) Clinical characteristics and outcomes of traditional Chinese medicine-induced liver injury: a systematic review. Exp Rev Gastroenterol Hepatol 12(4):425–434. https://doi.org/10.1080/17474124.2018.1427581
65. Wang J, Ma Z, Niu M et al (2015) Evidence chain-based causality identification in herb-induced liver injury: exemplification of a well-known liver-restorative herb Polygonum multiflorum. Front Med 9(4):457–467. https://doi.org/10.1007/s11684-015-0417-8
66. Zhu Y, Li YG, Wang JB et al (2015) Causes, features, and outcomes of drug-induced liver injury in 69 children from China. Gut Liver 9(4):525–533. https://doi.org/10.5009/gnl14184
67. Zhu Y, Niu M, Chen J et al (2016) Hepatobiliary and pancreatic: comparison between Chinese herbal medicine and Western medicine-induced liver injury of 1985 patients. J Gastroenterol Hepatol 31(8):1476–1482. https://doi.org/10.1111/jgh.13323
68. Zhu Y, Niu M, Wang JB et al (2019) Predictors of poor outcomes in 488 patients with herb-induced liver injury. Turk J Gastroenterol 30(1):47–58. https://doi.org/10.5152/tjg.2018.17847
69. Jing J, Teschke R (2018) Traditional Chinese medicine and herb induced liver injury: comparison with drug-induced liver injury. J Clin Transl Hepatol 6(1):57–68. https://doi.org/10.14218/JCTH.2017.00033
70. Jing J, Wang RL, Zhao XY et al (2019) Association between the concurrence of pre-existing chronic liver disease and worse prognosis in patients with an herb - *Polygonum multiflorum* thunb. induced liver injury: a case-control study from a specialised liver disease center in China. BMJ Open 9(1):e023567. https://doi.org/10.1136/bmjopen-2018-023567
71. Danan G, Teschke R (2018) Drug-induced liver injury: why is the Roussel Uclaf Causality Assessment Method (RUCAM) still used 25 years after its launch? Drug Saf 41(8):735–743. https://doi.org/10.1007/s40264-018-0654-2
72. Danan G, Teschke R (2019) Roussel Uclaf causality assessment method for drug-induced liver injury. Front Pharmacol 10:853. https://doi.org/10.3389/fphar.2019.00853
73. Teschke R (2019) Idiosyncratic DILI: analysis of 46,266 cases assessed for causality by RUCAM and published from 2014 to early 2019. Front Pharmacol 10:730. https://doi.org/10.3389/fphar.2019.00730
74. Frenzel C, Teschke R (2016) Herbal hepatotoxicity: clinical characteristics and listing compilation. Int J Mol Sci 17(5):588. https://doi.org/10.3390/ijms17050588
75. Larrey D, Faure S (2011) Herbal medicine hepatotoxicity: a new step with development of specific biomarkers. J Hepatol 54(4):599–601. https://doi.org/10.1016/j.jhep.2010.12.003
76. Larrey D, Vial T, Pauwels A et al (1992) Hepatitis after germander (*Teucrium chamaedrys*) administration: another instance of herbal medicine hepatotoxicity. Ann Intern Med 117(2):129–132. https://doi.org/10.7326/0003-4819-117-2-129

77. Lin G, Wang JY, Li N et al (2011) Hepatic sinusoidal obstruction syndrome associated with consumption of Gynura segetum. J Hepatol 54(4):666–673. https://doi.org/10.1016/j.jhep.2010.07.031
78. Gao H, Li N, Wang JY, Zhang SC, Lin G (2012) Definitive diagnosis of hepatic sinusoidal obstruction syndrome induced by pyrrolizidine alkaloids. J Dig Dis 13(1):33–39. https://doi.org/10.1111/j.1751-2980.2011.00552.x
79. Wang JY, Gao H (2014) Tusanqi and hepatic sinusoidal obstruction syndrome. Dig Dis 15(3):105–107. https://doi.org/10.1111/1751-2980.12112
80. Fugh-Berman A (2000) Herb-drug interactions. Lancet 355(9198):134–138. https://doi.org/10.1016/S0140-6736(99)06457-0
81. Fugh-Berman A, Ernst E (2001) Herb-drug interactions: review and assessment of reported reliability. Br J Clin Pharmacol 52(5):587–595. https://doi.org/10.1046/j.0306-5251.2001.01469.x
82. Kennedy DA, Seely D (2010) Clinically based evidence of drug-herb interactions: a systematic review. Expert Opin Drug Saf 9(1):79–124. https://doi.org/10.1517/14740330903405593
83. Horn JR, Hansten PD, Chan LN (2007) Proposal for a new tool to evaluate drug interaction cases. Ann Pharmacother 41(4):674–680. https://doi.org/10.1345/aph.1H423
84. Vohra S, Cvijovic K, Boon H et al (2012) Study of Natural Health Product Adverse Reactions (SONAR): active surveillance of adverse events following concurrent natural health product and prescription drug use in community pharmacies. PLoS One 7(9):e45196. https://doi.org/10.1371/journal.pone.0045196
85. Kelly WN, Arellano FM, Barnes J et al (2007) Guidelines for submitting adverse event reports for publication. Drug Saf 30:367–373. https://doi.org/10.2165/00002018-200730050-00001

Chapter 11
The Value of Complementary Approaches to Causality Assessment for Individual Case Safety Reports: The Example of *Artemisia annua* and Hepatotoxicity

Ruth L. Savage

11.1 Introduction

Drug safety signals may be generated from a variety of sources, including individual case safety reports (ICSRs) in pharmacovigilance databases, published case reports, clinical and pre-clinical trials, and observational studies. In the context of this chapter, the term "drugs" includes medicines, vaccines, herbal medicines and other natural health products. A signal, as defined by the Council for International Organizations of Medical Sciences (CIOMS), is "Information that arises from one or multiple sources (including observations and experiments) which suggests a new potentially causal association, or a new aspect of a known association, between an intervention and an event or set of related events, either adverse or beneficial, that is judged to be of sufficient likelihood to justify verificatory action" [1].

A core activity of pharmacovigilance is the generation of safety signals through causality assessment of reports of adverse effects to ascertain the likelihood that they represent a previously unknown or insufficiently documented adverse drug reaction (ADR). Verificatory action can range from gathering more information from prescribers and consumers, to undertaking large, formal studies. Clearly, from the CIOMS definition, causality assessment weighs up the evidence for and against a hypothesis of causality. It can rarely "prove" a causal association. In practice, this leads to a tension between publicizing a signal early enough to minimize the risk of harm, but not so early as to unnecessarily discourage use of a beneficial drug.

R. L. Savage (✉)
Uppsala Monitoring Centre, Uppsala, Sweden

New Zealand Pharmacovigilance Centre, Division of Health Sciences, University of Otago, Dunedin, New Zealand

Department of General Practice, University of Otago, Christchurch, New Zealand
e-mail: ruth.savage@otago.ac.nz

J. Barnes (ed.), *Pharmacovigilance for Herbal and Traditional Medicines*,
https://doi.org/10.1007/978-3-031-07275-8_11

11.2 Causality Assessment for Individual Case Safety Reports

The merits and demerits of various approaches and methods for causality assessment of ICSRs have been widely debated, all the more so because there is not—and cannot be—a "gold standard". The two most widely used approaches are the World Health Organization–Uppsala Monitoring Centre (WHO-UMC) System for Standardized Case Causality Assessment [2] and the Naranjo Algorithm [3]. The WHO-UMC system is a set of criteria that focuses on the data within the reports examined, while the Naranjo algorithm is a scoring system incorporating data from within reports, and some external data, such as previous similar reports. Both methods stratify reports of suspected adverse drug reactions into "Certain", "Probable", "Possible" and "Unlikely"/ "Doubtful" categories. The WHO-UMC system also includes "Unclassified" and "Unclassifiable" categories. The recently revised French method is a more extensive scoring system that uses both internal and external information and also includes a weighting for the informativeness of the report [4].

The WHO-UMC system (Table 11.1) has been described as a method employing "global introspection" and the Naranjo algorithm as a more objective and transparent system (see Chap. 10). However, there is only an incomplete dichotomy. "Global introspection" was a term used for the unstructured process by which clinicians assessed the likelihood of a causal association depending largely on their previous knowledge and experience. It was recognized that this approach was inadequate and that seven general categories of information needed to be addressed [5]:

1. Previous general experience with the suspect drug
2. Alternative aetiologic candidates
3. Prior history of the patient
4. Timing of events
5. Characteristics of the adverse event
6. Dechallenge
7. Rechallenge

Both the WHO-UMC and the Naranjo approaches incorporate these categories and, equally, require knowledge and experience or "global introspection", especially for considering co-morbidities, concomitant drugs and the indication for the suspect drug, as alternative aetiologies for the adverse effect under consideration. Nevertheless, the Naranjo scoring system does show external observers how a likelihood score was arrived at to some extent. However, it is still not explicit how, for example, the question regarding onset after start of a drug is assessed. The questions merely ask if the suspected reaction occurred after the drug was started and if it resolved when the drug was stopped. There is no apparent consideration given as to whether it is a reasonable time to onset and recovery. The WHO-UMC method does state that for a "certain" relationship, the times to onset and recovery should be

Table 11.1 WHO-UMC Causality Categories (Source: The use of the WHO-UMC system for standardized case causality assessment [2]). See reference for full explanatory details

Causality term	Assessment criteria[a]
Certain	• Event or laboratory test abnormality, with plausible time relationship to drug intake • Cannot be explained by disease or other drugs • Response to withdrawal plausible (pharmacologically, pathologically) • Event definitive pharmacologically or phenomenologically (i.e. an objective and specific medical disorder or a recognized pharmacological phenomenon) • Rechallenge satisfactory, if necessary
Probable/likely	• Event or laboratory test abnormality, with reasonable time relationship to drug intake • Unlikely to be attributed to disease or other drugs • Response to withdrawal clinically reasonable • Rechallenge not required
Possible	• Event or laboratory test abnormality, with reasonable time relationship to drug intake • Could also be explained by disease or other drugs • Information on drug withdrawal may be lacking or unclear
Unlikely	• Event or laboratory test abnormality, with a time to drug intake that makes a relationship improbable (but not impossible) • Disease or other drugs provide plausible explanations
Conditional/ unclassified	• Event or laboratory test abnormality • More data for proper assessment needed, or • Additional data under examination
Unassessable/ unclassifiable	• Report suggesting an adverse reaction • Cannot be judged because information is insufficient or contradictory • Data cannot be supplemented or verified

[a]All points should be reasonably complied with

"plausible", i.e. in keeping with the host's responses given the nature of the suspected adverse effect, and the pharmacology of the drug or herbal product. Since the latter information is not always known for a new drug or herbal product, the time to onset is required to be reasonable for a "probable" or "possible" classification, at the very least occurring after onset of the exposure. Both methods require the "certain" and "probable" categories to show that the exposure of interest is the most likely aetiology and that there was recovery when the patient was no longer exposed to the drug or herbal product implicated. The "certain" category in most cases also requires recurrence on re-exposure. It is apparent that what is needed for the best use of either method is adequate training for the assessor.

Both tools could also benefit from a degree of revision to increase clarity. The utility of both methods, and a comparison of their use in relation to a signal concerning a conventional medicine, comes from a study of over 500 published case reports of metformin-associated lactic acidosis, in which causality was assessed using both the WHO-UMC classification and the Naranjo algorithm [6]. The outcomes showed a considerable degree of agreement, but a larger group of "possible" reports arose

Table 11.2 Summary of the causality assessments of published case reports of metformin and lactic acidosis using the World Health Organisation–Uppsala Monitoring Centre (WHO-UMC) system and Naranjo adverse drug reaction (ADR) scale, including a sensitivity analysis where cases with poor completeness scores were excluded [6]

Causality assessment	Causality category	Cases ($n = 559$)	Sensitivity analysis excluding cases with a completeness score of ≤10 ($n = 386$)
WHO-UMC system	Certain	–	–
	Probably/Likely	17 (3%)	11 (2.8%)
	Possible	473 (84.6%)	338 (87.6%)
	Unlikely	2 (0.4%)	1 (0.3%)
	Conditional/ Unclassified	49 (8.8%)	24 (6.2%)
	Unassessable/ Unclassifiable	18 (3.2%)	12 (3.1%)
Naranjo ADR scale	Definite	–	–
	Probable	22 (3.9%)	16 (4.3%)
	Possible	536 (95.9%)	369 (95.6%)
	Doubtful	1 (0.2%)	1 (0.3%)

from using the Naranjo method (Table 11.2). This was largely attributable to the lack of an "unclassified" category in Naranjo. This is a WHO-UMC classification for reports considered to have insufficient data to assign the report to a causality category. The bar for a "possible" report is lower for Naranjo, and there is discussion about whether the Naranjo "possible" category should be subdivided [7].

11.3 Causality Assessment for Case Series

When reports constitute a case series, individual case causality assessment can be strengthened by application of the Bradford Hill guidelines, or criteria for causal inference, each of which, if fulfilled, provides additional support for causality; use of these criteria also allows external information to be considered which is not included in the WHO-UMC classification. These guidelines were developed by Sir Austin Bradford Hill when the causal link between tobacco smoking and lung cancer was first identified [8]. The guidelines were developed for epidemiological studies and have more recently been applied to pharmacovigilance [9]. The nine criteria are strength, consistency, specificity, temporality, biologic gradient (dose or duration response), plausibility, coherence, experiment and analogy. Table 11.3 shows how these might be applied to causality assessment of case series in pharmacovigilance. Published considerations of extrinsic and intrinsic mechanisms of adverse drug reactions and their application to causal inference have supported and informed the use of the Bradford Hill Guidelines [10–12].

Table 11.3 A suggested application of the Bradford Hill criteria to case series assessment for signal detection in pharmacovigilance

Bradford Hill criteria	Application to pharmacovigilance
Strength	Well-documented case reports of unexpected or incompletely documented suspected ADRs with no obvious confounders. Recurrence on rechallenge increases strength. Sometimes discovered through statistical disproportionality measures
Specificity	ADRs—Drugs generally cause ADRs through specific mechanisms—specific clinical conditions lend more weight than symptoms or diagnoses with many causes. Drugs—Many suspect drugs, rather than one or two, reduce the possibility of causality
Temporality	The time to onset (TTO) or recovery from starting the drug is consistent with the known pharmacology of the suspect medicine or the host response. If these are not known consistency of TTO is also supportive
Consistency	Reports with similar content from a range of reporters and/or geographic areas. OR similar reports from a specific location suggesting a product or administration problem.
Biologic gradient	High proportion of reports indicate use of maximum recommended doses. Suspected ADR occurring as dose increases or resolving with dose decrease. Duration of drug use also relevant.
Plausibility	The known pharmacology of the drug suggests a mechanism for the suspected ADR. Supportive if present but absence of a known mechanism does not preclude an ADR.
Coherence	A causal relationship does not usually contradict generally accepted medical/scientific knowledge
Experiment	Evidence for the suspected ADR or related events found in animal studies or pre-marketing data.
Analogy	Similar drugs, e.g. members of the same ATC group, are known to cause the suspected adverse reaction.

ADR adverse drug reaction, TTO time to onset, ATC anatomic, therapeutic, chemical

11.4 Focused Causality Assessment Methods

The methods discussed here are general. They can be applied to medicines, vaccines and medicinal herbs and used to assess suspected drug interactions. Some modifications have been suggested for particular types of product; for example, the WHO has published guidelines on the safety monitoring of herbal medicines in pharmacovigilance systems [13]. There is also the issue of organ-specific drug injury. One of the most important of these is drug-induced liver injury (DILI) since many drugs are implicated and there are specific characteristics that, if present, support the causality hypothesis, but are not given particular attention in the general causality assessment methods. These include time to onset and recovery of specific hepatic reactions and the alternative aetiologies that should be investigated. A generally accepted standardized, structured method for causality assessment of individual case reports of DILI is the Roussel Uclaf Causality Assessment Method (RUCAM) [14]. This is discussed more fully in Chap. 10 and by the recently published CIOMS consensus on DILI in relation to drug development and the post-market setting [15]. A modified version applied to herbal products has been published [16]. The CIOMS consensus notes some limitations of RUCAM for herbal and dietary supplements and liver injury and refers to an evidence-chained method, published by the China Food and Drug Administration, which is close to RUCAM in principle, but which incorporates verification of the product under suspicion and its quality [17].

RUCAM is designed for—and is a very useful tool for—prospective use when a patient with suspected DILI is under medical care with the possibility for detailed history taking and access to a range of diagnostic tests including radiological and serological investigations. A problem arises if the general methods for causality assessment are considered inadequate for suspected DILI, but the RUCAM method is limited in its usefulness for retrospective evaluation of case reports in pharmacovigilance, since detailed information is required that is often not present in ICSRs. However, if there is a suspicion of a safety signal, the dilemma is whether a delay in publication while waiting for reports that can fulfil all the details required for a full RUCAM assessment is acceptable.

11.5 *Artemisia annua* L. and Liver Injury: Linking Methodologies for Causality Assessment

In New Zealand, a cluster of case reports was received by the New Zealand Pharmacovigilance Centre between late 2017 and early 2019 describing hepatic disorders associated with the use of a supercritical carbon dioxide extract of *Artemisia annua* L. in grapeseed oil [18]. These reports were unexpected, since *A. annua* has a history of use as a medicinal herbal product over two millennia without being recognized as hepatotoxic [19]. Routine causality assessment was applied to each report using the WHO-UMC method. Consideration was also given to applying the RUCAM method as an alternative. A majority of the reports was of a high standard for reports in a pharmacovigilance database; however, it became obvious that it was very difficult to fulfil all the requirements for a RUCAM "probable" causal relationship retrospectively because of lack of information rather than unsupportive evidence. Nevertheless, the RUCAM criteria were invaluable in providing information on what were "reasonable" times to onset and recovery for different hepatic responses and which alternative aetiologies should be excluded. This allowed the identification of reports in which a causal association was clearly "unlikely".

One of the problems in applying RUCAM is testing for all possible aetiologies, since not all will be carried out routinely in patients presenting with acute hepatitis. For example, testing for hepatitis E when it is not endemic and there has not been overseas travel, or lack of comprehensive testing in resource-poor settings.

Furthermore, if patients recover quickly after stopping the suspect drug, further testing for alternative aetiologies may not be considered cost-effective.

The first approach to responding to limited information was to contact reporters to provide as much additional information as possible; this achieved an excellent response. However, given the limitations on the range of tests for alternative aetiologies carried out for the above reasons, the WHO-UMC method was retained for causality assessment of each ICSR while incorporating knowledge from RUCAM. This resulted in 12 of 29 reports being assessed as "probable" using the WHO-UMC criteria informed by RUCAM. After receiving the follow-up information requested, five of the reports were assessed as "probable" using the RUCAM method alone. Well-documented "probable" reports are the usual initial evidence for a drug safety signal, since the "certain" criteria are rarely fulfilled initially. The WHO-UMC "probable" category requires a reasonable time to onset and recovery from starting and stopping the drug and no obvious alternative aetiologies. The RUCAM method follows the same principle, but with more specific detail about onset and recovery times and alternative aetiologies.

To strengthen the causality assessment beyond individual report assessment, the case series was assessed using the Bradford Hill criteria for causal inference. This found: (1) *strength of evidence* through a cluster of similar and unexpected reports being submitted over a short period; (2) clear *consistency* in reporting with a range of health care professionals, including hepatologists, submitting similar reports from throughout New Zealand; (3) the expected *temporal relationship* with respect to the pharmacology of the product could not be known, but there was consistency in that the time to onset from starting the product was in most cases within three to four months and this was in keeping with the hepatotoxicity described; (4) the reports were very *specific* as hepatotoxicity was the suspected reaction and the *A. annua* product the only suspect medicine in all the reports, although some concomitant medicines were assessed as co-suspect in a small number of reports; (5) finally, there was some *experimental evidence*, as one of 28 participants assigned to the *A. annua* product in a randomized controlled study for osteoarthritis symptoms developed hepatitis, and one of 34 participants in an open-label extension of the study was found to have increased hepatic enzyme concentrations. *Biologic plausibility*, except that HILI is an accepted entity, could not be assessed, but this is an equally important aspect and investigations into the precise composition of the product were instigated.

In keeping with the CIOMS consensus [15] the New Zealand product was verified. The possibility remains that the extraction process may have contributed to the HILI observed. However, former lack of pharmacovigilance systems, lack of awareness of reporting and heavy promotion and uptake of the recently marketed product may also have been reasons for observations of hepatotoxicity only emerging now. The authors concluded that there was a safety signal of a causal association between the herbal product and an adverse reaction sufficient to be communicated and investigated further.

11.6 Conclusion

In conclusion, it is important not to dismiss the observations of concerned clinicians, or other healthcare professionals and patients themselves, especially where there is consistency of observations in the absence of obvious biases for reporting. In this context, it is true that the ideal pharmacovigilance analysis would be based on complete data sets to which sophisticated causality assessment methods have been applied, and it is true that poorly documented reports carry little weight. However, an appropriate assessment of a case series of partially incomplete reports, as is often the nature of ICSRs, may well be important to trigger an alert. Routinely collected and assessed ICSRs do make a major contribution to important regulatory action for medicines [20] and have the ability to do so for medicinal herbal products too.

Acknowledgements I would like to thank Dr. Birgitta Grundmark of the Uppsala Monitoring Centre for her peer review and helpful comments on this chapter.

Caveat
The opinions expressed in this chapter are those of the author and do not necessarily represent those of the Uppsala Monitoring Centre, the New Zealand Pharmacovigilance Centre or the Department of General Practice, Christchurch, University of Otago, New Zealand.

References

1. Report of the Council for International Organizations of Medical Sciences Working Group VIII (2010) Practical aspects of signal detection in pharmacovigilance. https://cioms.ch/wp-content/uploads/2018/03/WG8-Signal-Detection.pdf. Accessed 18 Jul 2019
2. Uppsala Monitoring Centre (2018) The use of the WHO-UMC system for standardised case causality assessment. https://www.who-umc.org/media/164200/who-umc-causality-assessment_new-logo.pdf. Accessed 22 May 2021
3. Naranjo CA, Busto U, Sellers EM et al (1981) A method for estimating the probability of adverse drug reactions. Clin Pharmacol Ther 30(2):239–245. https://doi.org/10.1038/clpt.1981.154
4. Arimone Y, Bidault I, Dutertre J-P et al (2013) Updating the French method for the causality assessment of adverse drug reactions. Therapie 68(2):69–76. https://doi.org/10.2515/therapie/2013016
5. Kramer MS (1986) Assessing causality of adverse drug reactions: global introspection and its limitations. Drug Inf J 20(4):433–437. https://doi.org/10.1177/009286158602000409
6. Kuan IHS, Savage RL, Dufull SB, Walker RJ, Wright DFB (2019) The association between metformin therapy and lactic acidosis. Drug Saf 42(12):1449–1469. https://doi.org/10.1007/s40264-019-00854-x
7. Théophile H, André M, Miremont-Salamé G, Arimone Y, Bégaud B (2013) Comparison of three methods (an updated logistic probabilistic method, the Naranjo and Liverpool Algorithms) for the evaluation of routine pharmacovigilance case reports using consensual expert judgement as reference. Drug Saf 36(10):1033–1044. https://doi.org/10.1007/s40264-013-0083-1

8. Hill AB (1965) The environment and disease: association or causation? Proc R Soc Med 58(5):295–300
9. Shakir SA, Layton D (2002) Causal association in pharmacovigilance and epidemiology: thoughts on the application of the Bradford-Hill criteria. Drug Saf 25(6):467–471. https://doi.org/10.2165/00002018-200225060-00012
10. Aronson JK, Ferner RE (2003) Joining the DoTS: new approach to classifying adverse drug reactions. BMJ 327(7425):1222–1225. https://doi.org/10.1136/bmj.327.7425.1222
11. Ferner RE, Aronson JK (2010) EIDOS: a mechanistic classification of adverse drug effects. Drug Saf 33(1):15–23. https://doi.org/10.2165/11318910-000000000-00000
12. Edwards IR (2017) Causality assessment in pharmacovigilance: still a challenge. Drug Saf 40(5):365–372. https://doi.org/10.1007/s40264-017-0509-2
13. World Health Organization (2004) WHO guidelines on safety monitoring of herbal medicines in pharmacovigilance systems. https://apps.who.int/iris/handle/10665/43034. Accessed 21 Jan 2021
14. Danan G, Teschke R (2015) RUCAM in drug and herb induced liver injury: the update. Int J Mol Sci 17(1):14. https://doi.org/10.3390/ijms17010014
15. Council for International Organizations of Medical Sciences (2020) Drug-induced liver injury: current status and future directions for drug development and the post-market setting. https://cioms.ch/wp-content/uploads/2020/06/CIOMS_DILI_Web_16Jun2020.pdf. Accessed 21 Jan 2021
16. Ahn BM (2004) Herbal preparation-induced liver injury. Korean J Gastroenterol 44(3):113–125
17. Xiao X, Tang J, Mao Y et al (2019) Guidance for the clinical evaluation of traditional Chinese medicine-induced liver injury Issued by China Food and Drug Administration. Acta Pharm Sin B 9(3):648–658. https://doi.org/10.1016/j.apsb.2018.12.003
18. Savage RL, Hill GR, Barnes J, Kenyon SH, Tatley MV (2019) Suspected hepatotoxicity with a supercritical carbon dioxide extract of *Artemisia annua* in grapeseed oil used in New Zealand. Front Pharmacol 10:1448. https://doi.org/10.3389/fphar.2019.01448
19. Chang HM, But PPH (eds) (1986) Pharmacology and applications of Chinese materia medica, vol 1. World Scientific Publishing Ltd, Singapore
20. McNaughton R, Huet G, Shakir S (2014) An investigation into drug products withdrawn from the EU market between 2002 and 2011 for safety reasons and the evidence used to support the decision making. BMJ Open 4(1):e004221. https://doi.org/10.1136/bmjopen-2013-004221

Chapter 12
Development of a Natural Health Product Active Surveillance Method in Outpatient Centers in Canada

Morgan Bharadia, Candace Necyk, and Sunita Vohra

12.1 Background

Natural health products (NHPs) are defined by Health Canada as vitamins, minerals, herbal medicines, homeopathic remedies, traditional medicines, probiotics, amino acids, and essential fatty acids [1]. Similar definitions are used worldwide with the addition of enzymes [2], aromatherapy [3], and plants [4]. NHPs are commonly used across the globe [1, 5–7]. Despite a high prevalence of use, the pre-market regulations for NHPs are less rigorous in comparison to those for prescription medications in some jurisdictions, and nearly non-existent in others [8]. Thus, pharmacovigilance, including post-marketing surveillance, is vital for a better understanding of NHP adverse events (AEs) and interactions with other NHPs and prescription drugs [8].

M. Bharadia (✉)
Faculty of Pharmacy and Pharmaceutical Sciences, University of Alberta, Edmonton, AB, Canada

Department of Medicine, Faculty of Medicine and Dentistry, University of Alberta, Edmonton, AB, Canada
e-mail: basiuk@ualberta.ca

C. Necyk
Faculty of Pharmacy and Pharmaceutical Sciences, University of Alberta, Edmonton, AB, Canada
e-mail: cnecyk@ualberta.ca

S. Vohra
Department of Pediatrics, Faculty of Medicine and Dentistry, University of Alberta, Edmonton, AB, Canada

Department of Psychiatry, Faculty of Medicine and Dentistry, University of Alberta, Edmonton, AB, Canada
e-mail: svohra@ualberta.ca

J. Barnes (ed.), *Pharmacovigilance for Herbal and Traditional Medicines*,
https://doi.org/10.1007/978-3-031-07275-8_12

Pharmacovigilance is "the science and activities relating to the detection, assessment, understanding and prevention of adverse effects or any other drug related problems" and its success is reliant on reporting [9]. Passive surveillance, the "mainstay of pharmacovigilance," is also known as a spontaneous reporting system [9]. Unsolicited adverse drug reports are generally initiated by healthcare professionals and/or patients and submitted to a national pharmacovigilance center [9, 10]. Information on the whole population is achieved, reflecting real-world use. Spontaneous post-marketing reports can detect signals of new, rare, and serious adverse events, a key reason to use this method of surveillance [9, 11–13].

Spontaneous reporting systems have many shortcomings. The lack of a true denominator (total number exposed) impedes incidence calculation [13, 14]. Moreover, under-reporting hampers the determination of a reliable numerator. Data suggest that only 6% of all drug AEs are reported to spontaneous reporting systems [15]. These issues are further exacerbated with regard to NHP use. Under-reporting is particularly problematic for NHPs, as patients are less likely to disclose NHP use to their healthcare provider or report NHP-related AEs than for prescription medications [16, 17]. Healthcare providers also under-report NHP AEs. Less than 2% of community pharmacists who had identified a potential NHP-drug interaction reported the AE to a regulatory agency in comparison to nearly 20% that had reported a drug-drug interaction [18]. While some countries aim to address under-reporting by mandating reports from drug manufacturers and hospitals, NHPs are typically excluded from this legislation [19, 20]. In the limited AE reports that are submitted, they are often characterized by poor and heterogeneous reporting, including lack of detailed information that is essential for assessment of causation [11–13]. It has become apparent that the activities and goals of pharmacovigilance need to evolve to more proactive and rigorous system [13, 21].

Active surveillance "seeks to ascertain the exact number of adverse events via a continuous pre-organized process" [10] where AEs are solicited systematically [9, 22]. While a few countries have implemented varying degrees of active surveillance for drug AEs, these types of systems remain quite rare [9, 23, 24]. Increased rates of AE reporting have been seen with active surveillance [25]. For example, in a pediatric primary care setting, the number of AEs identified through active reporting increased from 4 to 1510 per 100,000 children when compared with passive reporting [26]. In addition, the collection of AE data through active searching allows for estimation of incidence and prevalence [9] and more resourceful data for public health and policy makers. Typically, better quality and more comprehensive reports are generated allowing a more complete understanding of the AE [9, 25].

Evidence on AEs related to concurrent NHP-conventional drug use is still limited and more data are urgently needed, particularly in patients at high risk of clinically meaningful NHP-drug interactions [12, 27, 28]. An opportunity to enhance signal detection and improve patient care is obtained through the integration of AE reporting into clinical settings as part of routine patient assessment, making pharmacovigilance a core aspect of healthcare practices.

To address this knowledge and care gap, a new approach to NHP AE active surveillance, including causality assessment, was developed and implemented, including adaptations as needed based on real-world application. Methodology and major study results to date are discussed, as well as strengths, limitations, and future directions of this approach.

12.2 Development of the *S*tudy *o*f *N*atural Health Product *A*dverse *R*eactions (SONAR) Method

The main objective was to identify clinically relevant NHP adverse reactions (ARs), through the implementation of active surveillance and causality assessment developed specifically for NHPs.

SONAR began in 2009 in selected community pharmacies across Canada. Community pharmacies were chosen as the initial screening setting since often a large proportion of patients visiting pharmacies are taking prescription medications and it is also possible to purchase NHPs at these locations [29, 30]. Moreover, community pharmacists are well suited to identify potential AEs and drug interactions [30]. This population-based cross-sectional study was initially piloted in Ontario (ON) and then expanded to Alberta (AB) and British Columbia (BC), Canada [29, 31].

12.2.1 Active Surveillance

The initial phase of this novel approach to NHP AE screening focused on active surveillance. Consecutive patients presenting to community pharmacy counters to either drop off a prescription or pick up a medication were systematically screened (Table 12.1) to investigate the rate of prescription drug, NHP and concurrent NHP-drug use as well as the AE rates of each [29–31]. Trained pharmacists and pharmacy staff used a simple screening log to question patients on their prescription drug and NHP use in the last one month [29–31]. Patients were also asked about any AEs they had experienced in the last month [29–31]. If a patient was taking an NHP, with or without prescription drug use, and had experienced an AE, they were provided an

Table 12.1 General screening questions used in SONAR (adapted from [29–33])

Question 1	In the last month, have you taken any prescription medications? If yes, list the medications
Question 2	In the last month, have you taken any natural health products? If yes, list the products
Question 3	In the last month, have you experienced any undesirable effects? If yes, describe these effects

information package outlining the study and a consent form [29–31]. If the patient consented, a follow-up telephone interview was conducted by a member of the research team, the study pharmacist [29–31].

The telephone interview focused on gathering details required for causality assessment [29–31]. Data collected included demographics, medical history (medical conditions, hospital admissions, family history), details of drugs and NHPs used at the time of the AE and details regarding the AE (symptoms, timeframe, medical treatment sought) [29–31].

12.2.2 Causality Assessment and Laboratory Analysis

Causality assessment is required to determine the likelihood that the reported adverse event occurred from product exposure [20]. The information gleaned from causality assessments facilitates AE management at both a clinical practice and regulatory level [34], and an AE becomes an AR if a causal association is suspected [20, 30, 35].

Although a vitally important step of pharmacovigilance, there is currently no universally accepted gold standard for causality assessment of drug AEs [30, 34, 36–38]. Three broad categories of causality assessment tools exist: (1) expert judgment—an expert's assessment without the use of a standardized tool; (2) probabilistic—the calculation of the probability of drug causation; and (3) algorithmic—the use of a series of questions and step-by-step instruction to determine a likelihood score [37–39].

The SONAR adjudication process utilized a blended approach which combined both algorithmic and expert judgment processes for a complete assessment with less risk for biased conclusions (Fig. 12.1) [30]. To the best of the investigators' knowledge, there are no causality assessment tools developed specifically for AEs involving NHPs [30]. The difficulty exists in that NHPs have unique complexities including the possibility of contamination, adulteration, misidentified or mislabeled ingredients, and variable quality control, making causality assessment more challenging [40, 41].

The data collected from participants who consented to telephone interviews were summarized and both the summary and original text were submitted to two adjudicators: one clinical NHP expert and one basic science NHP expert [29]. Each expert independently assessed cases based on the three instruments that were adapted to assess causality of product-associated AE [30]. First, the World Health Organization-Uppsala Monitoring Centre (WHO-UMC) system for standardized case causality assessment was used [42]; it uses expert judgment to identify how likely it is that an AE was triggered by a drug [37]. This assessment system considers several factors including plausible temporal relationship and clinical pharmacology [30, 37, 42]. Only minor revisions were made by adapting the term "drug" to "health products" to encompass prescriptions, over-the-counter medications, and NHPs [30]. Contrarily, the Naranjo scale [43] and Horn Drug Interaction Probability Scale

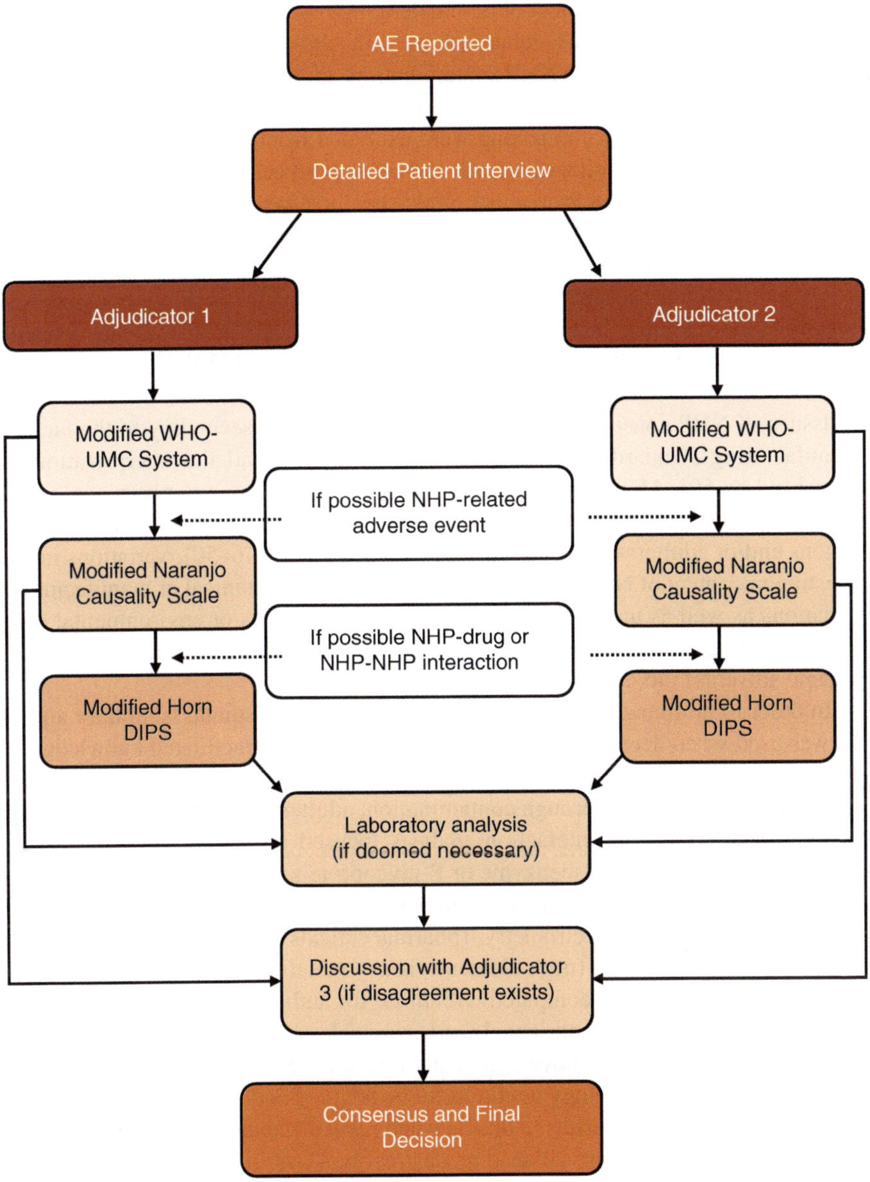

Fig. 12.1 Causality assessment process (adapted from [30])

(DIPS) [44] utilize an algorithmic approach and required more extensive revisions [30, 37]. The Naranjo scale assigns each AE a quantitative score based on answering "yes/no/do not know" to a series of questions [43]. The total score is converted to a qualitative descriptor of definite, probable, possible, or doubtful AR [43]. Echoing the Bradford Hill Criteria [45], Naranjo scores the consistency of previous reports, temporal relationship, the effect of de-challenging/re-challenging an agent, alternative causes, and dose-response relationships [43]. Alterations made were to broaden the scope of questions in that NHP contamination, adulteration, product quality, and manufacturing processes were considered [30]. Questions irrelevant for observational data regarding placebo testing were excluded from the scoring system [30]. Although it uses very similar questions and scoring processes as the Naranjo scale, the Horn Drug Interaction Probability Scale specifically estimates the likelihood of drug-drug interactions [44]. Again, modification was required to additionally address NHP-NHP and NHP-drug interactions [30].

Once individual causality assessment was complete, a consensus method was employed to strengthen the evaluation and reach a final decision [30]. Disagreements between experts were discussed until consensus was reached [30]. If consensus was not reached initially, a third expert was consulted [30].

Issues of NHP consistency, heterogeneity, and quality, secondary to the lack of manufacturing standards and regulations, further confound the interpretation of causality [46–50]. AEs identified during NHP use may be due to pharmacological actions of substances in NHPs alone, pharmacokinetic or pharmacodynamic interactions and/or adulteration or contamination of the NHP [46–50]. Variations in the constituent content of NHPs could be related to inherent botanical or manufacturing variations as well as undisclosed health product adulterants or environmental contaminants (e.g., heavy metals, pesticides, microbial toxins, micro-organisms, organic solvents) [46–50].

In order to minimize these challenges in causality assessment, laboratory analysis was used when deemed necessary by adjudicators; this facilitated knowledge of the mechanisms of action for AEs, confirmed the potential for interactions, and characterized the sample through contamination, adulteration, and constituent analysis [29–31]. NHP-drug interaction analysis focused on underlying mechanisms such as cytochrome P450 isoenzyme or P-glycoprotein inhibition and other shared metabolic routes. Adulterant and contaminant evaluation was performed using gas chromatography-mass spectrometry (pharmaceuticals) and inductively coupled plasma mass spectrometry (trace elements, heavy metals) [51]. Phytochemical constituent profiles were also compared with those for authentic products to determine product quality. This in-depth benchwork was able to inform causality decisions and enable the detection of novel, clinically relevant NHP ARs [29–31].

This SONAR methodology continues to be adapted and enhanced for a number of practice settings. The results to this date and related adaptations will be discussed to depict how the active surveillance process evolves over time.

12.3 Community Pharmacy SONAR Results

A total of 3733 patients were screened at 20 pharmacies in Alberta ($n = 7$), British Columbia ($n = 3$), and Ontario ($n = 10$) [29, 31]. The estimated national proportion suggests that nearly half (45.4%) of Canadians that present to community pharmacy counters to fill or pick up prescription medications, take NHPs and prescription drugs concurrently and, of those, about 7% report an AE [29, 31]. In the Western Canada analysis, it was determined that patients taking concurrent NHP-drug therapy are over six times more likely to experience an AE in comparison to people using prescription drugs alone [29]. In terms of causality assessment, one-quarter (6 of 24) of AEs among patients interviewed were adjudicated as "likely" or "probably" related to NHP use [29, 31]. Laboratory analysis was conducted on selected cases, which contributed to the causality assessment and adjudication [29, 31].

Several strengths were determined from this active surveillance process, including a marked increase in the number of AEs reported, compared to passive surveillance. For example, in the same timeframe that the study identified 54 AEs in 1118 patients screened (4.8%), spontaneous reporting captured 342 AE reports per approximately 30 million Canadians (0.0011%), representing a roughly 4000 times increase in reporting [29]. Moreover, the screening process and questioning was brief and overall well accepted by the pharmacy staff [29]. The meaningful and detailed information gathered from the interview process and laboratory analysis allowed for causality assessment [29, 31].

However, some limitations were also identified. One major barrier to adoption of active surveillance is that it involves the incorporation of methods into clinical workflow [32]. Time constraints, current practice environments, and community pharmacists' perceived lack of NHP knowledge made screening challenging at times; hence, only a fraction of presenting patients were screened [29, 32]. Sampling and recall bias are also possibilities [29]. Further, people attending community pharmacies may have different baseline AE risk than those in other settings, such as hospitals; this may potentially hinder generalizability [29, 31]. There was a significant loss to follow up for the detailed telephone interview, including difficulties reaching patients who had consented [29]. Patient screening and causality assessment continue to be refined and adjusted throughout SONAR progression.

Based on strengths and weaknesses of the process, the screening tools and assessment procedures have been adapted to better capture NHP AEs in the most efficient and effective way.

12.4 Specialty SONAR

The active surveillance SONAR methods have been expanded to selected special populations.

12.4.1 Mental Health

Mental health patients are at high risk of NHP AEs, including NHP-drug interactions, due to high prevalence of use of NHPs, lack of disclosure of NHP use, and polypharmacy [33, 52–54]. Many medications used in psychiatry, including antipsychotics and serotonergic medications, may lead to drug interactions due to their effects on, or the effects of, the cytochrome P450 enzyme system [55, 56]. The narrow therapeutic range of mood stabilizers also puts patients taking these medicines at high risk of AEs, including treatment failure [57–59].

In this study, prescription drug use, NHP use, and AEs were screened for at 6 outpatient mental health clinics in Edmonton [48]. Of 1466 adult patients with complete screening data available, 672 (45.8%) patients took prescription drugs only, 79 (5.4%) took NHPs only, 279 (19.0%) took NHPs and drugs concurrently, and 436 (29.7%) took neither [33]. In total, 147 patients reported an AE, representing 10.7%, 2.5%, 25.5%, and 0.5% of each group, respectively. Similar to the community pharmacy setting, patients who reported concurrent use of NHPs and prescriptions had an increased likelihood of experiencing an AE, in this case, nearly three times more likely compared to those taking prescription drugs alone [33]. These data reflect the pilot study completed during the telephone intake screening process for mental health clinics, which did not yet include causality assessment. After demonstrating that patients with mental health conditions are using NHPs at a highly prevalent rate and experiencing AEs more frequently than those taking prescription drugs alone, the research was expanded to include in-person active surveillance in additional adult clinics, pediatric clinics as well as causality assessment (this work is currently being completed).

12.4.2 Oncology

Patients with cancer similarly are at high risk of experiencing AEs associated with NHPs, including NHP-drug interactions, due to the high prevalence of use among this population, despite oncologists' requests to the contrary [52, 60–66]. Fear, the sense of losing control, managing cancer and medication side effects and poor prognosis may influence oncology patients' use of NHPs [67]. Anticancer medications have complex pharmacokinetic profiles with often narrow therapeutic range, making cancer patients particularly vulnerable to clinically important drug interactions [68]. With minimal differences between efficacious and toxic doses, a slight change in plasma concentrations due to an NHP-drug interaction may result in serious toxicity or treatment failure [69]. Moreover, cancer patients' care is complex, and they are often prescribed multidrug regimens [67, 70, 71].

Similar active surveillance methods, as previously described, are currently implemented at several cancer clinics, including integrative oncology centers, across Canada. Both pediatric and adult oncology centers are involved. With these data, the use and effects of NHPs in patients receiving anticancer medications will be determined.

12.4.3 *Evolution of SONAR*

Although the specialty clinic SONAR methods are very similar to those of previous work in community pharmacies, procedures have advanced in both phases. The most obvious changes include setting and patient population. First, the patients attending ambulatory specialty clinics, including integrative centers, are at a different baseline risk of AEs, and potentially NHP use, in comparison to community pharmacy patients. The team structure also differs at ambulatory care settings; the healthcare providers involved are no longer exclusively pharmacists and pharmacy staff; they are interdisciplinary teams of physicians, nurses, pharmacists, and others. Furthermore, the screening currently involves pediatric patients, who are high users of NHPs, but far less is known about NHP safety in this population [59, 72, 73]. The changes in setting and population have added new culture, complexities and challenges to methodology; it is also likely to contribute novel information to a very scarce body of evidence. The differing practice settings have required continual refinement of methods.

Improvements and changes have been made to the active surveillance tool. When patients identify that they have experienced an AE, they are now asked about what action they took regarding the AE, i.e., whether medical care was sought. This not only minimizes loss to follow-up for this information but gives us insight into the severity and seriousness of the AE. Previously, screening had solely relied on patient identification of AEs. Healthcare provider recognition and assessment of patient AEs is now additionally requested, which is especially important given underlying illness and complex treatment regimens. Healthcare providers may be able to identify AEs that the patient may be unaware of, such as potential changes in laboratory values or drug concentrations, or associated treatment delays. In this approach, both patient and healthcare provider may independently report an AE.

In Specialty SONAR, detailed patient follow-up and causality assessment is only performed when patients taking an NHP have experienced a serious and/or unexpected AE. This process change is particularly important in the vulnerable oncology population that are much more likely to experience AEs associated with anticancer regimens. By concentrating on the AEs deemed serious and/or unexpected we can enhance feasibility and focus on the most clinically relevant information.

12.5 Future Directions

The data collected to date have begun to demonstrate the impact of active surveillance on detecting NHP AEs. Active surveillance provides a means to collect high-quality, meaningful data on which causality assessment can be based, at a higher rate than does passive surveillance alone, perhaps increased by many thousand-fold

[29, 31].The feasibility of active surveillance has been demonstrated in several settings and the ability to be incorporated into the medical histories taken by clinicians [29, 31, 33].

There are plans to further study high-risk populations, such as those in whom product absorption, distribution, metabolism, and/or excretion is altered; patients taking medicines with a narrow therapeutic range are also at risk of clinically important AEs, including drug interactions, such as patients receiving treatment for HIV/AIDS, post-organ transplant, or those who require anticoagulation. Patients taking medicines with a narrow therapeutic range could develop life-threatening drug resistance [74], treatment failure [74–77], or toxicity [78] via NHP-drug interaction. There is also a plan to study patients with renal insufficiency or reduced liver function. As the kidney and liver are vital organs in the metabolism and clearance of medications and NHPs [79, 80], altered or reduced function may lead to the accumulation of toxic metabolites and serious AEs [81]. Expanding SONAR to these populations will provide crucial information needed to help guide policy and clinical decision-making.

Another important part of SONAR is knowledge translation and exchange. This will not only enhance post-marketing surveillance but make the information gathered very accessible and practical. Through scoping reviews and, now, systematic reviews, NHP-drug interaction grids have been developed that depict and distinguish NHP-drug interactions that are supported by clinical evidence vs. those that have only preclinical evidence or are postulated in theory. These grids are, or will be, published in open-access journals to enhance access [82, 83].

The development a population-based database of all reported NHP-drug combinations is also planned to complement existing vigilance programs and facilitate sharing and exchange of AE information between databases maintained by various national agencies. This will allow for identification of potential safety signals. Active surveillance can contribute to updating product monographs, safety alerts and inform standards, policy and regulation.

Healthcare practices are becoming more interconnected through electronic medical records, and the Best Possible Medication History (BPMH) has become a standard of practice [84]. The hope is that BPMHs and documentation of care becomes more inclusive and comprehensive by capturing all medicinal-type products being taken by a patient, including NHPs, and the presence or absence of AEs. As more NHP AE data are collected, the importance of merging SONAR questions and routine care becomes obvious.

Neither NHPs nor prescription medications are benign; enhanced pharmacovigilance is critical to improve accurate knowledge of product safety, including avoidance of potential NHP-drug interactions. Open communication about NHP use and AEs between patients and providers must be brought to the forefront of clinical practice to optimize patient safety and care.

References

1. Health Canada (2020) Natural health products. https://www.canada.ca/en/health-canada/services/drugs-health-products/natural-non-prescription.html. Accessed 16 Apr 2020
2. FDA (2015) FDA 101: dietary supplements. https://www.fda.gov/consumers/consumer-updates/fda-101-dietary-supplements. Accessed 16 Apr 2020
3. Australian Government Department of Health Complementary medicines. https://www.tga.gov.au/complementary-medicines. Accessed 16 Apr 2020
4. European Medicines Agency Procedures for monograph and list entry establishment. http://www.ema.europa.eu/ema/index.jsp?curl=pages/regulation/general/general_content_001149.jsp&mid=WC0b01ac0580033a9d. Accessed 13 Mar 2018
5. Nahin RL, Barnes PM, Stussman BJ, Bloom B (2009) Costs of complementary and alternative medicine (CAM) and frequency of visits to CAM practitioners: United States, 2007. Natl Health Stat Rep 18:1–14
6. Xue CCL, Zhang AL, Lin V et al (2007) Complementary and alternative medicine use in Australia: a national population-based survey. J Altern Complement Med 13(6):643–650. https://doi.org/10.1089/acm.2006.6355
7. Corazza M, Borghi A, Lauriola M, Virgili A (2009) Use of topical herbal remedies and cosmetics: a questionnaire-based investigation in dermatology out-patients. J Eur Acad Dermatol Venereol 23(11):1298–1303. https://doi.org/10.1111/j.1468-3083.2009.03314.x
8. Hassan MAG (2014) Need to incorporate pharmacovigilance of herbal medicine to the curriculum. Natl J Physiol Pharm Pharmacol 4(2):99–100. https://doi.org/10.5455/njppp.2014.4.080620141
9. WHO (2006) The safety of medicines in public health programmes: pharmacovigilance an essential tool
10. Black C, Tagiyeva-Milne N, Helms P, Moir D (2015) Pharmacovigilance in children: detecting adverse drug reactions in routine electronic healthcare records. A systematic review. Br J Clin Pharmacol 80(4):844–854. https://doi.org/10.1111/bcp.12645
11. Alvarez-Requejo A, Carvajal A, Begaud B et al (1998) Under-reporting of adverse drug reactions. Eur J Clin Pharmacol 54(6):483–488
12. Jacobsson I, Jönsson AK, Gerdén B, Hägg S (2009) Spontaneously reported adverse reactions in association with complementary and alternative medicine substances in Sweden. Pharmacoepidemiol Drug Saf 18(11):1039–1047. https://doi.org/10.1002/pds.1818
13. Begum SS, Mansoor M, Sandeep A et al (2013) Tools to improve reporting of adverse drug reactions - a review. Int J Pharm Sci Rev Res 23(1):262–265
14. Barnes J (2003) Quality, efficacy and safety of complementary medicines: fashions, facts and the future. Part II: efficacy and safety. Br J Clin Pharmacol 55(4):331–340
15. Hazell L, Shakir S (2006) Determinants of under-reporting of adverse drug reactions: a systematic review. Drug Saf 29(5):385–396. https://doi.org/10.2165/00002018-200932010-00002
16. Barnes J, Mills SY, Abbot NC et al (1998) Different standards for reporting ADRs to herbal remedies and conventional OTC medicines: face-to-face interview with 515 users of herbal remedies. Br J Clin Pharmacol 45(5):496–500. https://doi.org/10.1046/j.1365-2125.1998.00715.x
17. Vickers KA, Jolly KB, Greenfield SM (2006) Herbal medicine: women's views, knowledge and interaction with doctors: a qualitative study. BMC Complement Altern Med 6:40. https://doi.org/10.1186/1472-6882-6-40
18. Charrois TL, Hill RL, Vu D et al (2007) Community identification of natural health product-drug interactions. Ann Pharmacother 41(7):1124–1129. https://doi.org/10.1345/aph.1H463
19. Government of Canada (2014) Protecting Canadians from Unsafe Drugs Act (Vanessa's Law): questions/answers. https://www.canada.ca/en/health-canada/services/drugs-health-products/legislation-guidelines/questions-answers-regarding-law-protecting-canadians-unsafe-drugs-act-vanessa-law.html. Accessed 1 Apr 2020

20. Dal Pan GJ, Lindquist M, Gelperin K (2019) Postmarketing spontaneous pharmacovigilance reporting systems. In: Strom BL, Kimmel SE, Hennessy S (eds) Pharmacoepidemiology, 6th edn. Wiley-Blackwell, Hoboken, NJ, pp 165–201
21. Reform FG, Care P (2007) The Erice Manifesto: for global reform of the safety of medicines in patient care. Drug Saf 30(3):187–190. https://doi.org/10.2165/00002018-200730030-00001
22. Härmark L, Raine J, Leufkens H et al (2016) Patient-reported safety information: a renaissance of pharmacovigilance? Drug Saf 39(10):883–890. https://doi.org/10.1007/s40264-016-0441-x
23. Huang YL, Moon J, Segal JB (2014) A comparison of active adverse event surveillance systems worldwide. Drug Saf 37(8):581–596. https://doi.org/10.1007/s40264-014-0194-3
24. FDA (2017) FDA's sentinel initiative. https://www.fda.gov/Safety/FDAsSentinelInitiative/default.htm. Accessed 15 Mar 2018
25. ICH (2004) E2E Pharmacovigilance Planning
26. Menniti-Ippolito F, Raschetti R, Da Cas R et al (2000) Active monitoring of adverse drug reactions in children. Lancet 355(9215):1613–1614. https://doi.org/10.1016/S0140-6736(00)02219-4
27. Tabali M, Jeschke E, Bockelbrink A et al (2009) Educational intervention to improve physician reporting of adverse drug reactions (ADRs) in a primary care setting in complementary and alternative medicine. BMC Public Health 9:274. https://doi.org/10.1186/1471-2458-9-274
28. Jose J, Rao PGM, Kamath MS, Jimmy B (2009) Drug safety reports on complementary and alternative medicines (ayurvedic and homeopathic medicines) by a spontaneous reporting program in a tertiary care hospital. J Altern Complement Med 15(7):793–797. https://doi.org/10.1089/acm.2008.0128
29. Necyk C, Tsuyuki RT, Boon H et al (2014) Pharmacy study of natural health product adverse reactions (SONAR): a cross-sectional study using active surveillance in community pharmacies to detect adverse events associated with natural health products and assess causality. BMJ Open 4(3):e003431. https://doi.org/10.1136/bmjopen-2013-003431
30. Necyk C (2013) The Detection and Causality Assessment of Adverse Events Related to Natural Health Product Use in Community Pharmacies through the Implementation of Active Surveillance. Dissertation. University of Alberta
31. Vohra S, Cvijovic K, Boon H et al (2012) Study of natural health product adverse reactions (SONAR): active surveillance of adverse events following concurrent natural health product and prescription drug use in community pharmacies. PLoS One 7(9):e45196. https://doi.org/10.1371/journal.pone.0045196
32. Cvijovic K, Boon H, Jaeger W, Vohra S (2010) Pharmacists' participation in research: a case of trying to find the time. Int J Pharm Pract 18:377–383. https://doi.org/10.1111/j.2042-7174.2010.00067.x
33. Necyk C, Khamba B, Chue P et al (2016) Study of natural health product–drug adverse reactions (S.O.N.A.R.) in patients seeking mental health services. Curr Med Res Opin 32(8):1335–1343. https://doi.org/10.1185/03007995.2016.1174109
34. Jordan SA, Cunningham DG, Marles RJ (2010) Assessment of herbal medicinal products: challenges, and opportunities to increase the knowledge base for safety assessment. Toxicol Appl Pharmacol 243(2):198–216. https://doi.org/10.1016/j.taap.2009.12.005
35. European Medicines Agency (1995) Note for guidance on clinical safety data management: definitions and standards for expedited reporting (CPMP/ICH/377/95). ICH Topic E 2 A clinical safety data management: definitions and standards for expedited report. https://www.ema.europa.eu/en/documents/scientific-guideline/international-conference-harmonisation-technical-requirements-registration-pharmaceuticals-human-use_en-15.pdf. Accessed 16 Apr 2020
36. Théophile H, André M, Arimone Y et al (2012) An updated method improved the assessment of adverse drug reaction in routine pharmacovigilance. J Clin Epidemiol 65(10):1069–1077. https://doi.org/10.1016/j.jclinepi.2012.04.015
37. Agbabiaka TB, Savovi J, Ernst E (2008) Methods for causality assessment of a systematic review. Drug Saf 31(1):21–37

38. Khan LM, Al-Harthi SE, Osman AMM et al (2016) Dilemmas of the causality assessment tools in the diagnosis of adverse drug reactions. Saudi Pharm J 24(4):485–493. https://doi.org/10.1016/j.jsps.2015.01.010
39. Jones JK, Begaud B, Kingery E (2020) Assessing causation from case reports. In: Strom BL, Kimmel SE, Hennessy S (eds) Pharmacoepidemiology, 6th edn. Wiley-Blackwell, Hoboken, NJ, pp 723–745
40. Macedo AF, Marques FB, Ribeiro CF (2006) Can decisional algorithms replace global introspection in the individual causality assessment of spontaneously reported ADRs? Drug Saf 29:697–702. https://doi.org/10.2165/00002018-200629080-00006
41. Barnes J (2003) Pharmacovigilance of herbal medicines: a UK Perspective. Drug Saf 26(12):829–851. https://doi.org/10.2165/00002018-200326120-00001
42. The Uppsala Monitoring Centre (2005) The use of the WHO-UMC system for standardized case causality assessment
43. Naranjo C, Busto U, Sellers M et al (1981) A method for estimating the probability of adverse drug reactions. Clin Pharmacol Ther 30(2):239–245
44. Horn JR, Hansten PD, Chan L-N (2007) Proposal for a new tool to evaluate drug interaction cases. Ann Pharmacother 41(4):674–680. https://doi.org/10.1345/aph.1H423
45. Fedak KM, Bernal A, Capshaw ZA et al (2015) Applying the Bradford Hill criteria in the 21st century: how data integration has changed causal inference in molecular epidemiology. Emerg Themes Epidemiol 12:14. https://doi.org/10.1186/s12982-015-0037-4
46. Nelson MH, Cobb SE, Shelton J (2002) Variations in parthenolide content and daily dose of feverfew products. Am J Heal Pharm 59(16):1527–1531. https://doi.org/10.1093/ajhp/59.16.1527
47. Garrard J, Harms S, Eberly LE, Matiak A (2003) Variations in product choices of frequently purchased herbs: caveat emptor. Arch Intern Med 163(19):2290–2295. https://doi.org/10.1001/archinte.163.19.2290
48. Perharic L, Shaw D, Colbridge M et al (1994) Toxicological problems resulting from exposure to traditional remedies and food supplements. Drug Saf 11:284–294. https://doi.org/10.2165/00002018-199411040-00006
49. Gurley BJ, Gardner SF, Hubbard MA (2000) Content versus label claims in ephedra-containing dietary supplements. Am J Heal Pharm 57(10):963–969. https://doi.org/10.1093/ajhp/57.10.963
50. Ernst E (2000) Herb-drug interactions: potentially important but woefully under-researched. Eur J Clin Pharmacol 56:523–524. https://doi.org/10.1007/s002280000194
51. Cvijovic K, Boon H, Jaeger W, Vohra S (2011) Polypharmacy, multiple natural health products and hepatotoxicity. CMAJ 183(14):e1085–e1089
52. Miller MF, Bellizzi KM, Sufian M et al (2008) Dietary supplement use in individuals living with cancer and other chronic conditions: a population-based study. J Am Diet Assoc 108(3):483–494. https://doi.org/10.1016/j.jada.2007.12.005
53. Hazra M, Uchida H, Sproule B et al (2011) Impact of feedback from pharmacists in reducing antipsychotic polypharmacy in schizophrenia. Psychiatry Clin Neurosci 65:676–678. https://doi.org/10.1111/j.1440-1819.2011.02280.x
54. Poon SH, Sim K, Sum MY et al (2012) Evidence-based options for treatment-resistant adult bipolar disorder patients. Bipolar Disord 14:573–584. https://doi.org/10.1111/j.1399-5618.2012.01042.x
55. Manolopoulos VG, Ragia G, Alevizopoulos G (2012) Pharmacokinetic interactions of selective serotonin reuptake inhibitors with other commonly prescribed drugs in the era of pharmacogenomics. Drug Metabol Drug Interact 27(1):19–31. https://doi.org/10.1515/dmdi-2011-0033
56. Kruszkowksi M, Malti T, Modestin J (2003) Use of alternative therapy by psychiatric inpatients. Int J Psychiatry Clin Pract 7(3):161–166
57. Gitlin M, Frye MA (2012) Maintenance therapies in bipolar disorders. Bipolar Disord 14:51–65. https://doi.org/10.1111/j.1399-5618.2012.00992.x

58. Brzakoviać BB, Vezmar Kovačević SD, Vučiaćević KM et al (2012) Impact of age, weight and concomitant treatment on lamotrigine pharmacokinetics. J Clin Pharm Ther 37(6):693–697. https://doi.org/10.1111/j.1365-2710.2012.01351.x
59. Crepeau AZ, Moseley BD, Wirrell EC (2012) Specific safety and tolerability considerations in the use of anticonvulsant medications in children. Drug Healthc Patient Saf 4:39–54. https://doi.org/10.2147/DHPS.S28821
60. Zeller T, Muenstedt K, Stoll C et al (2013) Potential interactions of complementary and alternative medicine with cancer therapy in outpatients with gynecological cancer in a comprehensive cancer center. J Cancer Res Clin Oncol 139:357–365. https://doi.org/10.1007/s00432-012-1336-6
61. Boon HS, Olatunde F, Zick SM (2007) Trends in complementary/alternative medicine use by breast cancer survivors: comparing survey data from 1998 and 2005. BMC Womens Health 7:4. https://doi.org/10.1186/1472-6874-7-4
62. Greenlee H, Neugut AI, Falci L et al (2016) Association between complementary and alternative medicine use and breast cancer chemotherapy initiation the Breast Cancer Quality of Care (BQUAL) Study. JAMA Oncol 2(9):1170–1176. https://doi.org/10.1001/jamaoncol.2016.0685
63. Lee AH, Ingraham SE, Kopp M et al (2006) The incidence of potential interactions between dietary supplements and prescription medications in cancer patients at a Veterans Administration Hospital. Am J Clin Oncol 29(2):178–182. https://doi.org/10.1097/01.coc.0000209369.44100.25
64. Wilkinson JM, Stevens MJ (2014) Use of complementary and alternative medical therapies (CAM) by patients attending a regional comprehensive cancer care centre. J Complement Integr Med 11(2):139–145. https://doi.org/10.1515/jcim-2013-0048
65. Bonacchi A, Fazzi L, Toccafondi A et al (2014) Use and perceived benefits of complementary therapies by cancer patients receiving conventional treatment in Italy. J Pain Symptom Manag 47(1):26–34. https://doi.org/10.1016/j.jpainsymman.2013.03.014
66. Bishop FL, Prescott P, Chan YK et al (2010) Prevalence of complementary medicine use in pediatric cancer: a systematic review. Pediatrics 125(4):768–776. https://doi.org/10.1542/peds.2009-1775
67. Balneaves LG, Weeks L, Seely D (2008) Patient decision-making about complementary and alternative medicine in cancer management: context and process. Curr Oncol 15(Suppl. 2):s94–s100. https://doi.org/10.3747/co.v15i0.280
68. Alnaim L (2007) Therapeutic drug monitoring of cancer chemotherapy. J Oncol Pharm Pract 13(4):207–221. https://doi.org/10.1177/1078155207081133
69. Blix HS, Viktil KK, Moger TA, Reikvam A (2010) Drugs with narrow therapeutic index as indicators in the risk management of hospitalised patients. Pharm Pract (Granada) 8(1):50–55. https://doi.org/10.4321/S1886-36552010000100006
70. Statistics Canada (2015) Canadian cancer statistics. https://www.cancer.ca/~/media/cancer.ca/CW/cancer%20information/cancer%20101/Canadian%20cancer%20statistics/Canadian-Cancer-Statistics-2015-EN.pdf?la=en. Accessed 16 Apr 2020
71. Roberts D, Wilson C, Todd C et al (2013) Complementary therapies in cancer: patients' views on their purposes and value pre and post receipt of complementary therapy - a multi-centre case study. Eur J Integr Med 5(5):443–449. https://doi.org/10.1016/j.eujim.2013.06.005
72. Jean D, Cyr C (2007) Use of complementary and alternative medicine in a general pediatric clinic. Pediatrics 120(1):e138–e141. https://doi.org/10.1542/peds.2006-3105
73. Ang JY, Ray-Mazumder S, Nachman SA et al (2005) Use of complementary and alternative medicine by parents of children with HIV infection and asthma and well children. South Med J 98(9):869–875. https://doi.org/10.1097/01.smj.0000173089.51284.69
74. Piscitelli SC, Burstein AH, Chaitt D et al (2000) Indinavir concentrations and St John's wort. Lancet 355(9203):547–548. https://doi.org/10.1016/S0140-6736(99)05712-8
75. Sussman E (2002) Garlic supplements can impede HIV medication. AIDS 16(9):N5
76. Weigman D-J, Brinkman K, Franssen EJF (2009) Interaction of ginkgo biloba with efavirenz. AIDS 23(9):1184–1185. https://doi.org/10.1097/QAD.0b013e32832c412b

77. Light TD, Light JA (2003) Acute renal transplant rejection possibly related to herbal medications. Am J Transplant 3:1608–1609. https://doi.org/10.1046/j.1600-6135.2003.00270.x
78. Bolley R, Zülke C, Kammerl M et al (2002) Tacrolimus-induced nephrotoxicity unmasked by induction of the CYP3A4 system with St John's wort. Transplantation 73(6):1009
79. Dowling TC (2002) Drug metabolism considerations in patients with chronic kidney disease. J Pharm Pract 15(5):419–427. https://doi.org/10.1177/089719002237265
80. Anzenbacher P, Zanger UM (eds) (2012) Metabolism of drugs and other xenobiotics. Wiley-VCH
81. Dahl NV (2001) Herbs and supplements in dialysis patients: panacea or poison? Semin Dial 14:186–192. https://doi.org/10.1046/j.1525-139X.2001.00051.x
82. Kutt A, Girard L, Necyk C et al (2016) Natural health product – drug interaction tool: a scoping review. Can Pharm J 149(2):75–82. https://doi.org/10.1177/1715163516629156
83. Cvijovic K, Boon H, Barnes J et al (2009) A tool for rapid identification of potential herbal medicine – drug interactions. Can Pharm J 142(5):224–247. https://doi.org/10.3821/1913-701X-142.5.224
84. Accreditation Canada, the Canadian Institute for Health Information, the Canadian Patient Safety Institute, the Institute for Safe Medication Practices Canada (2012) Medication reconciliation in Canada: raising the bar - progress to date and the course ahead. https://www.ismp-canada.org/download/MedRec/20121101MedRecCanadaENG.pdf. Accessed 16 Apr 2020

Chapter 13
Ethnopharmacovigilance and Traditional Medicines

Eliana Rodrigues and Joanne Barnes

13.1 Ethnopharmacovigilance: Concepts and Methods

Ethnobotany is a discipline that has objectives and methodologies in common with ethnopharmacology. According to Harshberger [1], ethnobotany is the study of the utilitarian relationship between humans and the primitive plant environment in its entirety. Ethnopharmacology is defined as a subarea of ethnobotany, referring to the medical or pseudomedical use of plants and animals by pre-literate societies [2]. Current ethnobotanical and ethnopharmacological studies include, in addition to plants and animals, other natural resources, such as algae, fungi, minerals, and others. In an even more contemporary approach, work has been dedicated to unraveling the relationships between these substances in the composition of a traditional medicine "recipe"; for example, one study showed the mixed composition of a home remedy to involve a plant resin and the secretion of an amphibian [3], while another is the result of an insect-mineral-vegetable oil-interaction [4].

One of the applications of ethnobotany and ethnopharmacology is in the development of new drugs, and it is necessary for the ethnobotanist/ethnopharmacologist to collaborate with researchers in the fields of phytochemistry and pharmacology in this respect. Historically, around 6% of randomly collected samples sent to the USA National Cancer Institute (NCI) were bioactive, whereas ethnobotany- and ethnopharmacology-directed collections indicated 25% of bioactive plants [5].

E. Rodrigues (✉)
Center for Ethnobotanical and Ethnopharmacological Studies—Institute of Environmental, Chemical and Pharmaceutical Sciences, Universidade Federal de São Paulo, Sao Paulo, Brazil
e-mail: e.rodrigues@unifesp.br

J. Barnes
School of Pharmacy, Faculty of Medical and Health Sciences, University of Auckland, Auckland, New Zealand
e-mail: j.barnes@auckland.ac.nz

J. Barnes (ed.), *Pharmacovigilance for Herbal and Traditional Medicines*,
https://doi.org/10.1007/978-3-031-07275-8_13

Ethnobotanical and ethnopharmacological studies can take at least two different approaches. One approach is based on fieldwork, in which the researcher co-exists with a certain human group in order to understand and record their knowledge. In these studies, some requirements are necessary to obtain results worthy of further pharmacological and phytochemical investigations. To do so, one must use methods from at least two areas of knowledge: anthropology and biology. Specific literature [6–10] details the methods and techniques of cultural anthropology that provide a favorable interviewer-interviewee interaction. Such a relationship is based, above all, on reciprocal trust, and, for that, special care is needed during these studies. Another approach, using biological methods, involves the collection of substances (plants, animals, fungi, algae) indicated by the local medicine experts. During fieldwork, the use of these substances, their parts used, doses, durations of use, routes of administration, and contraindications are recorded in detail. The more detailed this information, the greater the chance of developing a pharmacological activity in the laboratory, consistent with the traditional knowledge, making feasible studies of efficacy on these plants, and facilitating the development of phytotherapeutic agents.

In addition to creating this record or "register," ethnobotany and ethnopharmacology have the task of interpreting local terms (termed "emic"), used by traditional / popular medicine, to the terms "ethic," those used by official medicine [11]. This is not always possible, since different cultures use their own perceptions to identify and denominate the diseases of their daily lives. In the Amazon rainforest, for example, some riverside people refer to mumps (ethic term) as *"papeira"* (emic). Sometimes such a correlation is totally impaired by the difficulty of translation and can be compared to a "puzzle," especially when there is not a medical doctor in the fieldwork team.

The second approach starts from the knowledge about medicinal plants published in ancient or contemporary literature: the literature or knowledge specific to a certain human group (past or present), or that which is widely diffused among populations of different regions. The reason for the diffusion of this knowledge is due to the fact that some uses for traditional medicines are very old and, therefore, have been passed on among people throughout history in their movements through various regions. There is growing recognition of the value of this type of research, where potential bioactive compounds are investigated from historical and current recordings; examples from all over the world can be cited, including in Brazil [12–16]. Thus, the extraction of the alkaloid pilocarpine from several species of the plant *"jaborandi"* (*Pilocarpus* sp.), native to Brazil, was possible from reports of its uses in ancient literature. In 1648, Piso and Macgrave reported in the work Historia Naturalis Brasiliae the use of this plant by the Guarani Indians to cause sweating. Centuries later, pilocarpine was isolated from the plant, and became a substance of great use worldwide and which is currently cited in approximately twenty patent applications [17, 18]. This method of selecting plants of interest is very useful for guiding pharmacological studies aimed at the development of new drugs and is used by up to 80% of pharmaceutical laboratories [19]. This is due, in part, to the difficulties inherent in fieldwork: obtaining data from the literature almost always requires fewer financial resources, is faster, and, above all, does not involve authorizations to access traditional/popular knowledge, nor establishing benefits distributions with the community that would have provided their traditional knowledge during fieldwork studies. A limitation in relation to the use of these historical data is the fact that hardly any quality data—with details of quantities of the substance(s) used in the traditional "prescription," dose, duration of use, special cautions, restrictions on use, adverse reactions, toxicity, and contraindications—are available in the literature, yet these types of information are necessary for assessing efficacy and safety.

Between 1995 and 2020, one of the authors (ER) conducted or supervised eighteen fieldwork ethnobotanical and ethnopharmacological studies among several traditional populations occupying different Brazilian biomes: caboclo river-dwellers of the Rio Unini and Rio Jaú (biome Amazon forest); *Quilombolas* (pantanal wetlands); Krahô Indians and migrants (cerrado brushlands); Guarani Indians, migrants and *Quilombolas* (Atlantic rain forest); and *sertanejos* (caatinga semi-arid lands). From these eighteen surveys, 1602 plant and 106 animal species were indicated, collected, and identified by taxonomy. For three of these surveys—whose traditional communities show great geographical isolation in relation to access to conventional medical care—366 plant species were indicated by members of these populations (82 by the *Quilombolas*, 164 by the Krahô Indians, 120 by the caboclo river-dwellers population). Of these, 57 (15.6%) species presented at least one of the 6 restrictions for use: plants with abortifacient effects; plants with contraceptive effects; plants contraindicated during pregnancy; plants that should be used/prescribed at lower doses for children and older people; plants used to aid childbirth; and plants known as poisonous to animals and/or humans. This work illustrated how ethnobotanical and ethnopharmacological studies can contribute not only to questions related to efficacy but also to safety by recording the following data for traditional prescriptions during fieldwork studies: composition of the prescription; therapeutic use(s); preparation and storage; route(s) of administration; dose(s), dosage(s), and duration(s) of administration; adverse/undesirable effects of the prescription and its ingredients; use in special patient groups: children and older patients; cautions and contraindications; and food and/or sexual taboos relating to this prescription [20, 21]. Thus, several examples emerged from this work of traditional knowledge relating to safety issues (see Table 13.1) [21]. From these examples, and from experiences with fieldwork in ethnobotany and ethnopharmacology, a new (sub)discipline within the ethnosciences was proposed: "ethnopharmacovigilance" [21]. A definition for ethnopharmacovigilance is that it is a branch of the ethnosciences that is concerned with the collection, collation, interpretation, and analysis of traditional knowledge relating to traditional medicines derived from plants, animals, and other natural resources to enhance understanding of the safety and harms profile of traditional medicines, including in relation to their use by traditional knowledge holders and other indigenous groups. Alongside this, a guided interview tool containing several aspects related to those described above was developed. This tool requires field testing and aims to make possible the collection and investigation of information on the harmful aspects of plants, and, in some cases, how these can be mitigated, for example, through specific preparation methods and/or restrictions on dose or route of administration, according to the

Table 13.1 Some examples of traditional knowledge on use, preparation, and route of administration of plant species relating to safety issues [21]

Population	Plant species (Family)	Traditional knowledge on use, preparation, and route of administration	Traditional knowledge relating to unsafe administration
Krahô Indians	*Chrysolaena herbacea* (Vell.) H.Rob (Compositae)	They scarify their children's legs with the root juice, three times a day for one week per month, as a leg fortifier.	*"this prescription cannot be ingested due to its toxicity"*
	Clitoria simplicifolia (Kunth) Benth. (Leguminosae)	The grated root is used topically for leg pain	*"the ingestion is not allowed because it is suspected to be poisonous"*
	Cissampelos ovalifolia DC. (Menispermaceae)	They place grated tubercles on snake bites claiming that an analgesic effect is evident within half an hour.	*"if the water of the tubercles is ingested, someone can die,"* since the plant is an *"human poison"*
	Tephrosia sinapou (Buc'hoz) A Chev. (Leguminosae) and *Serjania* spp. (Sapindaceae)	Leaves are used in cigarettes to alter perception, and also it has ichthyotoxic effects.	*"The ingestion must be contraindicated in humans; moreover, pregnant are not allowed to consume fish that has been caught using these plants"*
Quilombolas	*Rudgea viburnoides* (Cham.) Benth. (Rubiaceae)	Leaves are ingested as tea to treat insomnia	*"although it acts as a medicine, it diminish blood pressure and also provoke loss of sexual desire."* The interviewee also explained that naturally hypotensive individuals should avoid this plant and find a substitute.
	Ouratea spp. (Ochnaceae); *Ayenia* spp. (Malvaceae) and *Heteropterys aphrodisiaca* O. Mach (Malpighiaceae)	Leaves and/or roots are ingested as tea for adaptogenic-like effects.	They are said to be contraindicated in people with kidney problems.

perceptions of the practitioners of traditional healing using local medicinal plants [21].

Other authors have also proposed models on how data on safety and effectiveness can be obtained from traditional medicine practitioners. One such model—although involving conventionally trained health professionals rather than indigenous traditional healers—is based on methods developed by an association of anthroposophic physicians in Europe, a system of integrative medicine that includes the use of botanicals (herbal medicines) and certain other substances, and is practiced mostly by medical doctors. This method comprised a questionnaire seeking information, mostly for single-ingredient botanical or anthroposophical remedies, on the medical diagnosis or condition for which the botanical or anthroposophic remedy was considered effective, the patient's perspective on the remedy's effectiveness, how the practitioner measured or determined effectiveness, dose and duration of treatment [22]. This exercise was focused on collecting data on experiences relating to effectiveness of remedies, not their safety; practitioners were also asked to provide information on observed adverse effects (if any), but how these data are used and whether they could be useful from a pharmacovigilance perspective is not yet clear. It would appear that there is at least the opportunity for practitioners involved in this kind of data collection, albeit focused on effectiveness, to be encouraged to submit spontaneous reports of suspected adverse reactions associated with these remedies to their respective national pharmacovigilance center where appropriate.

A first step towards achieving comprehensive information on the experiences of traditional medicine practitioners and individual users of traditional medicines is to collect clinical data, including a patient's current and previous medical and medicines history and health outcomes—beneficial and adverse—following traditional medicine treatment during ethnopharmacological field studies [23]. Such data could contribute to developing clinical research exploring effectiveness and safety of traditional medicines, and could be of interest to users of traditional medicines and traditional medicine practitioners in their healthcare choices and practice, respectively [23].

13.2 Ethnopharmacovigilance in Published Literature: An Overview

Several ethnobotanical/ethnopharmacological fieldwork studies have dedicated their objectives to what was conceptualized earlier as ethnopharmacovigilance—a meeting of ethnobotany/ethnopharmacology and pharmacovigilance. Table 13.2 summarizes aspects relevant to ethnopharmacovigilance included in these studies, which were conducted among different cultures globally.

One study [24] recorded knowledge on one hundred and twenty-five toxic species collected from 80 interviewees (aged 40 to 70 years, most of whom had experience with folk medicine, and a long-standing relationship with the local area), interviewed from the southern part of Jordan. Examples of toxic effects described by the interviewees for commonly cited species included fatigue, skin irritation, and gastric and abdominal disturbances, as well as abortion, sterility, and neuralgic pains [24].The study shows that respondents hold substantial traditional knowledge on the toxicity of local plants.

Other studies also indicate the wealth of traditional knowledge relating to safe use of traditional medicines held by indigenous communities. Residents of the

Table 13.2 Examples of ethnopharmacological/ethnobotanical studies involving one or more items of data on: substance(s); therapeutic use; part used; mode of preparation; route of administration; dose; duration of use; contraindications; adverse reactions/adverse effects; and toxicity

First author(s) (year of publication); country [reference number]	Substance(s)	Therapeutic use	Part used	Mode of preparation	Route of administration	Dose	Duration of use	Contraindications	Adverse reactions/ adverse effects	Toxicity
Noumi and Tchakonang (2001); Cameroon [25]	Plants	x	x	x	x	x	x	–	x	–
Al-Qura'n (2005); Jordan [24]	Plants	–	x	–	x	x	x	–	–	x
Coelho-Ferreira (2009); Brazil [26]	Plants	x	x	x	x	–	–	–	x	x
Gbolade (2009); Nigeria [27]	Plants	x	x	x	x	x	x	–	x	–
Andel and Westers (2010); Netherlands [28]	Plants/ animals	x	x	x	x	–	–	–	x	–
Ogbole and Ajaiyeoba, (2010); Nigeria [29]	Plants	x	x	x	–	–	–	–	x	–
Maroyi and Maesen (2011); several (review) [30]	Plants	x	x	x	x	–	–	–	–	x
Olorunnisola et al. (2013); Nigeria [31]	Plants	x	x	x	x	x	x	–	x	x
Bahassan et al. (2014); Yemen [32]	Plants	x	x	x	x	–	–	–	–	x

Nergard et al. (2015); Mali [33]	Plants	x	x	x	x	–	–	–	x	–
Ngezahayo et al. (2015); Republic of Burundi [34]	Plants	x	x	x	x	x	x	x	–	–
Dey et al. (2017); India [35]	Plants/ animals	x	x	x	x	x	x	–	–	x

Amazonian coastal community of Marudá, Brazil, were found to possess knowledge on how to avoid adverse reactions of home remedies [26]. Interviewees were described as having a meticulous approach to harvesting, preparing, and using traditional medicines, and patients also took the same care [26]. For example, some interviewees described practices for boiling leaves of certain species before use to prevent the occurrence of specific adverse reactions. Other recommendations included specific instructions to collect plant material in a certain state (such as dry, withered, or yellowing leaves), or to collect plant material at a specific time of day, or in a particular season [26]. This knowledge, together with the practice of using small quantities of material for certain plants, may imply an understanding of dose-response and its relevance for toxicity [26]. Dose adjustment as a risk reduction strategy is also an approach used in other studies. The plant *Gloriosa superba* L. (Colchicaceae), for example, is used for a wide range of effects among many cultures worldwide, including those from tropical African and Asian countries [30]. According to the authors, it is used as a remedy for the treatment of urinary and reproductive systems, respiratory disorders, skin diseases, cardiovascular problems, and other disorders, as well as for toxic purposes, such as for head lice, as an abortifacient, as an antidote for snake bite, scorpion sting, skin diseases (antiparasitic), and as a poison. Apparently, healers manage the doses of this plant in order to promote healing or poisoning. Thus, they prescribe dose minimization to their patients to avoid toxic symptoms, and high doses when toxic effects are desired. Colchicine is the compound mainly responsible for these effects [30].

Burundian traditional healers claim to be able to adjust doses of traditional medicines used for microbial diseases based on the patient's age (child or adult) and/or his/her physiological state (e.g., pregnancy) using various measurements for quantifying medicinal plants (e.g., handfuls, pinches, teaspoons, tablespoons, cups, bottles) [34]. This claim requires further investigation, but, if substantiated, is important for herbal medicines since synergy depends not only on the combinations of ingredients used but also on the proportions (ratios) of extracts or combined products. As an example, a previous study conducted by us [36, 37] demonstrates the use of a recipe composed of seven plants [guiné (*Petiveria alliacea* L., Phytolaccaceae), rosemary (*Salvia rosmarinus* Spenn. (synonym: *Rosmarinus officinalis* L.), Lamiaceae), myrrh (*Commiphora* spp., Burseraceae), incense (*Pittosporum* spp., Pittosporaceae), benzoin (*Styrax* spp., Styracaceae), lavender (*Lavandula dentata* L., Lamiaceae), and rue (*Ruta graveolens* L., Rutaceae)], utilized as a smoke during Umbanda rituals—one of the Afro-Brazilian religions —in which the smoke is known as *defumador*, aiming balance and harmony in the environment, "cleansing, harmony, diseases of the spiritual plane" and "to calm," according to some Umbanda's priests. A priest explained that among the seven herbs, the relative proportion of three of them (guiné, rosemary, and rue) must always be higher in the recipe. This indication suggests that, in addition to the synergistic aspect of the seven plants, there is a dependence on their proportion in the recipe, giving the desired final effect. The priest further explained that, among these three species, *P. alliacea* is the one of the "strongest" substances with regard to its "harmonizing effects" [36, 37].

Another study evaluated which of the recipes involving different proportions of three plants from traditional knowledge (*Pterospartum tridentatum* (L.) Willk., *Gomphrena globosa* L., and *Cymbopogon citratus* (DC) Stapf.) had a greater antioxidant effect [38]. The authors concluded that the infusion obtained with 40% of *P. tridentatum* and 60% of *C. citratus* gave the highest antioxidant properties. This study shows how much the proportion between plants in a given recipe can amplify its pharmacological effect; therefore, it is extremely important to record the different proportions of the substances that make up a given recipe during ethnopharmacological studies.

Another study collected information on a total of 23 herbal recipes used by the indigenous people of Ogbomoso, Southwest Nigeria, for the treatment of malaria infection [31]. Around half (53%) of the recipes included in this survey were described as being associated with specific adverse effects, such as sweating, frequent urination, vomiting, stomach/intestinal pain, dizziness, weight loss, and allergic reactions; some preparations were described as having "no side effect" [31]. Data for this survey were collected from traditional healers, herbalists, and residents of rural communities; these respondents had detailed knowledge of the plants and plant parts used for malaria treatment, their mode(s) of preparation and administration, and possible adverse effects [31].

Similarly, in a study conducted among traditional medicine men and women of the indigenous population living in Bengal part of the Manbhum region in India, interviewees were found to hold substantial knowledge on the use of traditional medicine formulations for neurological and psychological conditions, including their possible toxicity and adverse effects [35]. For example, numerous animal parts, such as the whole body of cellar spiders (*Crossopriza lyoni*), skin water extract from cows (*Bos taurus*) and goats (*Capra aegagrus hircus*), and fox (*Vulpes bengalensis*) stools, were reported as ingredients of some traditional medicine formulations. In some instances, interviewees described being aware that, for some of these animal ingredients, the time and method of collection of animal excreta and products were considered important when preparing these traditional medicines [35].

Considering women's health conditions, one study [33] described the medicinal use of plants by Malian women during pregnancy, as well as their perception regarding safety in this practice. They observed that almost 80% of them had used medicinal plants when pregnant; but only 8.5% of them had received any orientation from a traditional healer; and 30% did not believe in the side effects of plants for the mother. Several studies describe the use of traditional medicines for their poisonous effects, including for use as abortifacients. An ethnobotanical survey in the Sangmelima region of Southern Cameroon (among the Bulu, Fang, and Maka, three ethnic groups of Sangmelima) recorded 20 plants used as abortifacients [25]. Information on adverse effects associated with these plants was collected from individuals with experience of using the plant. For several plants, adverse reactions were reported, several of them serious; examples included vaginal and/or vulval irritation, redness; vaginal and vulval burns and other wounds, including lacerations; vaginal bleeding; abdominal pains; menstrual cycle disorders; and fever [25].

Several of these effects were likely due to the method of administration of the plant material: fresh plant parts (such as leaves, pieces of bark) were used as a pad or formulated as a paste using saliva and applied intravaginally; some of the adverse effects described may be consequences of infections. It is clear that there is substantial potential for women using these practices to achieve abortion are at risk of harms [25]. These studies show that the knowledge of adverse effects from medicinal plants is limited among these women, and their beliefs around safety are more of a rule than an exception worldwide. Such beliefs persist beyond use in people's country of origin. Surinamese migrants in the Netherlands continue to use traditional medicinal herbs from Surinam for several reasons, including the belief that traditional herbal remedies have fewer adverse effects than do conventional medicines [28].

In conclusion, the studies described above have brought additional data to those normally obtained during ethnobotanical and ethnopharmacological fieldwork, but important information on one or more aspects of medicinal plant use, including therapeutic uses, plant parts used, mode(s) of preparation, route(s) of administration, dose and dosage, duration of use, contraindications, adverse reactions, and toxicity, are often lacking. In part, the lack of these data may be a reflection of the limited knowledge regarding safety aspects of traditional medicines among interviewees; conversely—and perhaps more likely—it may be due to methodological limitations of ethnobotanical and ethnopharmacological studies in that many do not comprehensively collect all relevant data and, in particular, do not adequately consider and explore safety aspects.

Against this background, and as pointed out previously by some authors [21, 22], it is necessary to test and validate the ethnobotanical tools that have been developed with the aim of advancing data collection for ethnopharmacovigilance.

References

1. Soejarto DD, Fong HHS, Tan GT et al (2005) Ethnobotany/ethnopharmacology and mass bioprospecting: issues on intellectual property and benefit-sharing. J Ethnopharmacol 100(1–2):15–22. https://doi.org/10.1016/j.jep.2005.05.031
2. Schultes RE (1988) Ethnopharmacological conservation: a key to progress in medicine. Acta Amazon 18(1–2):393–406
3. Rodrigues E, de Santos JFL, Souza SM, Lago JHG (2012) The mystery of the 'resin-of-canuaru': a medicine used by caboclos river-dwellers of the Amazon, Amazonas, Brazil. J Ethnopharmacol 144(3):806–808. https://doi.org/10.1016/j.jep.2012.10.026
4. Rodrigues E, Lago JHG, de Santos JFL, Bitencourt ALV (2018) Nests of caba-leão wasps (Sceliphron sp., Sphecidae) used in traditional medicine by riverine communities of the Jaú and Unini Rivers, Amazon, Brazil: ethnopharmacological, chemical and mineralogical aspects. Rev Bras Farmacogn 28:352–357. https://doi.org/10.1016/j.bjp.2018.04.005
5. Balick MJ (1990) Ethnobotany and the identification of therapeutic agents from the rainforest. Ciba Found Symp 154:22–31. https://doi.org/10.1002/9780470514009.ch3
6. Foote-Whyte W (1980) Treinando a observaçao participante. In: Guimaraes AZ (ed) Desvendando máscaras sociais. Francisco Alves, Rio de Janeiro, pp 77–86

7. Bernard HR (1988) Research methods in cultural anthropology. Sage Publications, London
8. Malinowski B (1990) O objeto, método e alcance dessa pesquisa. In: Guimaraes AZ (ed) Desvendando máscaras sociais, 3rd edn. Francisco Alves, Rio de Janeiro, pp 39–61
9. Minayo MCdeS (1992) O desafio do conhecimento: pesquisa qualitativa em saúde. Hucitec, São Paulo
10. Alexiades MN (1996) Collecting ethnobotanical data: an introduction to basic concepts and techniques. In: Alexiades MN, Sheldon JW (eds) Selected guidelines for ethnobotanical research: a field manual. The New York Botanical Garden, New York, pp 53–94
11. Pagani E, Santos JFL, Rodrigues E (2017) Culture-bound syndromes of a Brazilian Amazon Riverine population: tentative correspondence between traditional and Western medicine terms and possible ethnopharmacological implications. J Ethnopharmacol 203:80–89. https://doi.org/10.1016/j.jep.2017.03.024
12. Rodrigues E, Mendes FR, Negri G (2006) Plants indicated by Brazilian Indians for disturbances of the central nervous system: a bibliographical approach. Curr Med Chem – Central Nervous System Agents 6(211–244). https://doi.org/10.2174/187152406778226725
13. Giorgetti M, Negri G, Rodrigues E (2007) Brazilian plants with possible action on the central nervous system - a study of historical sources from the 16th to 19th century. J Ethnopharmacol 109(2):338–347. https://doi.org/10.1016/j.jep.2006.08.003
14. Mendes FR, Carlini EA (2007) Brazilian plants as possible adaptogens: an ethnopharmacological survey of books edited in Brazil. J Ethnopharmacol 109(3):493–500. https://doi.org/10.1016/j.jep.2006.08.024
15. Brandão MGL, Zanetti NNS, Oliveira P, Grael CFF, Santos ACP, Monte-Mór RLM (2008) Brazilian medicinal plants described by 19th century European naturalists and in the official pharmacopoeia. J Ethnopharmacol 120(2):141–148. https://doi.org/10.1016/j.jep.2008.08.004
16. Otsuka RD, Lago JHG, Rossi L, Galduróz JCF, Rodrigues E (2010) Psychoactive plants described in a Brazilian literary work and their chemical compounds. Cent Nerv Syst Agents Med Chem 10(3):218–237. https://doi.org/10.2174/187152491100603021
17. European Patente Office (ESPACENET) (2009). http://www.ep.espacenet.com/. Accessed 15 Aug 2009
18. Instituto Nacional de Propriedade Industrial (2009). http://www.inpi.gov.br/. Accessed 13 Aug 2009
19. ten Kate K, Laird SA (1999) Natural products and the pharmaceutical industry. In: ten Kate K, Laird SA (eds) The commercial use of biodiversity: access to genetic resources and benefit-sharing. Earthscan Publications, London, pp 34–77
20. Rodrigues E (2007) Plants of restricted use indicated by three cultures in Brazil (Caboclo-river dweller, Indian and Quilombola). J Ethnopharmacol 111(2):295–302. https://doi.org/10.1016/j.jep.2006.11.017
21. Rodrigues E, Barnes J (2013) Pharmacovigilance of herbal medicines: the potential contributions of ethnobotanical and ethnopharmacological studies. Drug Saf 36(1):1–12. https://doi.org/10.1007/s40264-012-0005-7
22. Elsas SM (2015) A model on how to obtain data from botanical practitioners. Epilepsy Behav 52(Part B):333–337. https://doi.org/10.1016/j.yebeh.2015.07.006
23. Graz B (2013) What is "clinical data"? Why and how can they be collected during field surveys on medicinal plants? J Ethnopharmacol 150(2):775–779. https://doi.org/10.1016/j.jep.2013.08.036
24. Al-Qura'n S (2005) Ethnobotanical survey of folk toxic plants in southern part of Jordan. Toxicon 46(2):119–129. https://doi.org/10.1016/j.toxicon.2005.04.010
25. Noumi E, Tchakonang NYC (2001) Plants used as abortifacients in the Sangmelima region of Southern Cameroon. J Ethnopharmacol 76(3):263–268. https://doi.org/10.1016/S0378-8741(01)00252-5
26. Coelho-Ferreira M (2009) Medicinal knowledge and plant utilization in an Amazonian coastal community of Marudá, Pará State (Brazil). J Ethnopharmacol 126(1):159–175. https://doi.org/10.1016/j.jep.2009.07.016

27. Gbolade AA (2009) Inventory of antidiabetic plants in selected districts of Lagos State, Nigeria. J Ethnopharmacol 121(1):135–139. https://doi.org/10.1016/j.jep.2008.10.013
28. Van Andel T, Westers P (2010) Why Surinamese migrants in the Netherlands continue to use medicinal herbs from their home country. J Ethnopharmacol 127(3):694–701. https://doi.org/10.1016/j.jep.2009.11.033
29. Ogbole OO, Ajaiyeoba EO (2009) Traditional management of tuberculosis in Ogun State of Nigeria: the practice and ethnobotanical survey. Afr J Tradit Complement Altern Med 7(1):79–84. https://doi.org/10.4314/ajtcam.v7i1.57270
30. Maroyi A, Van der Maesen LJG (2011) *Gloriosa superba* L. (family Colchicaceae): remedy or poison? J Med Plants Res 5(26):6112–6121. https://doi.org/10.5897/JMPR11.913
31. Olorunnisola OS, Adetutu A, Balogun EA, Afolayan AJ (2013) Ethnobotanical survey of medicinal plants used in the treatment of malaria in Ogbomoso, Southwest Nigeria. J Ethnopharmacol 150(1):71–78. https://doi.org/10.1016/j.jep.2013.07.038
32. Bahassan A, Zidane L, El Yacoubi H, Rochdi A (2014) Ethnobotanique et taxonomie des plantes médicinales utilisées pour le traitement des pathologies du système digestif à Hadramaout (Yémen). Phytothérapie 12:399–416. https://doi.org/10.1007/s10298-014-0897-8
33. Nergard CS, Ho TPT, Diallo D, Ballo N, Paulsen BS, Nordeng H (2015) Attitudes and use of medicinal plants during pregnancy among women at health care centers in three regions of Mali, West-Africa. J Ethnobiol Ethnomed 11:73. https://doi.org/10.1186/s13002-015-0057-8
34. Ngezahayo J, Havyarimana F, Hari L, Stévigny C, Duez P (2015) Medicinal plants used by Burundian traditional healers for the treatment of microbial diseases. J Ethnopharmacol 173:338–351. https://doi.org/10.1016/j.jep.2015.07.028
35. Dey A, Gorai P, Mukherjee A, Dhan R, Modak BK (2017) Ethnobiological treatments of neurological conditions in the Chota Nagpur Plateau, India. J Ethnopharmacol 198:33–44. https://doi.org/10.1016/j.jep.2016.12.040
36. Garcia D, Domingues MV, Rodrigues E (2010) Ethnopharmacological survey among migrants living in the Southeast Atlantic Forest of Diadema, Sao Paulo, Brazil. J Ethnobiol Ethnomed 6:29. https://doi.org/10.1186/1746-4269-6-29
37. Alves TC, Rodrigues E, Lago JHG, Prado CM, Giradi CEN, Hipólide DC (2019) Petiveria alliacea, a plant used in Afro-Brazilian smoke rituals, triggers pulmonary inflammation in rats. Rev Bras Farmacogn 29(5):656–664. https://doi.org/10.1016/j.bjp.2019.06.005
38. Roriz CL, Barros L, Carvalho AM, Santos-Buelga C, Ferreira ICFR Scientific validation of synergistic antioxidant effects in commercialised mixtures of Cymbopogon citratus and Pterospartum tridentatum or Gomphrena globosa for infusions preparation. Food Chem 185:16–24. https://doi.org/10.1016/j.foodchem.2015.03.136

Chapter 14
Pharmacovigilance for Herbal Medicines: A Perspective from the Herbal Medicines Industry

Phil Rasmussen

14.1 Introduction

Pharmacovigilance is a crucial science for all medicines. This includes phytomedicines, which are those that are plant-based, or herbal, in origin. With the high and increasing use of herbal medicines for human and veterinary health, companies producing and selling these products need to have good understanding of, and engagement with, activities related to detecting, assessing, understanding and preventing adverse effects and other safety concerns that may arise in relation to their products.

The global natural health products (NHPs) or complementary medicines industry is extremely diverse, both from a commercial business model and size perspective, as well as in relation to the types of products manufactured and sold. It incorporates a huge spectrum of companies, ranging from 'one-person' operators making and selling 'home herbal remedies' at their local farmers' market each weekend, to multi-billion dollar turnover network marketing companies employing several thousands of people and working with millions of distributors in multiple countries, to manufacturers of well-researched and patented herbal extracts sold as ingredients and used by numerous brands in thousands of different marketed products globally.

From a public health perspective, the risks and pharmacovigilance considerations arising from each of these types of operations are very different. Inevitably, the capacity and resources available to, or allocated by, companies to the post-market surveillance, monitoring, analysis and reporting of suspected adverse reactions involving their herbal medicinal products is also highly variable. This ranges from operations that have no capacity and/or allocate no resource to pharmacovigilance activities, to those that employ a team of

P. Rasmussen (✉)
Phytomed Medicinal Herbs Ltd., Auckland, New Zealand
e-mail: phil@phytomed.co.nz

J. Barnes (ed.), *Pharmacovigilance for Herbal and Traditional Medicines*,
https://doi.org/10.1007/978-3-031-07275-8_14

pharmacovigilance experts and have a qualified person in pharmacovigilance (QPPV) within the company, and who frequently liaise with third parties such as toxicologists, public health officials, medicines' regulators, researchers and clinical specialists.

Herbal medicines are a complex and disparate group of products. In practice, most countries have more than one regulatory agency (generally those responsible for foods and for medicines) involved in overseeing compliance with what can be complicated and, all too often, outdated legislative frameworks. The regulatory classification of and level of therapeutic efficacy, or 'health' claims permitted for herbal medicinal products, also varies in different jurisdictions [1–3].

Even within different European Union (EU) member states, for example, the legal status of herbal products varies, and it is possible to find the same product classified in several different categories. These include registered medicinal products, including prescription only medicines, traditional herbal medicinal products, well-established use herbal medicinal products, food supplements, medical devices, homoeopathic/anthroposophical medical products, and cosmetics [3]. This is the situation despite ongoing political and regulator attempts to harmonize key attributes.

As a result, substantive differences exist internationally in terms of what can and cannot be legally stated on product packaging and marketing materials. This is also the case for regulatory pharmacovigilance obligations, and how these are interpreted and managed by the agencies concerned. An often challenging landscape is sometimes complicated further by inequities in the political will, regulatory resource and ability to ensure regulations are properly complied with. Clearly, these factors will have a pivotal impact on the level of effective product safety and post-marketing monitoring and review activity taking place in practice.

Despite these difficulties, a majority of operators in the NHPs and herbal medicines industry want to see good and appropriate legislation and regulatory requirements in place, both to help ensure product end users benefit from the desired health outcomes, and that the need for product safety crisis management and product recalls is minimized.

Herbal medicines are chemically rich substances with pharmacological activities, and their history, contemporary usage, and a growing body of scientific data demonstrate this. However, as with all medicines, in order to be able to make and impart the therapeutic and health-enhancing claims for which consumers and practitioners rely upon herbal medicines, systems must be in place to ensure that labelled and actual ingredients align, and that there are adequate safeguards against overtly false claims and other quality, effectiveness and safety concerns [4, 5]. This can only be achieved through fit-for-purpose legislation and regulations that are developed and regularly reviewed, by a partnership between industry, scientists and governments.

14.2 Under-Reporting of Adverse Reactions Associated with Herbal Medicines

Reports of adverse reactions associated with herbal medicines appear to be relatively infrequent, but, as with conventional medicines, under-reporting is of concern [6–8]. Reasons for this are multiple, as they are for under-reporting of suspected ADRs associated with medicines in general. An additional issue with respect to herbal medicines is that the term pharmacovigilance is not one that is familiar to, nor well understood by, all those involved in the herbal medicines industry.

There are unfortunately still elements within the herbal medicines industry that for various different, including historical, reasons regard the need for an effective pharmacovigilance system as somehow a 'threat' to their livelihood, and which has an agenda to restrain manufacturers from being able to continue selling their products and operating a profitable business. While only a small minority of the industry, the influence and reach that these views can have on public and consumer perceptions, particularly in a world in which social media sometimes supersedes science in affecting people's behaviour, is a concern. It is important that this minority should see the overarching public health need for a reasonable and evidence-based pharmacovigilance system, and related regulatory requirements.

Contributing to these views and the low level of reporting of adverse reactions associated with herbal medicines, both by end users, health professionals and industry itself, is that there is sometimes a perception that, as herbal medicines come from natural resources, they must be safe. This is a false and potentially dangerous view, which industry itself needs to challenge, discuss and address when it arises. However, in order to help facilitate the required shift in thinking, it is important that medicines regulators recognize the many differences between single chemical entity (drug)-based products and herbal medicinal products and, again, seek to actively engage with industry prior to and during legislation development.

14.3 The Need for Pharmacovigilance for Herbal Medicines

Ensuring adequate pharmacovigilance systems are in place for all medicines, including herbal products, helps to foster the fundamental public health requirement to protect populations against avoidable harm from medicines.

While most herbal medicines are generally very safe when manufactured to high pharmaceutical quality standards and used appropriately, without a pharmacovigilance programme and accompanying regulatory requirements in operation, it is not possible to be certain that a particular product is, in fact, without risk of harm.

There are multiple reasons for industry to actively embrace a good pharmacovigilance system. Apart from enabling detection and study of adverse events, it can help anticipate any rare, but serious, safety concerns, help improve the end user experience, and enable valuable data to be gathered that may help better understand and measure product effectiveness.

If industry, or the regulator, lacks sufficient or robust data, it becomes more difficult to make a case in defence of moves to reschedule or prohibit a particular herbal product or substance. Where there is evidence to support a favourable benefit versus harm and cost profile, other approved indications and potential sales channels may become possible. Adjunctive use alongside conventional medicines is an emerging area of opportunities for patients, companies and government health budgets [9], but understanding the nature of any clinical interactions between drugs and herbal medicines is very important.

There are also things that can go wrong within the herbal medicine industry. As with all under-regulated industries, some participants lack integrity. It is, therefore, crucial that unethical or dangerous activities undertaken by a minority of companies are identified by regulatory authorities, and that appropriate action is taken early on, before product safety issues cause harm to human or animal health, and such misadventures damage the reputation of the entire industry.

It is inappropriate, for instance, for herbal medicines manufactured without quality control and quality assurance systems in place, and with no guarantee that the ingredients are authentic or true to label, to make unsubstantiated claims and be sold for the treatment of a serious health condition. Based upon product efficacy claims with little if any evidence base, an individual with, for example, cancer, or severe depression, may purchase and rely upon self-medication with herbal products as a substitute for other interventions. This is a situation that has ethical, safety and other implications, including the need for improved regulations that address different scopes of practice for trained herbal practitioners, versus those with little or no training.

Unfortunately intentional adulteration of raw materials and finished products with unlabelled extraneous or synthetic substances added to confuse analytical techniques, and the addition of active synthetic drugs continues to taint the herbal medicines industry. This can present serious safety concerns, and is still a problem, particularly for products aimed at improving athletic performance, sexual functioning or achieving bodyweight loss [10–13]. Product contamination with bacterial or fungal pathogens can also cause harm to patients.

These particular types of issues relate more to the need for an appropriate overarching regulatory system, and requirements, such as Good Manufacturing Practice (GMP) manufacturing and quality assurance systems, to be adopted and implemented. They clearly also have important impacts for pharmacovigilance.

When these requirements are properly managed and trained out to all relevant staff and divisions within the company, and adverse event reporting mechanisms for end product users are made easily accessible and user friendly, clear benefits derive for patients and end users. Further, as with a good regulatory environment in relation to the manufacture and sale of herbal medicines, a strong pharmacovigilance system is a powerful tool to protect the herbal medicines industry's reputation, and ensure that companies produce good quality products, then market and sell them into jurisdictions and market channels and for purposes that are appropriate. The overarching need to comprehensively understand a product's safety parameters is intrinsic to many other business objectives and, ultimately, it also provides benefits to company staff and management, shareholders and regulators.

14.4 Regulatory Obligations for Pharmacovigilance in the Herbal Medicines Industry

Regulatory requirements for industry in relation to pharmacovigilance are set out in the relevant country's regulations, such as the Australian Regulatory Guidelines for Complementary Medicines (ARGCM), and the MHRA's recently published pharmacovigilance procedures, following its departure from EU membership [14]. These describe the obligations on sponsors, or marketing authorisation holders, of authorized/registered/listed medicines, including herbal medicines, to undertake pharmacovigilance-associated activities.

Herbal medicine companies need to provide information to product users to optimize the safe and effective use of their medicines, and, at the very least, keep and retain records of reports they receive of adverse reactions associated with the company's products. Also, depending on the country and corresponding regulatory framework, sponsors/marketing authorisation holders of authorized herbal medicines should promptly notify the relevant pharmacovigilance agency (e.g. Adverse Drug Reactions Unit, TGA, or European Medicines Agency) should a serious adverse event report be received, and report all adverse events received to this centralized agency on a regular basis. National and international pharmacovigilance agencies require adequate data in order to undertake proper and meaningful causality assessments and signal detection, and ensuring companies themselves submit adverse event reports is a key part of this.

There is also a requirement for sponsors/marketing authorisation holders of authorized herbal products to monitor, or be regularly notified of, the international pharmacovigilance and other relevant scientific literature that might relate in some way to the company's products. This ensures the company is promptly informed of any adverse event reports, interaction reports, or other safety concerns for products containing the same or similar herbal substances/ingredients. Again, this is important in an environment where under-reporting and a relative paucity of data in general are serious deficits.

Most companies, particularly large enterprises, or those selling into multiple markets, should undertake some kind of signal detection activities themselves. Engagement and retention of a QPPV to help ensure key pharmacovigilance obligations and processes are properly pursued is a valuable asset and often a mandatory requirement.

The overarching document that stipulates each company's procedures and protocols in relation to pharmacovigilance is the Pharmacovigilance System Master File

(PSMF). This and the associated SOPs, as well as having clear and transparent internal processes and comprehensive staff training for dealing with all pharmacovigilance matters, leads to a good pharmacovigilance culture and beneficial outcomes for all stakeholders. A clear crisis management process and flow chart that stipulate the different stages of emerging safety concerns, and whether and how to escalate, are also important.

Another integral component is the need for an easily accessible adverse event report form, and awareness and training of company and third-party sales and distribution staff around how to identify, collect and report data on any potential adverse events. This helps foster an environment whereby reporting such adverse events is actively encouraged and viewed as something that is good for the company and its products, rather than being considered unimportant or perhaps even a threat.

While covered more under GMP and marketing authorisation requirements, the testing of raw materials for adulterants, manufacturing artefacts or mycotoxins, such as aflatoxins and ochratoxins, is sometimes required. The presence of these compounds is fortunately a rare, but potentially very serious, quality concern that can have a major impact on product safety and thus pharmacovigilance, and places a higher compliance burden on the industry. Understanding and appropriately managing the level of risk with these types of contaminants, and other quality concerns for herbal medicines, is something that a good (medicines) regulator will understand and monitor in partnership with industry, in a measured and risk-appropriate manner.

14.5 Challenges for the Herbal Medicines Industry

The relative shortage of comprehensive data on efficacy and safety aspects for many herbal medicines and the huge diversity in product types and their composition are major and universal challenges for all involved with the herbal medicines industry. The ability to apply an appropriate stance to the pharmacovigilance of herbal medicines is impacted by these factors as well as by the low level of reporting of adverse reactions associated with herbal medicines to national pharmacovigilance agencies.

Just as regulators have many challenges to navigate when applying appropriate legislative obligations and compliance expectations upon industry, and considering how to address many of the inherent complexities of phytochemically rich herbal medicines, so industry itself needs to manage a wide variety of challenges on a regular basis.

There are substantial differences in legislative obligations and regulatory interpretations across different countries. This situation—discussed earlier in this chapter—relating to whether herbal products are dietary/food supplements, or traditional or other forms of herbal medicines, can lead to lack of ownership for all parties involved, including health professionals. As there are different thresholds and regulations for 'dietary supplements', compared with those for herbal medicines, many companies elect to pursue 'dietary supplement' status for their products. 'Dietary supplement' regulations typically stipulate more food—rather than medicine-oriented manufacturing and testing requirements, yet some of those same companies still seek to make overtly medicinal- or therapeutic-type claims in the marketplace. This leads to unfair disparities in the regulatory and cost burden versus market access incurred by different company's products and can unintentionally give a competitive advantage to companies who are also less likely to apply sufficients integrity and resource to safety and pharmacovigilance matters.

Confusion about whether, in a given country, it is the medicines regulatory agency or the food regulatory agency who is the regulator of 'dietary supplements' can also lead to the serious situation in which responsibility for pharmacovigilance matters, and the ability to enforce when required in a timely, appropriate, and sometimes urgent manner, can be compromised. Another outcome of this is that herbal medicine companies can also experience difficulties in obtaining export certificates for their products if the local legislation and regulatory requirements are materially different to those of the country to which the product is being exported.

From a safety and pharmacovigilance perspective, an herbal extract manufactured using methanol as a solvent and by a supercritical carbon-dioxide method, incorporated into a capsule or tablet dose form, and marketed by the sponsor for the treatment of arthritis, is surely a medicine, not a 'dietary supplement', and should therefore be required to meet medicines-based pharmacovigilance (and other) obligations. Even where products legitimately meet definitions for 'dietary supplements', mandatory recall powers may occasionally be required, in relation to emerging safety concerns, and this ability needs to be stipulated in legislation.

Better engagement, and thus regard by industry, for the importance of having effective pharmacovigilance systems in place, will become apparent once more fit-for-purpose regulatory frameworks are implemented by regulatory agencies. These should enable reasonable therapeutic claims to be made for herbal medicinal products, better differentiate herbal medicines and dietary supplements, and require GMP-based manufacturing systems to be employed by herbal medicine manufacturers.

Other industry challenges stem from supply chain inconsistencies and disruption due to climatic factors or crop failures, global pandemics, or herbal materials experiencing sudden increases in demand. These can impact very seriously and sometimes suddenly on both quality parameters and price, and potentially pharmacovigilance.

The highly competitive landscape of the herbal medicines market, consumers and patients increasingly sourcing herbal medicinal products online, and frequently from other countries apart from that they live in and with no health professional involved at the point of dispensing or purchase, are further impediments to gathering post-marketing surveillance data and mitigating risk of harm. A requirement to hold a pregnancy registry (which records usage of the product(s) by pregnant or breastfeeding women) is something that in an ideal world would apply to all medicinal products, including herbal medicines, but for various reasons is challenging and unrealistic for most herbal medicines companies to achieve.

Regulatory authorities themselves often seem to be impacted by insufficient resource or specialist expertise in order to make appropriate rulings in relation to safety issues for herbal medicines. Information sources can also be of variable quality [15]. As a consequence, industry has encountered several instances of what it (the herbal medicines industry) considers to have been poor decision-making, which has resulted in products being unjustifiably removed from the market, or sponsors obliged to add warning statements to all packaging and information sources, that some regard as lacking a strong evidence-base.

To help minimize such instances and ensure a higher level of industry engagement and collaboration, with the ultimate objective of creating and operating a robust pharmacovigilance system, adequate consultation between the regulator and industry, and the employment of staff with relevant specialist training and expertise in what are essentially complex and challenging products, is critical.

Concerns relating to extracts of kava (*Piper methysticum* G.Forst) rhizome and hepatotoxicity became a protracted situation in which pharmacovigilance experts, epidemiologists, hepatologists, toxicologists and regulators spent many years investigating potential aetiological factors behind the cases of liver damage associated with use of products containing extracts of this plant. While several contributory factors have since been identified for these serious adverse events, inconsistencies occurred and, to some extent, are ongoing, in how these safety concerns are viewed and addressed in different countries [16].

Alleged hepatotoxicity associated with extracts of black cohosh (*Actaea racemosa* L.) root/rhizome is another instance where there was a delay in regulators and industry applying the required evidence-based standpoint and considering potential contributory factors, such as the incorrect plant species being used, and sometimes lack of robust GMP manufacturing and testing requirements [17, 18].

Science is vital in regulatory settings and the scientific challenges in relation to herbal medicines and their regulations provide new opportunities for scientists and regulators to work together both nationally and internationally. In addition to learning from each other, such co-operation has huge benefits, including harmonization of assessment and regulatory approaches when appropriate, to improve public health [4].

Some of the claimed adverse events associated with particular herbal medicines, and associated mandatory warnings, also remain contentious and poorly substantiated according to evidence-based principles. Sun sensitivity when taking St John's wort (*Hypericum perforatum* L.) preparations, for example, may be due to hypericism, a rare sensory nerve hypersensitivity experienced after ingesting forms of St John's wort that are high in the compound hypericin [19, 20]. Allergic reactions associated with echinacea preparations frequently relate to products made using flowering aerial parts where high pollen content is a factor, rather than those made using root-only extracts. Research suggests echinacea (*Echinacea purpurea* (L.) Moench) root exhibits mast cell stabilizing and anti-inflammatory properties that make it potentially indicated, rather than contraindicated, in conditions, such as allergic rhinitis and eczema [21–23].

14.6 The Future

The ongoing growth of the herbal medicines industry, and its ability to continue to make a major contribution to human and animal health, is dependent on having achievable, accessible and reasonable pharmacovigilance processes and legal requirements in place. This is more likely to be embraced by industry in a regulatory environment that adequately addresses the broader regulatory needs and market access requirements for herbal medicinal products in general. Effective implementation is more achievable, and compliance and enforcement more likely, when a fit-for-purpose regulatory framework is in place. This should enable reasonable therapeutic claims to be made in exchange for adherence with GMP-based manufacturing systems that address the particular quality and safety issues intrinsic to herbal medicines, and better differentiate a herbal medicine from a dietary supplement.

To optimize the approach to dealing with safety concerns associated with herbal medicines, there is a need for a regulatory framework that is well-resourced, evidence-based, risk-based, and appropriate. Additionally, and very importantly, the regulatory framework should provide for consultative and open discussions with industry. Herbal medicines are phytochemically very complex, their sourcing, growing and processing methods differ enormously, and product manufacturing and presentation parameters will become even more diverse as new technologies are embraced by industry. There is, therefore, an increasing need for a high level of specialist skills in a range of different areas, for any regulator who aims to optimally oversee the manufacture, sale, use and pharmacovigilance considerations for herbal medicines.

In parallel with the above, better regulations for the large, diverse and important complementary–/natural-health and traditional medicine practitioner sector is needed. Many traditional and complementary medicine professions operate without little accountability and regulatory oversight for maintaining public protection, despite playing a key role in providing herbal medicines to end users globally. The World Health Organization has called for improved regulations for traditional practitioners under its Traditional Medicine Strategy 2014–2023 [24] and broad support for this exists [25]. Promoting evidence-based traditional and complementary medicine as self-care modalities has also been identified as an efficient way of reducing the healthcare economic burden and promoting healthy ageing in many countries with lower income levels. This, and the popularity of more traditional forms of herbal products in many parts of the world, mandates policy makers to implement appropriate risk-based regulation and quality assurance, and to establish pharmacovigilance systems to detect potential harm [26].

The internet is also disrupting traditional purchasing channels and the ability of regulators to execute regulatory inputs in relation to product safety, and the herbal medicines industry overall is rapidly changing and continuing to evolve. Cumulatively, these factors mean that there will be increasing reliance on industry itself in the future to take more responsibility for ensuring pharmacovigilance needs are properly addressed.

14.7 Conclusion

Understanding and actively embracing the public health need for pharmacovigilance is an important requirement for an engaged herbal medicine manufacturer, or brand, in the rapidly changing health landscape of the twenty-first century. Gathering comprehensive product safety data is crucial to the herbal medicines industry's ongoing steady growth in sales and reach, and its further maturity. Changing consumer attitudes towards safety and efficacy, and values about what is important in food and medicine, will also determine future needs for the post-market surveillance and monitoring, and scientific evaluation of herbal medicines.

However, to further enable an effective and robust pharmacovigilance system for herbal medicines and its active embracement by industry, the overarching regulatory environment and its underpinning legislation needs to be deemed to be 'fit for purpose' by most of the industry itself, and more cross-border harmonization needs to be achieved. Finally, it is critical that scientists, regulators and industry in many countries work together and learn from each other in both identifying and debating issues and developing ways to address them, in order to continue to drive the science and positive outcomes of pharmacovigilance for herbal medicines into the future.

References

1. Alostad AH, Steinke DT, Schafheutle EI (2018) International comparison of five herbal medicine registration systems to inform regulation development: United Kingdom, Germany, United States of America, United Arab Emirates and Kingdom of Bahrain. Pharmaceut Med 32(1):39–49. https://doi.org/10.1007/s40290-018-0223-0
2. Li J, Zhu J, Hu H, Hu H et al (2018) Internationalization of traditional/complementary medicine products: market entry as medicine. Chin Med 13:50. https://doi.org/10.1186/s13020-018-0209-6
3. Bilia AR, Costa MDC (2021) Medicinal plants and their preparations in the European market: Why has the harmonization failed? The cases of St. John's wort, valerian, ginkgo, ginseng, and green tea. Phytomedicine 81:153421. https://doi.org/10.1016/j.phymed.2020.153421
4. Dwyer JT, Coates PM, Smith MJ (2018) Dietary supplements: regulatory challenges and research resources. Nutrients 10(1):41. https://doi.org/10.3390/nu10010041
5. Sardella M, Belcher G, Lungu C et al (2021) Monitoring the manufacturing and quality of medicines: a fundamental task of pharmacovigilance. Ther Adv Drug Saf 12:20420986211038436. https://doi.org/10.1177/20420986211038436
6. Shin HK, Jeong SJ, Huang DS, Kang BK, Lee MS (2013) Usage patterns and adverse experiences in traditional Korean medicine: results of a survey in South Korea. BMC Complement Altern Med 13:340. https://doi.org/10.1186/1472-6882-13-340
7. Park JH, Choi SM, Moon S et al (2017) Development of an adverse events reporting form for Korean folk medicine. Pharmacoepidemiol Drug Saf 26(5):498–508. https://doi.org/10.1002/pds.4077
8. Xu Y, Patel DN, Ng SP (2018) Retrospective study of reported adverse events due to complementary health products in Singapore from 2010 to 2016. Front Med (Lausanne) 5:167. https://doi.org/10.3389/fmed.2018.00167

9. Rasmussen PL (2016) Potentially positive herb-drug interactions. Phytomed Medicinal Herbs Ltd, Auckland, New Zealand. Practitioner Seminar, 19 Jul 2016
10. Rocha T, Amaral JS, Oliveira MBPP (2015) Adulteration of dietary supplements by the illegal addition of synthetic drugs: a review. Compr Rev Food Sci Food Saf 15(1):43–62. https://doi.org/10.1111/1541-4337.12173
11. Yéléhé-Okouma M, Pape E, Humbertjean L et al (2021) Drug adulteration of sexual enhancement supplements: a worldwide insidious public health threat. Fundam Clin Pharmacol 35(5):792–807. https://doi.org/10.1111/fcp.12653
12. Firozian F, Nili-Ahmadabadi A, Moradkhani S, Moulaei M, Fasihi Z, Ahmadimoghaddam D (2021) Adulteration of the herbal weight loss products by the illegal addition of synthetic anti-obesity medications: a pilot study. J Obes 2021:9968730. https://doi.org/10.1155/2021/9968730
13. Koncz D, Tóth B, Roza O, Csupor D (2021) A systematic review of the European rapid alert system for food and feed: tendencies in illegal food supplements for weight loss. Front Pharmacol 11:611361. https://doi.org/10.3389/fphar.2020.611361
14. Medicines and Healthcare Products Regulatory Agency (2021) Guidance on pharmacovigilance procedures. Available from: https://www.gov.uk/government/publications/guidance-on-pharmacovigilance-procedures. Accessed 13 Sept 2021
15. Ng JY, Munford V, Thakar H (2020) Web-based online resources about adverse interactions or side effects associated with complementary and alternative medicine: a systematic review, summarization and quality assessment. BMC Med Inform Decis Mak 20(1):290. https://doi.org/10.1186/s12911-020-01298-5
16. Bian T, Corral P, Wang Y et al (2020) Kava as a clinical nutrient: promises and challenges. Nutrients 12(10):3044. https://doi.org/10.3390/nu12103044
17. Teschke R, Bahre R, Fuchs J, Wolff A (2009) Black cohosh hepatotoxicity: quantitative causality evaluation in nine suspected cases. Menopause 16(5):956–965. https://doi.org/10.1097/GME.0b013e31819d3904
18. Teschke R (2010) Black cohosh and suspected hepatotoxicity: inconsistencies, confounding variables, and prospective use of a diagnostic causality algorithm. A critical review. Menopause 17(2):426–440. https://doi.org/10.1097/gme.0b013e3181c5159c
19. Connor HE (1951) The poisonous plants of New Zealand. Government Printer, Wellington, New Zealand
20. Schempp CM, Winghofer B, Müller K et al (2003) Effect of oral administration of Hypericum perforatum extract (St. John's Wort) on skin erythema and pigmentation induced by UVB, UVA, visible light and solar simulated radiation. Phytother Res 17(2):141–146. https://doi.org/10.1002/ptr.1091. PMID: 12601676
21. Gulledge TV, Collette NM, Mackey E et al (2017) Mast cell degranulation and calcium influx are inhibited by an Echinacea purpurea extract and the alkylamide dodeca-2E,4E-dienoic acid isobutylamide. J Ethnopharmacol 212:166–174. https://doi.org/10.1016/j.jep.2017.10.012
22. Rasmussen PL (2021) Herbs for seasonal rhinitis. Pharmacy Today, The Health Media Ltd, Auckland, New Zealand
23. Oláh A, Szabó-Papp J, Soeberdt M et al (2017) Echinacea purpurea-derived alkylamides exhibit potent anti-inflammatory effects and alleviate clinical symptoms of atopic eczema. J Dermatol Sci 88(1):67–77. https://doi.org/10.1016/j.jdermsci.2017.05.015
24. World Health Organization (2013) WHO traditional medicine strategy: 2014–2023. Available from: https://www.who.int/publications/i/item/9789241506096. Accessed 13 Sept 2021
25. Care J, Steel A, Wardle J (2021) Stakeholder attitudes to the regulation of traditional and complementary medicine professions: a systematic review. Hum Resour Health 19(1):42. https://doi.org/10.1186/s12960-021-00579-y
26. Chung VCH, Wong CHL, Zhong CCW, Tjioe YY, Leung TH, Griffiths SM (2021) Traditional and complementary medicine for promoting healthy ageing in WHO Western Pacific Region: Policy implications from utilization patterns and current evidence. Integr Med Res 10(1):100469. https://doi.org/10.1016/j.imr.2020.100469

Chapter 15
Pharmacovigilance and Risk Communication for the Safe Use of Herbal and Traditional Medicines: How to Promote Evidence and Keep People Safe

Bruce Hugman and Souad Skalli

15.1 Background

Although there are few precise estimates of prevalence in the use of herbal medicines [1], the market for them continues to expand rapidly and has grown into a multibillion-dollar industry across the world [2]. Market research data indicate increasing sales of licensed and unlicensed products [3–5], including international trading, suggesting that very large numbers of people are users, on every continent. The influence of religious, sociocultural, and socioeconomic issues, traditional practices and belief in the use of herbal medicines is evident, particularly in Chinese, Indian and African societies where traditional remedies remain predominant. Documented use of herbals in Western societies is also high [6, 7]: there are 100 million users in Europe, according to the World Health Organization (WHO) [8]. On every continent there are millions of practitioners whose professional and economic survival depends on the demand for herbal and alternative medicines and services. Worldwide, most herbal medicines are unregulated and can be obtained from multiple sources without prescription.

As with all medicines, herbals have the potential to cause adverse effects that may be mild, serious or fatal. Professionals in the field know that these are related to a variety of causes, including: inherent properties such as the presence of toxic constituents; adulteration; mistaken use of the wrong plant species or plant part; incorrect dosing; errors in indication and use; contamination; herb–drug interactions [9], caused by effects on the pharmacokinetic and pharmacodynamic properties of conventional drugs. This range of risk factors is poorly understood by the public and often understated or absent in labelling. Benefits are commonly

B. Hugman (✉)
Uppsala Monitoring Centre (at the time of writing), Uppsala, Sweden

S. Skalli
Faculty of Science, Mohammed V University in Rabat, Rabat, Morocco

J. Barnes (ed.), *Pharmacovigilance for Herbal and Traditional Medicines*,
https://doi.org/10.1007/978-3-031-07275-8_15

promoted colourfully; the safety issues, including that of concomitant use, require equally vivid and transparent communication at the point of purchase or use, whether in a shop, at a market stall, on the internet, or in consultation with a prescriber or dispenser.

For all these reasons, there is an increasing awareness of the need to improve public communication and to maintain and develop the practical usefulness of regulation and pharmacovigilance for herbal medicines (phytovigilance). Advice on how to prevent harm caused by herbals and traditional practices is a decisive challenge for regulatory authorities, phytovigilance and all health professionals. Several factors make communicating phytovigilance information challenging. These include: serious limitations in the quantity and quality of evidence on safety and effectiveness; the slow advance of research in phytotherapy and traditional practices; the belief that herbs as plants, are natural and safe. There is a host of other variables that profoundly influence public attitudes and choices: differential perceptions of risk; the power of commercial and social pressure through advertising, blogs, social media (including high profile influencers), mass media; levels of knowledge, literacy and numeracy, risk literacy [10] in the population, amongst them.

In essence, billions of people use herbal and traditional medicines; the challenge to educate people about rational use and protect them from harm is gigantic. Official efforts to minimize the risk of harm must contend with the weight of history and belief and compete with vested interests and sophisticated and widespread promotion. Risk minimization projects require sociological insight, empathy, creativity and substantial human and financial resources. The impact of some of these issues on the safe use of conventional drugs has also been extensively discussed [11].

15.2 The Radical Problems of Risk Communication in all Aspects of Traditional and Modern Medicine

Public communication about complex scientific issues has always been problematic and is especially so today [12]. It is even more so when the aim is to influence attitudes and behaviour, as is the case with all risk communication in medicine, and especially in herbal medicine; the problems arise in the provision of specific medicines information and guidance as well as from the operation of the whole vigilance process.

In a modern world, where evidence and expert opinion are being questioned, even mocked or vilified the difficulties multiply. In a post-truth era of 'alternative facts' and perspectives, it is feelings, beliefs and allegiances that have become potent drivers of opinion and choice, while scientific consensus and evidence are losing popular and political influence and credibility. For example, a UK Academy of Medical Science report 'How can we all make better decisions about medicines?' revealed that only about one third of UK citizens trusted medical research, while two thirds reported that their family and friends were their preferred sources of medical opinion and advice [13].

In our field of interest, vaccine and climate change denial or scepticism, homoeopathy, naturopathy, and undiscriminating commitment to natural remedies and all kinds of health fads are vivid examples of the process of popular opinion and wishful thinking side-lining scientific evidence. In Japan, and other countries, for example, resistance to vaccination for Covid-19 reached alarming proportions [14]. We must, nevertheless, avoid complacency about modern medicine's pure rationality: the practice of modern medicine also has many irrational and inconsistent elements, where best evidence plays little or no part. For example, regional variation in care and outcomes in the UK and the USA cannot be explained by disease patterns, patient preferences or economic arguments and are, therefore, unwarranted [15]. This and other deficiencies demand a measure of humility that should moderate claims to having all the answers, which so antagonize those with alternative perspectives. The question of 'ownership' in contemporary medicine and systems of traditional medicine, and its implications for communication, is considered in Box 15.1.

Herbal medicines, in their many and varied forms, with all their potential for benefits and harms, have been integral to most cultures for thousands of years. While there may be little strictly scientific evidence of safety and effectiveness, we must assume that both were (and are) perceived to be at socially acceptable levels when measured against the occurrence of harm: no traditional healer would survive the fact of most patients dying, and no community would tolerate such a threat to its survival. The evidence for use in such societies comes from traditional practice and, primarily, a consensus based on oral history and experience. An implicit understanding of risk would be integral to the constellation of beliefs and perceptions underlying behaviour; each society would have its own levels of expectation and tolerance and these would be very different from modern Western opinion. For example, a common belief in Ghana that vomiting provoked by ingestion of a medicine is a sign of a strong and effective remedy is quite contrary to the tenets of scientific medicine. Such fundamental differences across cultures must deeply affect any discussion of risk communication and require a range of distinct, differential, targeted approaches. In sub-Saharan Africa it is estimated that around 60% of the population relies on herbal and traditional remedies [16], so the communication project is on a grand scale there and throughout the world.

Ancient traditions of herbal remedies exist in all advanced countries too; while they have been largely displaced by modern medicine, they still have their adherents and practitioners, numbers now greatly enlarged by revival of old beliefs and by the importation and adoption of novel philosophies, substances and practices from distant places. With no tradition or oral history of use to rely on, these foreign adopters of indigenous remedies are acting largely on the basis of faith, sometimes prejudice against aspects of modern medical practice, on anecdotal evidence, in the footsteps of celebrities or on the advice of commercially driven gurus or entrepreneurs, some whose motives may be anything but altruistic. These influences provide only a fragile basis, at best, for safe, healthy choices. The knowledge delivered by pharmacovigilance and risk communication have essential parts to play in reducing the threat to health that these factors represent, by broadening and deepening the pool of evidence, and disseminating it widely and vividly.

Along with the assumption that 'natural' means safe, the beliefs held, often passionately, by the proponents of herbal medicines, are not easily amenable to influence or change. The prejudices *against* herbal medicines are equally stubborn, of course, and such entrenched positions militate against productive rapprochement. That herbal medicines hold enormous benign potential is hardly to be disputed, but we are very far from having good, specific evidence about safety and effectiveness for most of the thousands of products available. There is evidence about harms from limited research and clinical trials in human populations, and when patients suffer adverse effects (as they have from preparations containing *Aristolochia fangchi*

Box 15.1 The Question of Ownership and Its Implications for Communication

There is a profound difference in the popular status of modern and traditional medicines. It is the issue of *ownership*, in the sociological (not financial) sense: Are authority and decision-making external or internal? Are others the principal actors or am I in charge? Is this a system in which I am an object or a member and creator? The widespread paternalism of modern medicine, with patients as objects of benign, but complex and mysterious purposes, contrasts starkly with the subjectivity and authenticity of belief in herbal and traditional medicines, the experience of *agency* in the solving of problems. Modern medicine is *theirs*; traditional medicines are *mine* and *ours*. To access modern medicine, I must submit myself to gatekeepers; for access to herbal and traditional medicines, I am my own gatekeeper (at the simplest level, I collect my own herbs from nature); if I submit myself to a local expert, it is to one who already shares my values, who is familiar and trusted.

Patients have fought to achieve a greater degree of involvement in modern medicine, to assert a greater degree of ownership in decision-making at all levels, moving along the spectrum from *theirs* to *ours*. Quite the opposite happens when medicines regulators try to assert greater control over herbal and traditional medicines: that is an enforced move along the spectrum from *ours* to *theirs*; from me as independent agent to me as unwilling victim of official control. (Similar issues and emotions are echoed by vaccine deniers.) Patients who feel excluded from decisions about their medical therapy, and those citizens who feel their freedom to make choices about alternative therapy is under threat, readily become hostile to those in authority, perceive real or imaginary conspiracies to manipulate and deceive them, circulate stories that allege harm, and find champions to promote and consolidate their causes. Scientists may be impugned because of their alleged conflicts of interests or political allegiances, because of differences of opinion among them, or evidence and guidance changing over time.

None of these groups responds well to the assertive repetition of official wisdom or evidence, or to the recitation of data, however scientifically authoritative. Any group under what it perceives to be external attack, is likely to retrench and consolidate. Evidence will not survive where group identity is determined by belief. Risk communication cannot succeed if these profound issues are not fully considered; it stands a chance only if it engages accurately and empathetically with them.

Y.C.Wu ex L.D.Chow & S.M.Hwang root or *Bupleureum chinense* DC root, for example), but only, of course, when such events are reported or come to public notice in some way.

Among the risks associated with herbal medicines, hepatobiliary disorders (HBD) have been associated with numerous products [17]. A recent study on the analysis of hepatobiliary disorder reports associated with the use of herbal medicines in the WHO global ADR database Vigibase has shown that the most commonly reported herbal drugs (after exclusion of non-specified herbal products) among serious reports with ADRs coded to MedDRA HBD system organ class (SOC) are *Cimicifuga racemosa, Valeriana officinalis, Camellia sinensis, Hypericum perforatum, Serenoa repens, and Pelargonium sidoides* [18].

In short, the use of herbal and traditional medicines is driven largely by belief and not by scientific evidence; this has enormous implications for risk minimization planning. Alongside this, increased investment in systematic research into the benefits and harms of herbals and the detection of adverse effects are high priorities.

15.3 Challenges of Risk Communication in Adverse Effect Reporting

Risk communication is among the most complex aspects of the practice of professional communication, even with a willing and attentive audience sharing common principles and committed to rational discourse. If we consider that our popular audience for information about herbal medicines may be largely faithful followers, some intelligent and well educated but with a low opinion of scientific processes and the credibility of statistics—at the same time distracted by many other daily preoccupations, sceptical (even cynical) about establishment motives and opinions on the subject, and making choices on the basis of belief—we can see how the difficulties multiply. Given the skill, creativity and enthusiasm with which herbal and other natural remedies are persistently promoted, notes of scientific or regulatory caution are very easily overshadowed or drowned out. Official risk communication has shown a deficiency in creative innovation and, especially, an unwillingness to use stories as a powerful tool of influence, so effectively used, for example, by the proponents of alternative remedies and vaccine sceptics.

Physicians, pharmacists, herbalists and traditional practitioners also need access to the best, up-to-the-minute information about efficacy, effectiveness and safety of herbal medicines, where it exists. Gaining the attention of health professionals, often under great pressure of time and administrative duties, as well as patient care, is a notoriously difficult project. In addition, there may be significant prejudices within the modern medical community against herbal and traditional medicines and alternative practices that may close the doors on valuable resources that do exist. Herbal and traditional practitioners have generally been outsiders to the enterprise of modern medicine and regulation; the safety of patients requires that they should be partners involved, and active, in communications and risk management.

It is commonly accepted that only a very small percentage of adverse events is reported through pharmacovigilance systems (for example, MedWatch in the USA; the Yellow Card scheme in the UK). Currently, it is an even smaller proportion of herbal safety concerns that is received through such systems; stimulating reporting is a major communications challenge in itself. We need vigilance systems that will record and expose harm, but we also need creative methods for facilitating reporting, promoting its benefits, and for making sure that information is widely shared and actually influences what people think and do. (We should note that there is little research into the impact of risk minimization measures in all fields.)

Importantly, key audiences, including all providers and patients, need to be identified and segmented, their complex character researched in order that they can be targeted and engaged differentially in productive dialogue about risk. Systems for reporting the adverse effects of all medicines, including herbals and related products, need to be simple, accessible, dynamic and familiar; there are current systems that fail by some, or all, of these criteria. Risk communication practitioners must understand the rapidly evolving complexity of modern communications and master novel methods and channels; they must be in touch, connected and media-savvy.

15.4 Communicating Herbal Medicine Safety Concerns

The challenge of effective communication about the risks of health products, including herbals, is not a new issue. Patients and patient organizations have long complained about the inadequacies of the official means of communication—principally patient information leaflets (package inserts)—as well as the inadequacy of much risk communication during clinical consultations. Consumers and patients have always resorted to alternative sources of information—relatives, friends, websites, blogs, the outpourings of ideologues/proponents, social media, and stories about celebrities and their lifestyle habits. Sources such as these seem particularly influential in relation to alternative remedies of all kinds, as many people in the West may be reluctant to seek the advice of their healthcare professionals about choices that might provoke disapproval. The situation is very different in countries such as China, South Korea or Switzerland, where medical services are almost completely integrated, or in Japan, where there are nearly 300,000 medical doctors who are sole prescribers of kampo medicines (Japanese traditional herbal medicines) [8]. The influence of anecdotes of effective treatment with herbal products and preparations is probably especially strong in social networks of all kinds outside countries of traditional use.

Official communications with health professionals from regulators and pharmacovigilance centres have been severely criticized in the past as bureaucratic and ineffective. There is evidence, for example, that multi-page Dear

Healthcare Professional letters and comparable documentary tools have variable effects, sometimes none at all [19, 20]. National formularies, printed and electronic, independent or integrated into prescribing and dispensing systems, apps and websites, seem to hold the greatest promise for health professionals in all areas of medicine, including herbal and traditional medicine. The exploitation of mobile devices and social media may hold the greatest promise for communication in both directions for populations in many parts of the world. Mobile apps for medicines information and adverse event reporting have been introduced in several countries, including, Kenya, South Africa, the UK, and the USA; a pan-European solution has grown out of the inventive WEB-RADR project [21].

In summary, traditional paper-based methods of risk communication have been shown to be often ineffective for professional and popular audiences. Solutions driven by the latest technology and the widespread ownership of mobile devices, even in low- and middle-income countries, appear to hold the greatest potential for influence. Risk communication must also move into the (for regulators) uncomfortable territory of storytelling, theatre and creative engagement with audiences, even with celebrities, to match the creative power of those promoting herbal and traditional medicines who are sceptical of evidence.

15.5 Multiple Conditions for Herbal Medicine Use

The context within which herbal medicines are accessed and used varies greatly around the world. In considering communication in general, and risk communication in particular, historical, social, cultural and other factors will influence public needs and perceptions in any given location. The purposes of risk communication are described in Box 15.2. In the WHO monographs on selected medicinal plants [22], information on medicinal use has been divided into three categories: use supported by clinical data; uses described in pharmacopoeias and in traditional systems of medicine; uses described in traditional medicine, not supported by experimental or clinical data ('folk-medicine').

Communications have to take account of whether a country's approach to herbal and traditional medicine is:

- *Integrative*, where herbals and traditional medicines are the primary sources of healthcare.
- *Inclusive*, where there is the use and prescription of herbals and traditional medicines in conjunction with conventional medicine due to cultural/historical influence (China, Vietnam and the Republic of Korea, India, for example).
- *Tolerant*, where the use of herbals and traditional medicines is primarily in a complementary or alternative role with conventional medicine (as in North America and many European countries).

These aspects provide just a few of the many variables that set the scene for communications; they demand sociocultural, demographic and psychological knowledge, understanding, empathy and extraordinary inventiveness in methods and approaches.

Box 15.2 The Aims of Risk Communication

The purpose of risk communication is to inform and protect: to support wise, balanced and rational decisions that match patients' and consumers' wishes and needs. The European Union Guideline on Good Pharmacovigilance Practice [23] offers clear objectives that are relevant to all medicines therapy communication, Western, herbal or traditional:

- Providing timely evidence-based information on the safe and effective use of medicines.
- Facilitating changes to healthcare practices (including self-medication practices) where necessary.
- Improving attitudes, decisions and behaviour in relation to the use of medicines.
- Supporting risk minimisation behaviour.
- Facilitating informed decisions on the rational use of medicines.

In addition, safety communication should support public confidence in the regulatory system.

The guidelines do not specify *how* these aspirations are to be achieved; that question can be answered only by highly skilled communications specialists working with intimate knowledge of their local audiences.

15.6 Professional Collaboration and Education

To manage risks with herbal medicines and traditional practices adequately to reduce consumer/patient harm, phytovigilance professionals must communicate actively about safety issues with all participants in the phytovigilance system, locally and globally. Phytovigilance stakeholders include patients, patient organizations, carers, media personnel (including social media influencers), healthcare professionals, health policy makers, researchers and academics, providers of phytotherapy products, herbalists, traditional practitioners and manufacturers. The multiplicity of systems, beliefs, practices and information across the world, especially when local medicines are exported to foreign places, increases the need for a common language, ideally common standards, definitions, taxonomy and procedures. Safety information must be shared and accessible to all [24]. Significant efforts towards this ideal have been made among a handful of countries outside the European Union (EU), but there is a long way to go.

Avoidable adverse effects and medication errors cause morbidity and mortality [25], and there are urgent patient needs for information to facilitate safe and rational use of herbal medicines. Risk communication is an important part of risk minimization plans, and is one of the major roles of pharmacovigilance in general and, in the present case, of phytovigilance in particular [26]. Co-ordination between all stakeholders is necessary to determine best practice in the presentation of herbal safety data in pharmacopoeias, prescribing information and labelling. Such co-operation should take into account universal ethical principles and the substantial amount of research related to the proper communication of risk information [27]. WHO, the International Conference of Drug Regulatory Authorities (ICDRA), and other international bodies and collaborations, such as CIOMS [28], have a great responsibility in this area; useful work has been underway for many years, for example, agenda items and recommendations from ICDRA conferences and meetings.

Various methods can be considered to reach all relevant target audiences, such as: involvement of the mass media (radio is a valuable tool in many places) and patient/consumer associations (including translating information into local languages); education of health professionals via the delivery of adverse event or poisoning print or digital bulletins/articles; meetings and education about the implications for herbal medicine providers, academics, researchers/scientists, journalists and the pharmaceutical and herbal medicine industries; skilled use of social media platforms. Communication must take place in a dynamic network, well structured, targeted, interactive, collaborative and finely adapted to the local and cultural situation [29] if there is to be any hope of changing beliefs and behaviours regarding herbal medicines, traditional practice and irrational use. The development of communication materials for targeted audiences (young and old, male and female, low literacy or numeracy, remote rural populations, rich and poor, for example) with the use of apps, texting, graphic materials, comics or videos as means of communication is desirable. Nothing, however, is likely to be superior to the impact of face-to-face engagement: training for members of local communities, teachers, community leaders, traditional birth-attendants, and others, who can provide finely calibrated, dispersed communications, provides a potent resource for change. All communication plans must include tools for recipient feedback, evaluation and impact assessment, without which nothing can be proved or learnt.

Information on herbal medicines should be included in academic programmes. Pharmacological aspects of phytotherapy and training on the need for suspected adverse drug reaction reporting for herbal products, and its practical aspects, should be included in all medical and pharmacy curricula [30]. Improvements in the education of all healthcare professionals should be made with regard to the principles and practice of phytovigilance communication. The addition of such topics to the curriculum of both conventional (medical) and alternative (e.g. naturopathic, herbal medicine) schools would inform these practitioners of the importance of in-depth discussion of the use of herbal medicines—and their benefits and harms—with their patients, as well as how to recognize and report suspected adverse effects. These proposals mirror many of those made in regard to the establishment of pharmacovigilance as a whole as a core discipline in medical training.

Key points in respect of the above are that fragmented scientific and healthcare communities will not have optimal impact on consumer and patient safety;

engagement and integration across the world may bring major rewards and results, but these require careful planning, monitoring and evaluation. Dispersed, local communications networks may offer powerful ways to educate and influence populations. Wherever herbal medicines are used, they must have a place in the education of health professionals.

15.7 Obstacles and Solutions in Risk Communication for Herbal Medicines

Successful communications, including the timing, content and method of delivery of messages regarding safety concerns of herbal medicines, mirror those for conventional medicines (such as: availability at the point of need; primary information on quality, indication, dose, adverse effects, contraindications, and so on; methods of delivery tailored to the habits and preferences of multi-segment audiences). Such messages '...should deliver relevant, clear, accurate and consistent messages and reach the right audiences at the right time' [28].

There are many excellent guidelines on these kinds of communications, including those produced for conventional medicines by the European Medicines Agency [31]. However, guidelines alone are insufficient for generating good communications: there must be a belief that communications matter, a recognition that they are difficult, as well as the talent, creativity, resources and vision to create and deliver them and follow them up [32, 33].

There are quite different variables to be considered in relation to herbal medicines, not least the relatively low level of regulatory control and the extent to which herbal medicines are perceived as safe and frequently taken without any expert diagnosis, prescription or advice. This relates to the question of ownership and self-determination raised earlier. The nature and risks of herbals must be part of general public health awareness and education; at the same time, individual users must have complete and specific point-of-sale information available to them on the products they may be browsing in a shop or market or on the internet. They must have the evidence available, even though we recognize that they may ignore it on rational or irrational grounds. (We must remember that public health communication of all kinds, for example, with regard to smoking, or obesity, faces immense obstacles, and may take decades to have significant effects).

There are three problems specific to herbal medicines that make communication difficult [34]: great difficulty in assessing the causality of any suspected adverse effect; commentary often confined to theory or unvalidated anecdotes; poor estimates of the probability of harm and of its likely seriousness and intensity. There is the additional problem that healthcare professionals are unlikely to know which of their patients are using herbals. It is obviously of great importance that prescribers of conventional medicines should always ask the specific question ('Are you taking any other medicines, herbal products, dietary supplements, vitamins or anything else?'), aware that herbal products may not always be categorized or perceived as medicines, or their consumption seen as relevant to standard healthcare. Health professionals must know where to find authoritative information about the risks of serious drug interactions once they learn their patient is taking herbal medicines and where there is a need to prescribe conventional medicine. This issue is being addressed, where herbal and traditional medicines are both included in medical training (South Africa, for example [35]). It may also reduce the risk of an herbal medicine being used in circumstances of particular risk, such as renal insufficiency.

Health professionals have a great responsibility for the detection and reporting of harm. This requires (a) a simple, accessible reporting system and, in the case of herbal medicines, (b) accurate identification of the suspected products and substances. This is an issue very far from resolution; it may, in some respects, be quite beyond ultimate resolution, but first steps need to be taken and there is much that can be done, even if it falls short of the ideal we might hope possible for modern medicines. Herbal and traditional practitioners must also be engaged with the adverse reaction reporting system.

In brief, there are many excellent guidelines for effective communications, but the time, energy, resources and skills to fulfil them on the ground are often insufficient. Public health awareness of the nature and risks of herbals and traditional medicines must develop alongside best practice point-of-sale and point-of-use labelling. Health professionals and herbal and traditional practitioners need good reporting systems and good sources of information for the identification of substances suspected of causing harm.

15.8 Product Labelling and Safety Information

There have been giant steps in the practice of food labelling in many Western countries, in Asia and other places, where completeness and transparency of information have been popularly demanded. All sources of herbals safety information (trials, other research, pharmacovigilance) should contribute to the information carried by herbal products. Comprehensive, product-specific information should be readily available to prescribers and retailers, as well as to consumers and patients. Unlike conventional medicines, herbal products often lack the summary of product characteristics (SmPC) that specifies the information essential for safe use. For traditional herbal medicinal products authorized in the European Union, European Union (EU) herbal monographs (formerly known as Community herbal monographs) are available and contain the scientific opinion of the Committee on Herbal Medicinal Products (HMPC) on safety and efficacy data about a herbal substance and its preparations intended for medicinal use in an SmPC format [36].

Alongside indication and putative benefits, labelling and prescribing information for herbal products should list the potential harmful effects of each ingredient together with the following information about each potential adverse effect: relation to dose; time course; factors altering an individual's susceptibility; seriousness; remedial action; and probability of occurrence, at least in the population, preferably in the individual.

There are several reasons that, in practice, this information is rarely available. First, most studies on phytotherapy products and herbal medicines focus on benefits and do not, or cannot, adequately explore harms; larger studies are required to provide useful data. In addition, there is often no regulatory or other legal requirement to list all the ingredients of every (multi-ingredient) herbal preparation on the product label; even listed ingredients may not be adequately described with the accepted name of the plant species, botanical authority, plant part, and type of extract. In some cases, only selected ingredients are listed on labels, omitting others that could be harmful.

In some countries there is no requirement to state the quantity of active ingredients contained in herbal preparations precisely [37]. Also, there are problems with the use of common names of herbal medicines, which vary from country to country and within regions of the same country. Latin names are rarely mentioned in labelling; for a similar common name, the active principle may be different and, in such cases, comprehensive causality assessment relating to individual herbal ingredients is impossible.

There are other, more traditional, means for sharing information about herbal safety: adverse reaction newsletters, information bulletins and specific risk messages. These are issued by many regulators though we know relatively little about their influence and impact. Newsletters and bulletins may be specific to herbals or may cover both herbals and conventional medicines. In any case, the open, transparent, timely and efficient knowledge-transfer of safety information is critical to inform all stakeholders, locally and internationally. Information may also be shared via articles in professional journals, conferences, courses, mass media, social media platforms, targeted messages for consumers and through the websites of regulators or pharmacovigilance centres [38]. Direct communication with consumers is also important, given high levels of self-medication and the likelihood that healthcare professionals may not know that their patients are using herbals. Such communication may include alerts and warnings on the websites of government agencies (which are often picked up by the mass media), information articles issued by regulators and information provided at points-of-sale. We should emphasize again, however, that data alone may not influence risk perception driven by belief, especially from sources that are perceived as biased or untrustworthy. It is hard to overstate the importance of the expert knowledge and advice of pharmacists or retailers, and their empathetic engagement at the point of sale.

Recommendations have been made on communication in health products safety and what targeted, clear messages mean in practice:

- Messages should address, in clear terms, the information needs of the various audiences, in an appropriate language and by suitable methods and, in particular, fill the respective information gaps.

- Messages should describe the action desired as an outcome from the communication, where that action has been agreed between the communicating parties on the basis of shared understanding and purpose.
- Messages prepared for targeted populations should easily enable further individualization for one-to-one communications, taking into account the literacy level, psychology and social environment of the given patient [39].

All of these admirable principles require substantial, sensitive information about the needs, preferences, values, expectations and behaviour of target audiences, and processes for continuous research, feedback and impact assessment. They also require an understanding that information itself is not necessarily a guarantee of influence between parties who have contrasting values and biases. Measures of uncertainty must be acknowledged and unrealistic claims avoided.

Pharmacovigilance and phytovigilance have a major part to play in the discovery and provision of safety information about herbal medicines. Herbal medicines should be labelled with comprehensive and accurate information about their constituents, benefits and harms. Further, a range of methods should be used to share information and promote safety to all audiences, and messages should be explicit about their purposes and expected outcomes. Such messages must be respectful and sophisticated, taking account of maybe wide differences in perception of science and risk.

15.9 Phytovigilance in the Age of Covid-19 (SARS-CoV-2)

Herbal medicines are the oldest form of healthcare known to mankind and have always been used in the treatment of viral diseases, especially in times of pandemic such as Covid-19 (SARS-CoV-2).

In general, the benefit, when it exists, of using herbal medicines in viral respiratory infections, is thought to be associated with stimulation of the immune system and/or anti-inflammatory effects [40]. Review of current literature on herbal medicines and Covid-19, reveals that no high-quality clinical trials have been published. Furthermore, all published articles on this topic mention that good evidence is still required through controlled clinical trials to examine safety and efficacy. This deficiency is also underlined by Yichang Yang [41]. Indeed, when we apply herbal medicines to a novel disease like Covid-19, especially in combination with antivirals or steroids, comprehensive research into safety is urgently needed.

When populations are overwhelmed by a disease and official and medical responses are uncertain and problematic, there are at least two significant public responses: doubt about the competence and openness of the authorities; a search for explanations and remedies outside of science and official frameworks. Covid has revealed a powerful sub-culture of suspicion and scepticism about everything from the existence of the virus itself to the safety and integrity of vaccines designed to protect against Covid-19 infection. There has been an outpouring of unscientific recommendations, experimentation and speculation. In such a context of anxiety and turmoil, the challenges of risk communication are all but insurmountable. However, the basic principles of clarity, consistency, transparency, empathy, targeting and ingenuity hold good and represent the only hopeful path forward.

15.10 Conclusion

Herbal and traditional remedies play a large part in the management of the health of billions of people. Those billions embrace populations of enormous variety in every dimension of social, economic, cultural and historical characteristics. About the safety and efficacy of the thousands of products available, and about their adverse effects, we know relatively little. There is accurate and useful information about only a handful of products, resulting in consumers and patients using products of uncertain benefit and unknown risk. Systems for collecting information about the adverse effects of herbal and traditional products are immature and insufficient.

Major reform and innovation are required in:

- The operation of pharmacovigilance and phytovigilance
- Research into the safety of herbal and traditional remedies
- Labelling and point-of-sale and point-of-use information
- Education and risk communication for multiple, complex audiences
- The time, energy, resources and creativity for risk communication in general

References

1. Fugh-Berman A, Ernst E (2001) Herb-drug interactions: review and assessment of report reliability. Br J Clin Pharmacol 52(5):587–595. https://doi.org/10.1046/j.0306-5251.2001.01469.x
2. De Smet PA (2002) Herbal remedies. N Engl J Med 347(25):2046–2056. https://doi.org/10.1056/NEJMra020398
3. Mintel International Group Ltd (2003) Complementary medicines, UK, Mintel International Group Limited, London
4. Eisenberg DM, Davis RB, Ettner SL et al (1998) Trends in alternative medicine use in the United States, 1990–1997: results of a national follow-up survey. JAMA 280(18):1569–1975. https://doi.org/10.1001/jama.280.18.1569
5. MacLennan AH, Wilson DH, Taylor AW (2002) The escalating cost and prevalence of alternative medicine. Prev Med 35(2):166–173. https://doi.org/10.1006/pmed.2002.1057
6. Barnes J, Abbot NC, Harkness EF, Ernst E (1999) Articles on complementary medicine in the mainstream medical literature: an investigation of Medline, 1966 through 1996. Arch Intern Med 159(15):1721–1725. https://doi.org/10.1001/archinte.159.15.1721
7. Fisher P, Ward A (1994) Complementary medicine in Europe. BMJ 309(6947):107–111. https://doi.org/10.1136/bmj.309.6947.107
8. WHO (2013) Traditional medicines strategy 2014–2023, Geneva. http://www.who.int/medicines/publications/traditional/trm_strategy14_23/en/

9. Skalli S, Zaid A, Soulaymani R (2007) Drug interactions with herbal medicines. Ther Drug Monit 29(6):679–686. https://doi.org/10.1097/FTD.0b013e31815c17f6
10. Spiegelhalter D (2020) The art of statistics: learning from data. Pelican Books
11. Waller PC, Evans SJW, Beard K (2005) Drug safety and the media. Br J Clin Pharmacol 61(2):123–126. https://doi.org/10.1111/j.1365-2125.2005.02528.x
12. Bucchi M (2017) Credibility, expertise and the challenges of science communication 2.0. Public Underst Sci 26(8):890–893. https://doi.org/10.1177/0963662517733368
13. The Academy of Medical Sciences. How can we all best use scientific evidence? http://acmedsci.ac.uk/policy/how-can-we-all-best-use-evidence
14. de Figueiredo A, Simas C, Karafillakis E, Paterson P, Heidi J, Larson HJ (2020) Mapping global trends in vaccine confidence and investigating barriers to vaccine uptake: a large-scale retrospective temporal modelling study. Lancet 396(10255):898–908. https://doi.org/10.1016/S0140-6736(20)31558-0
15. Wennburg J (2011) Time to tackle unwarranted variations in practice. BMJ 342:d1513. https://doi.org/10.1136/bmj.d1513
16. Real M, Barnhill MS, Higley C, Rosenberg J, Lewis JH (2019) Drug- induced liver injury: highlights of the recent literature. Drug Saf 42(3):365–387. https://doi.org/10.1007/s40264-018-0743-2
17. van Hunsel F, van de Koppel S, Skalli S et al (2019) Analysis of hepatobiliary disorder reports associated with the use of herbal medicines in the global suspected ADR database vigibase. Front Pharmacol 10:1326. https://doi.org/10.3389/fphar.2019.01326
18. James PB, Wardle J, Steel A, Adams J (2018) Traditional, complementary and alternative medicine use in sub-Saharan Africa: a systematic review. BMJ Glob Health 3(5):e000895. https://doi.org/10.1136/bmjgh-2018-000895
19. Thompson CA (2014) 'Dear healthcare professional' letters may not be effective REMS communication tool. Am J Health-Syst Pharm 71(3):177–178. https://doi.org/10.2146/news140012
20. Dusetzina SB, Higashi AS, Dorsey ER et al (2012) Impact of FDA drug risk communications on health care utilization and health behaviors: a systematic review. Med Care 50(6):466–478. https://doi.org/10.1097/MLR.0b013e318245a160
21. WEB-RADR: Recognising adverse drug reactions. https://web-radr.eu/
22. WHO (1999) WHO monographs on selected medicinal plants—volume 1 (introduction). World Health Organization, Geneva. http://apps.who.int/medicinedocs/en/d/Js2200e/2.html
23. European Medicines Agency (2018) Guidelines on good pharmacovigilance practices (EMA/722239/2018). https://www.ema.europa.eu/en/documents/regulatory-procedural-guideline/guidelines-good-pharmacovigilance-practices-gvp-introductory-cover-note-last-updated-chapter-piv_en.pdf
24. Berry DC, Knapp PR, Raynor DK (2002) Is 15% very common? Informing people about the risks of medication side effects. Int J Pharm Pract 10(3):145–151. https://doi.org/10.1111/j.2042-7174.2002.tb00602.x
25. Hakkarainen KM, Hedna K, Petzold M, Hägg S (2012) Percentage of patients with preventable adverse drug reactions and preventability of adverse drug reactions: a meta-analysis. PLoS One 7(3):e33236. https://doi.org/10.1371/journal.pone.0033236
26. Brown EG, Goldman SA (2004) The future is now: current concepts in risk management [topical edition]. Drug Saf 27:489–617
27. Kesselheim AS, Franklin JM, Avorn J, Duke JD (2013) Speaking the same language? International variations in the safety information accompanying top-selling prescription drugs. BMJ Qual Saf 22(9):727–734. https://doi.org/10.1136/bmjqs-2012-001704
28. CIOMS The Council for International Organizations of Medical Sciences. www.cioms.ch
29. Skalli S, Jordan SA (2017) Herbal and traditional medicines, now and future. In: Edwards IR, Lindquist M (eds) Pharmacovigilance critique and ways forward. Springer International Publishing, Switzerland, pp 145–159
30. Firenzuoli F, Gori L, Neri D (2005) Fitoterapiaclinica: opportunita' e problematiche. Ann Ist Super Sanità 41(1):27–33

31. European Medicines Agency (2013) Guideline on good pharmacovigilance practices—Module XV—Safety communication (EMA/118465/2012). http://academy.gmp-compliance.org/guidemgr/files/GVP%20MODULE%20XV%20%E2%80%93%20SAFETY%20COMMUNICATION.PDF
32. WHO (2017) Strategic communications framework for effective communications. https://www.who.int/mediacentre/communication-framework.pdf
33. CDC (2015) Health communication strategies and resources. https://npin.cdc.gov/pages/health-communication-strategies
34. Ferner RE, Aronson JK (2006) Communicating information about drug safety. BMJ 333(7559):143–145. https://doi.org/10.1136/bmj.333.7559.143
35. Chitindingu E, George G, Gow J (2014) A review of the integration of traditional, complementary and alternative medicine into the curriculum of south African medical schools. BMC Med Educ 14:40. https://doi.org/10.1186/1472-6920-14-40
36. European Medicines Agency. European Union monographs and list entries. https://www.ema.europa.eu/en/human-regulatory/herbal-products/european-union-monographs-list-entries
37. Skalli S, Soulaymani Bencheikh R (2012) Safety monitoring of herb-drug interactions: a component of pharmacovigilance. Drug Saf 35(10):785–791. https://doi.org/10.1007/BF03261975
38. Yamamoto M, Kubota K, Okazaki M et al (2015) Patients views and experiences in online reporting adverse drug reactions: findings of a national pilot study in Japan. Patient Prefer Adher 9:173–184. https://doi.org/10.2147/PPA.S75761
39. Bahri P, Mol P, Théophile H, Edwards IR, Hugman BPJ (2010) Communication in drug safety: a report from an interactive debate held at the 10th annual meeting of the international society of pharmacovigilance (ISoP). Drug Saf 34(10):881–882. https://doi.org/10.2165/11594430-000000000-00000
40. Yarnell E (2018) Herbs for viral respiratory infections. Altern Complement Ther 24(1):35–43
41. Yang Y (2020) Use of herbal drugs to treat COVID-19 should be with caution. Lancet 395(10238):1689–1690. https://doi.org/10.1016/S0140-6736(20)31143-0

Part II
International Perspectives in Pharmacovigilance for Herbal Medicines

Chapter 16
Reports for Herbal Medicines in the Global Suspected ADR Database VigiBase

Florence van Hunsel, Lucie M. Gattepaille, Camilla Westerberg, and Joanne Barnes

16.1 Background

VigiBase is the World Health Organization (WHO) global database of individual case safety reports (ICSRs), reports of suspected adverse drug reactions (ADRs) of medicines. VigiBase is maintained by the Uppsala Monitoring Centre (UMC), an independent centre for drug safety and scientific research. By June 2020, VigiBase contained over 21 million de-duplicated [1] ICSRs, submitted since 1968 by member countries of the WHO Programme for International Drug Monitoring (PIDM). Initially, the WHO PIDM members comprised ten countries. Currently (2020), 140 countries are members of the programme, and another 31 associate member countries are in the early stages of establishing their pharmacovigilance systems in preparation for full membership [2]. This chapter provides a broad, descriptive overview of the reports involving herbal medicines (HM) present in VigiBase.

VigiBase contains ICSRs that were reported, collected, coded and assessed locally, i.e. in their country of origin. Healthcare professionals and, in most countries, patients/consumers, and pharmaceutical companies, report suspected ADRs to national pharmacovigilance centres. These ICSRs are reviewed and analysed locally and may lead to regulatory action in that country if there are grounds for concern.

F. van Hunsel (✉)
Netherlands Pharmacovigilance Centre Lareb, 's-Hertogenbosch, The Netherlands
e-mail: f.vanhunsel@lareb.nl

L. M. Gattepaille · C. Westerberg
Uppsala Monitoring Centre, Uppsala, Sweden
e-mail: Lucie.Gattepaille@who-umc.org; Camilla.Westerberg@who-umc.org

J. Barnes
School of Pharmacy, Faculty of Medical and Health Sciences, University of Auckland, Auckland, New Zealand
e-mail: j.barnes@auckland.ac.nz

J. Barnes (ed.), *Pharmacovigilance for Herbal and Traditional Medicines*,
https://doi.org/10.1007/978-3-031-07275-8_16

The reports are then sent to VigiBase by member countries of the WHO PIDM either directly or, for European countries in recent years, indirectly from the European database Eudravigilance®. All EU countries share data via Eudravigilance, which also uses the VigiBase application programming interface (API) for transfer, i.e. sending E2B (R3) files directly from any national database.

National competent authorities of the WHO PIDM member countries in other regions can send ICSRs to UMC as e-mail attachments (preferably as encrypted files) [3]. UMC also provides VigiFlow, which national pharmacovigilance centres can elect to use for sending reports to VigiBase. VigiFlow is an ICSRs management system that is widely used by WHO PIDM member countries. It enables them to manage their national pharmacovigilance data and supports the processes of collection, collation and sharing of ICSRs to enable data analysis in a global context [4]. NotiFACEDRA is another regional database, similar to Eudravigilance (i.e. also using the VigiBase API), which is used by many of the WHO PIDM member countries in Central America.

Member countries are expected to share their PV data on a regular basis—at least quarterly and, ideally, more than once per month—to keep VigiBase as up-to-date as possible [3]. Data transfer from the European Eudravigilance® database to VigiBase occurs electronically on a daily basis [5].

Most national centres review case reports before they are sent to VigiBase. However, it is important to note that the information in VigiBase comes from a variety of sources, and the likelihood that the suspected ADR is drug-related is not the same in all cases [6]. Once sent to VigiBase, ICSRs can be viewed in VigiLyze, a web-based advanced analytics tool developed by UMC that can be used to browse VigiBase data and support signal detection; in VigiLyze, there is the option to choose between a national focus or a global focus for analysis. VigiLyze is available to the national competent authorities of the WHO PIDM member countries, including affiliated independent pharmacovigilance centres, and, on request, to other stakeholders [7]. All suspected ADRs in VigiBase are coded using MedDRA® (Medical Dictionary for Regulatory Activities) terminology, which is the international medical terminology developed under the auspices of the International Council for Harmonisation of Technical Requirements for Pharmaceuticals for Human Use (ICH). By coding data using this standardized terminology, analyses can be performed on individual medical events (e.g. 'influenza') or issues involving a system, organ or aetiology (e.g. infections) using its hierarchical structure. There are five levels to the MedDRA hierarchy, arranged from very specific to very general. The most specific level is 'lowest level terms' (LLTs). Each member of the next level, 'preferred terms' (PTs), is a distinct descriptor (single medical concept) for a symptom, sign, disease diagnosis, therapeutic indication, investigation, surgical or medical procedure, and medical social or family history characteristic. Related PTs are grouped together into 'high level terms' (HLTs) based upon anatomy, pathology, physiology, aetiology or function. HLTs, related to each other by anatomy, pathology, physiology, aetiology or function, are in turn linked to 'high level group terms' (HLGTs). Finally, HLGTs are grouped into 'system organ classes' (SOCs), which are groupings by aetiology (e.g. infections and infestations), manifestation site (e.g. gastrointestinal disorders) or purpose (e.g. surgical and medical procedures). In addition, there is an SOC containing issues pertaining to products and one containing issues relating to social circumstances [8].

Correspondingly, all medicinal products listed in the ICSRs are coded using the WHODrug dictionary [9]. Developed and maintained by UMC, WHODrug is a global medicinal information dictionary that contains medicinal products and active ingredients intended for human use, such as active chemical substances, biotherapeutics, vaccines, dietary supplements, herbal medicines, radiopharmaceuticals and diagnostic agents. The information on the medications in WHODrug includes proprietary (trade/brand) name, ingredient(s), pharmaceutical form, strength, country of sales and marketing authorisation holder. Some of this is not applicable where the item is solely an ingredient, e.g. an herbal ingredient or herbal substance [9].

In general, herbal ingredients/substances are properly described using their scientific binomial with botanical authority, plant family and plant part (e.g. *Ginkgo biloba* L. (Ginkgoaceae) leaf). Reports of ADRs involving herbal ingredients/substances may describe the products/preparations concerned using proprietary names, botanical names, synonyms, genus name only and/or vernacular (common) names. More on botanical names and their importance is provided in Chaps. 8 and 9. Also, there are many different ways in which different national pharmacovigilance centres code ICSRs in their respective databases after receipt of reports. For example, a national pharmacovigilance centre may choose to record only the proprietary name of a (local) product, without including its specific ingredients, or, to give another example, record only the common name of the herbal ingredient/substance involved (e.g. 'ginseng'). These items may be recorded in local language, although centres might also use English to summarize the reports and ingredients. The UMC receives data on herbal products/preparations with different levels of coding, for instance, ingredient names and proprietary names. More on coding of herbal medicines in pharmacovigilance databases is provided in Chap. 9.

In VigiBase, all substances must be assigned a 10-digit ID number, besides a serial substance ID. Where there is an official CAS (chemical abstracts service) number [10] for a substance, the 10-digit ID is the CAS number. Non-plant/allopathic substances, including ingredients of animal origin, e.g. deer velvet, and substances such as vitamins/minerals, without an official CAS number have an autogenerated 10-digit ID starting with '8'; plant substance ID numbers start with '9'.

To perform the descriptive analysis presented in this chapter, a data extraction was performed by UMC. Substances were retrieved from the WHODrug database on 2020-05-03, filtered in Excel software based on their 10-digit ID number (8 or 9) and curated by one author (JB). Substances with an ID starting with '9' that were not herbal were manually identified and excluded. Based on the characterization of the substances, products were classified as follows: one herbal substance and nothing else = single; more than one herbal substance and nothing else = multi herbal; at least one herbal substance with non-herbal = mixed; no herbal substance at all = non-herbal. Each reported herbal medicine can be described as a 'suspected', 'interacting' or 'concomitant' drug; for the descriptive analysis presented here, only reports with 'suspected' or 'interacting' herbal medicines are included. Reports dated between '1968-01-01' and '2019-12-31' were extracted from a frozen version of VigiBase, dated May 3rd, 2020.

16.2 Reports Involving Herbal Medicines

Figure 16.1 shows the total number of ICSRs in VigiBase and in subsets of ICSRs relating to herbal medicines, including the subset of reports with a sole, suspected or interacting, single-ingredient herbal medicine for which the herbal ingredient is

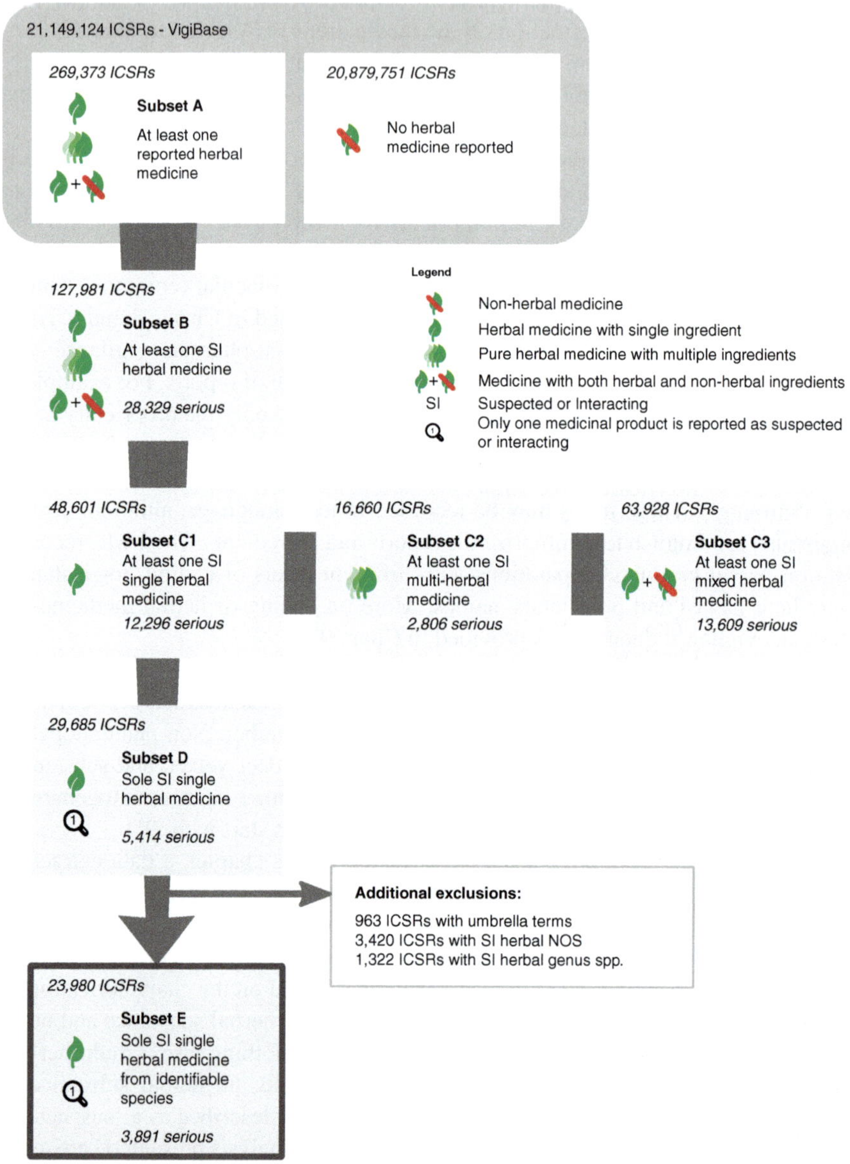

Fig. 16.1 Flowchart indicating the total number of ICSRs in VigiBase, numbers of reports involving herbal ingredients listed as suspected or interacting agents for various subsets, and number of serious reports for each subset

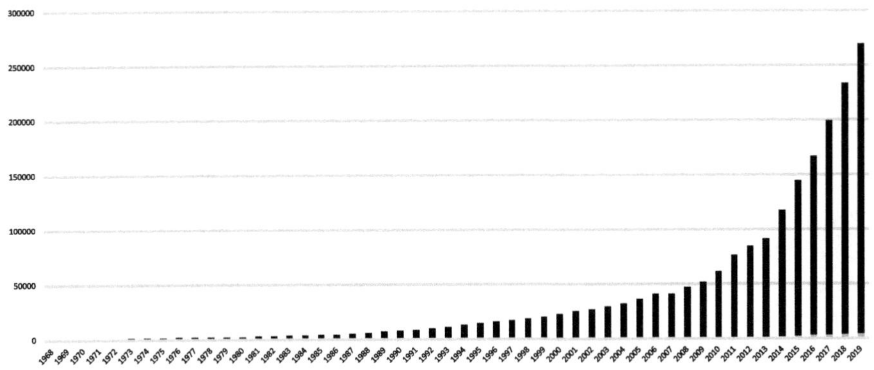

Fig. 16.2 Cumulative number of ICSRs held in VigiBase with at least one herbal substance, ingredient, or product/preparation (including sole-herbal-ingredient products, multi-ingredient herbal products, and multiple-mixed-ingredient products/preparations) listed as suspected, interacting or concomitant agents

identifiable at the species level (subset E). 'Sole' in the sentence above meaning that the specified herbal medicine was the *only* suspect medicine in the report.

From the set of substances captured with the 10-digit ID starting with '9', substances that were not herbal were manually identified by one author (JB) and excluded. Of the 3283 'herbal' substances, 3067 (93.4%) were classified as being from plants or plant parts (including resin); 216 substances related to animal species, including reptiles, and insects, or were fungi, algae or other substances. Thus, suspected or interacting products remaining for further description in this chapter are purely plant-based.

Figure 16.2 shows the cumulative number of reports in VigiBase per year since 1968 (subset A in Fig. 16.1) that have at least one herbal ingredient/substance listed (pure or mixed) regardless of characterization (suspected, interacting or concomitant). These products can contain a single herbal ingredient, or a 'mixed' product/preparation containing one or more herbal ingredients. There has been a substantial increase in the number of reports received by WHO-UMC since the early 2000s, alongside an increase in the number of countries contributing reports to the WHO-UMC. The total number of reports with at least one herbal product/preparation listed as the suspected, interacting or concomitant agent (subset A, 269,373 ICSRs in total) is shown in Table 16.1 by WHO region for member countries contributing reports. This is approximately 1% of the total number of reports in VigiBase.

In Fig. 16.1, subset B represents the subset of reports (n = 127,981) for which at least one herbal product was reported as the suspected or interacting agent. This herbal product could contain multiple herbal ingredients. The reports were submitted by physicians (n = 33,272), consumers/non-health professionals (n = 30,356), pharmacists (n = 22,537), other health professionals (n = 10,756) and lawyers (n = 308); one report (i.e. relating to the same incident) can be submitted by multiple reporters. The type of reporter is unknown for 43,476 reports.

Table 16.1 Total number of ICSRs in VigiBase and number of reports listing products/preparations containing at least one herbal ingredient/substance regardless of classification (suspected, interacting or concomitant) in VigiBase for the period 1968 to 31 Dec 2019, by WHO region for WHO Programme for International Drug Monitoring member countries contributing reports, ordered for highest proportion of herbal reports

Region	Number of countries contributing reports (2019)	Total number of reports	Number of herbal reports (% of total number of reports)
Eastern Mediterranean	13	137,936	4043 (2.9)
Western Pacific Region	14	3,961,758	95,243 (2.4)
European Region	47	5,055,487	82,474 (1.6)
Region of the Americas	21	11,086,963	82,565 (0.7)
South-East Asia Region	8	783,404	4654 (0.6)
African Region	33	123,576	394 (0.3)

The three subsets C1, C2 and C3 are not mutually exclusive, as one ICSR may list products from one or more of these categories of products/preparations. In these reports patients may or may not have been taking conventional medicines as a suspect drug in addition to a herbal product. The subset D contains reports with a single herbal ingredient being the sole suspected agent. Subset E is the main subset of interest for this analysis as it contains reports involving a sole-suspected/interacting single herbal ingredient and where the species name is available (n = 23,980). Characteristics of ICSRs in subset E are summarized in Table 16.2.

Data quality for ICSRs in subset E was scored using vigiGrade [11], a multidimensional measure of the quantity of information on reports, which gives a score for the completeness of ICSRs. The maximum score is 1.0. In this subset, 4707 ICSRs (19.6%) had a vigiGrade score ≥0.8, meaning that the reports were well documented, i.e. the data fields had a high degree of completion. The remaining 19,273 reports (80.4%) had a vigiGrade score <0.8.

For subset E, reports involving sole-suspected/interacting single herbal ingredients and where the species name is available (n = 23,980), the ten most frequently reported herbal substances—representing 45% of the herbal substances named in these reports—are listed in Table 16.3.

The ten most frequently reported adverse events by MedDRA PT (preferred term) level for subset E, stratified for reports with non-serious ADRs and serious ADRs, are listed in Tables 16.4 and 16.5. The seriousness of the cases is based on international criteria [12].

16.3 Considerations

This chapter gives a broad descriptive overview of ICSRs involving herbal medicines available in VigiBase. From the full database involving all reports available, this descriptive analysis navigated through a series of filters to focus on a tightly

Table 16.2 Characteristics of reports involving sole-suspected/interacting single herbal ingredients where the species name was available (subset E; $n = 23{,}980$)

Characteristic		*N* reports
Reporter type[a]	Physician	7123
	Consumer/non-health professional	7882
	Pharmacist	5394
	Other health professional	1502
	Lawyer	1
	Not stated	3846
Patient age group[b]	0–27 days	101
	28 days–23 months	333
	2–11 years	1011
	12–17 years	622
	18–44 years	5102
	45–64 years	5736
	65–74 years	2365
	≥75 years	1896
	Not stated	6814
Patient sex	Female	15,445
	Male	7165
	Not applicable	2
	Not stated	1368
Completeness score	vigiGrade ≥0.8	4707
	vigiGrade <0.8	19,273

[a] The total is greater than 23,980 as one report (i.e. relating to the same incident) can be submitted by multiple reporters

[b] Reports were not individually checked to see if the reported age was correct, for instance, in the neonate group

Table 16.3 The ten most frequently reported herbal substances in subset E (reports involving sole-suspected/interacting single herbal ingredients and where the species name is available ($n = 23{,}980$))

Reported herbal (scientific name)	Examples of reported herbal (common name(s))	Number of reports
Ginkgo biloba	Ginkgo, maidenhair tree	2198
Phleum pratense	Timothy grass	1574
Hypericum perforatum	St. John's wort	1311
Actaea racemosa (synonym: *Cimicifuga racemosa*)	Black cohosh, black bugbane, black snakeroot	1125
Vitis vinifera	Grapevine, grape	1055
Vitex agnus-castus	Vitex, chaste tree, chasteberry, Abraham's balm, monk's pepper	982
Plantago ovata	Blond plantain, desert Indian wheat, blond psyllium, ispaghula	894

(continued)

Table 16.3 (continued)

Reported herbal (scientific name)	Examples of reported herbal (common name(s))	Number of reports
Silybum marianum	Milk thistle, blessed milk thistle, Marian thistle, Mary thistle, Saint Mary's thistle	676
Viscum album	Common mistletoe, European mistletoe	603
Hedera helix	Common ivy, English ivy	515
Total		10,933

Table 16.4 The ten most frequently reported adverse events on MedDRA preferred term level for non-serious reports in subset E (reports involving sole-suspected/interacting single herbal ingredients and where the species name is available (n = 20,089))

Reported adverse events (MedDRA PT)	Number of reports
Nausea	1550
Pruritus	1447
Rash	1241
Diarrhoea	1223
Headache	1094
Dizziness	1023
Vomiting	986
Abdominal pain	929
Urticaria	815
Abdominal discomfort	770
Total	11,078

Table 16.5 The ten most frequently reported adverse events on MedDRA preferred term level for serious reports in subset E (reports involving sole-suspected/interacting single herbal ingredients and where the species name is available (n = 3891))

Reported adverse events (MedDRA PT)	Number of reports
Dyspnoea	276
Nausea	241
Pruritus	218
Vomiting	209
Dizziness	168
Diarrhoea	162
Hypersensitivity	162
Rash	143
Anaphylactic reaction	142
Headache	128
Total	1849

defined subset of reports involving sole-suspected/interacting single herbal ingredients for which the species name was provided. When investigating a signal between a herbal substance(s) and a reported adverse event(s), these latter reports are particularly useful since the assessment of causality is more straightforward if only one suspected herbal substance is implicated [13]. However, cases with multiple suspected herbal substances, or mixed products containing herbal and other ingredients, should also be considered in signal analysis. It could also be that the reported cases only contain reports with multiple herbal substances or mixed products. Furthermore, this broad description of the VigiBase reports involving herbal substances does not consider which plant parts were implicated, reports relating to different manufacturers' products containing the same herbal ingredients, different extraction methods and administration route of the reported herbals (or, indeed, whether these items of information were provided in the reports). In signal analysis for herbal medicines, it is important to consider these issues.

The analysis for this chapter was limited to herbal substances only, thus, other natural health products of non-herbal origin, which are also widely used, such as 'fish oils', glucosamine, and chondroitin, are not represented here. Also, cannabis-containing products are present in subset E of the database, but some cannabis-containing products could be missing from this analysis due to international differences in coding products/preparations containing different types of cannabis and/or cannabis chemical compounds. More on this topic is provided in Chap. 20, which discusses pharmacovigilance for cannabis products in Canada.

An estimate of the frequency with which ADRs occur in association with use of herbal medicines is not possible based on analyses of spontaneous reporting data. It is likely that only a small proportion of ADRs experienced in association with use of herbal medicines is represented in VigiBase [14].

The 'V90' code in VigiBase is used for 'unspecified herbal and traditional medicine' and, thus, includes substances that are not of plant origin; these were filtered manually for this analysis since this chapter is focused on herbal medicinal products. In addition, not all non-plant traditional medicines are included in the V90 ATC code, so this code does not comprehensively capture all herbal and traditional medicines and substances. Even within the subset of products containing only herbal substances, further issues with coding arise: many reports were coded with an 'umbrella term' for non-otherwise-specified (NOS) products (e.g. 'herbal drug NOS'), or specified only the plant genus (e.g. *Echinacea* spp.) without the species epithet. This lack of specification is problematic when investigating signals of suspected ADRs associated with herbal medicines as the reports could relate to several different species (with different chemical profiles) belonging to the same genus [15]. Also, some herbal substances might be used as ingredients in conventional medicines and, therefore, will have a regular ATC code. For instance, an extract of *Phleum pratense* (Timothy-grass) pollen is used as sublingual immunotherapy for seasonal allergic rhinitis ('hayfever') [16], and *Plantago ovata* (psyllium, ispaghula) is an ingredient used in some laxative products that may be considered by users to be conventional medicines [17] even though the ingredient(s) is an herbal substance by definition.

A similar descriptive analysis of ICSRs in VigiBase involving herbal medicines was undertaken around 20 years ago [18]. At that time (1999), there were around two million ICSRs in VigiBase, of which around 0.5% involved herbal medicines, with 55 countries contributing to the programme [12]. The ten most frequently reported sole-suspected single herbal substances reported up to 1997 have some similarities with the present analysis (Table 16.3): *Ginkgo biloba*, *Viscum album*,

Silybum marianum and *Plantago ovata* were among the most frequently reported herbal ingredients in the earlier analysis, along with 'total opium alkaloids', *Oenothera biennis* (evening primrose) oil, *Mentha* × *piperita* (peppermint) oil and 'senna'. The most frequently reported non-serious reactions (up to 1997) included pruritus, rash (including erythematous), urticaria, gastrointestinal ADRs, such as nausea, vomiting, diarrhoea, and headache. Serious reactions included anaphylactic reaction/shock and other terms related to anaphylaxis or allergy, hallucination and intestinal obstruction [12]. The current ten most frequently reported ADRs also include hypersensitivity reactions and gastrointestinal ADRs. There is close similarity between the ten serious and non-serious ADRs in the present analysis. This could be due to the fact that the same coding might be used for a mild reaction and those that require medical attention. It should be noted that for reports not submitted using the E2B-R3 format for report submission (mostly older reports), seriousness could only be applied to the whole report, not to each reaction.

This analysis has identified that there is relative under-representation of reporting from some regions, such as the African region, where there is a strong tradition in the use of herbal (and other traditional) medicines, as well as an ongoing reliance on these traditional medicines as the main source of healthcare for many people [19]. Some regions, such as the African region, have lower ADR reporting rates overall [20], which is explained in part by the fact that countries in those regions have a shorter history of having national pharmacovigilance systems and involvement in the WHO PIDM. Also, many countries in these regions face different challenges, including in achieving patient reporting, and reporting by other potential reporter groups, such as traditional healers [21–24]. There is likely much to gain from increasing the number of reports submitted by countries in the African and other under-represented regions: this could provide important insights into the safety of herbal medicines used locally in those regions. This could include enhancing knowledge on well-known herbal medicines that are used by different populations for different health reasons in those regions, as well as building knowledge on the safety profiles of herbal medicines that are endemic in those regions. The ten most frequently reported herbal substances overall reflects the patterns of use of herbal products in regions with higher reporting rates (i.e. typically western countries).

A higher proportion of ICSRs involving herbal medicines in VigiBase related to females (rather than males) who had experienced suspected ADRs. Use of orally administered herbal medicines may be more common among females than among males [25, 26], thus, it could be expected that this population is also overrepresented in VigiBase. A general analysis of ICSRs in VigiBase found that of the reports with information on sex, 60.1% concerned females and 39.9% males, across children and adults. More ADR reports involving females were submitted in all regions of the world and by all types of reporters [27]. A higher ADR reporting rate for and by females could be a factor contributing to such a pattern, although important underlying sex-related differences in the occurrence of ADRs are also likely [27, 28].

To conclude, the VigiBase database maintained by the Uppsala Monitoring Centre contains almost 128,000 reports listing at least one herbal ingredient as the suspected or interacting drug. These reports represent around 0.6% of the total number of reports in VigiBase. There are almost 24,000 reports listing a sole-suspected/interacting single herbal substance for which the herbal ingredient is identifiable at the species level. Even though the level of documentation in VigiBase is heterogeneous, the database can provide a rich source of information concerning pharmacovigilance for herbal medicines.

References

1. Norén GN, Orre R, Bate A, Edwards IR (2007) Duplicate detection in adverse drug reaction surveillance. Data Min Knowl Disc 14(3):305–328. https://doi.org/10.1007/s10618-006-0052-8
2. The Uppsala Monitoring Centre (2019) What is VigiBase? https://www.who-umc.org/vigibase/vigibase/. Accessed 7 Jul 2020
3. The Uppsala Monitoring Centre (2020) Sharing pharmacovigilance data in the WHO Programme for International Drug Monitoring 2020. https://www.who-umc.org/media/3236/reporting-fact-sheet.pdf. Accessed 7 Jul 2020
4. Vogler M, Ricci Conesa H, de Araujo Ferreira K et al (2020) Electronic reporting systems in pharmacovigilance: the implementation of VigiFlow in Brazil. Pharmaceut Med 34(5):327–334. https://doi.org/10.1007/s40290-020-00349-6
5. European Medicines Agency (EMA) (2020) Access to EudraVigilance data 2020. https://www.ema.europa.eu/en/human-regulatory/research-development/pharmacovigilance/eudravigilance/access-eudravigilance-data. Accessed 7 Jul 2020
6. Uppsala Monitoring Centre (2019) Caveat document - statement of reservations, limitations and conditions relating to data released from VigiBase, the WHO global database of individual case safety reports (ICSRs). https://www.who-umc.org/media/164610/umc_caveat.pdf. Accessed 7 Jul 2020
7. Lindquist M (2008) VigiBase, the WHO global ICSR database system: basic facts. Drug Inf J 42(5):409–419. https://doi.org/10.1177/009286150804200501
8. The International Conference on Harmonisation of technical requirements for registration of pharmaceuticals for human Use (2017) Welcome to MedDRA. https://www.meddra.org/. Accessed 7 Jul 2020
9. Lagerlund O, Strese S, Fladvad M, Lindquist M (2020) WHODrug: a global, validated and updated dictionary for medicinal information. Ther Innov Regul Sci 54(5):1116–1122. https://doi.org/10.1007/s43441-020-00130-6
10. CAS Registry - The gold standard for chemical substance information. https://www.cas.org/support/documentation/chemical-substances. Accessed 7 Jul 2020
11. Bergvall T, Noren GN, Lindquist M (2014) vigiGrade: a tool to identify well-documented individual case reports and highlight systematic data quality issues. Drug Saf 37(1):65–77. https://doi.org/10.1007/s40264-013-0131-x
12. CIOMS Working Group VIII (2010) Practical aspects of signal detection in pharmacovigilance: report of CIOMS Working Group VIII. CIOMS Working Group, Geneva
13. Barnes J (2003) Pharmacovigilance of herbal medicines: a UK perspective. Drug Saf 26(12):829–851. https://doi.org/10.2165/00002018-200326120-00001
14. Pokladnikova J, Meyboom RH, Meincke R, Niedrig D, Russmann S (2016) Allergy-like immediate reactions with herbal medicines: a retrospective study using data from VigiBase(R). Drug Saf 39(5):455–464. https://doi.org/10.1007/s40264-016-0401-5
15. van Hunsel F, van de Koppel S, Skalli S et al (2019) Analysis of hepatobiliary disorder reports associated with the use of herbal medicines in the global suspected ADR database Vigibase. Front Pharmacol 10:1326. https://doi.org/10.3389/fphar.2019.01326

16. Kay AB (2007) An extract of Timothy-grass pollen used as sublingual immunotherapy for summer hay fever. Drugs Today (Barc) 43(12):841–848. https://doi.org/10.1358/dot.2007.43.12.1162079
17. Alsalimy N, Madi L, Awaisu A (2018) Efficacy and safety of laxatives for chronic constipation in long-term care settings: a systematic review. J Clin Pharm Ther 43(5):595–605. https://doi.org/10.1111/jcpt.12721
18. Farah HM, Edwards R, Linquist M, Leon C, Shaw D (2000) International monitoring of adverse health effects associated with herbal medicines. Pharmacoepidemiol Drug Saf 9(2):105–112. https://doi.org/10.1002/(SICI)1099-1557(200003/04)9:2<105::AID-PDS486>3.0.CO;2-2
19. Skalli S, Bencheikh RS (2015) Pharmacovigilance of herbal medicines in Africa: questionnaire study. J Ethnopharmacol 171:99–108. https://doi.org/10.1016/j.jep.2015.05.033
20. Ampadu HH, Hoekman J, de Bruin ML et al (2016) Adverse drug reaction reporting in Africa and a comparison of individual case safety report characteristics between Africa and the rest of the world: analyses of spontaneous reports in VigiBase(R). Drug Saf 39(4):335–345. https://doi.org/10.1007/s40264-015-0387-4
21. Matos C, Harmark L, van Hunsel F (2016) Patient reporting of adverse drug reactions: an international survey of national competent authorities' views and needs. Drug Saf 39(11):1105–1116. https://doi.org/10.1007/s40264-016-0453-6
22. James PB, Wardle J, Steel A, Adams J (2018) Traditional, complementary and alternative medicine use in Sub-Saharan Africa: a systematic review. BMJ Glob Health 3(5):e000895. https://doi.org/10.1136/bmjgh-2018-000895
23. Dodoo ANO, Appiah-Danquah A (2006) Safety of herbal medicines: the practitioners' view. Drug Saf 29(4):350
24. Dodoo ANO, Appiah-Danquah A, Gyansa-Lutterodt M, Duwiejua M (2006) Safety monitoring of herbal medicines in Ghana: challenges and opportunities. Drug Saf 29(4):352
25. Eardley S, Bishop FL, Prescott P et al (2012) A systematic literature review of complementary and alternative medicine prevalence in EU. Forsch Komplementmed 19(Suppl 2):18–28. https://doi.org/10.1159/000342708
26. Kennedy J (2005) Herb and supplement use in the US adult population. Clin Ther 27(11):1847–1858. https://doi.org/10.1016/j.clinthera.2005.11.004
27. Watson S, Caster O, Rochon PA, den Ruijter H (2019) Reported adverse drug reactions in women and men: aggregated evidence from globally collected individual case reports during half a century. EClinicalMedicine 17:100188. https://doi.org/10.1016/j.eclinm.2019.10.001
28. de Vries ST, Denig P, Ekhart C et al (2019) Sex differences in adverse drug reactions reported to the National Pharmacovigilance Centre in the Netherlands: an explorative observational study. Br J Clin Pharmacol 85(7):1507–1515. https://doi.org/10.1111/bcp.13923

Chapter 17
The Italian Phytovigilance Spontaneous Reporting Scheme

Francesca Menniti-Ippolito and Fabio Firenzuoli

17.1 Introduction

The Phytovigilance system, coordinated by the Italian National Institute of Health, collects spontaneous reports of suspected adverse reactions associated with the use of food supplements, and compounded preparations containing plant ingredients. Adverse reactions associated with traditional or "well-established use" herbal medicines, registered in accordance with the European Union Traditional Herbal Medicinal Products Directive, are not included in this surveillance, and are collected within the usual pharmacovigilance system. The Phytovigilance system activities are conducted separately from the medicines pharmacovigilance system, which is coordinated by the Italian Medicines Agency.

The surveillance system was activated in 2002 as a research project and, in 2012, became a national system to support the Ministry of Health in monitoring the safety of products in its regulatory competence. Anybody observing or experiencing a suspected adverse reaction associated with the above-mentioned products (i.e. food supplements) can report the reaction. Online reporting has been possible through the website Vigierbe (www.vigierbe.it) since December 2018; all reports previously received on an ad hoc form by fax or mail are included in a unique database.

Following receipt of a report, the coordinating centre conducts the initial check activities. If information is incomplete, the reporter is contacted for filling, if available, the missing data. Adverse reactions are coded according to the Medical

F. Menniti-Ippolito (✉)
National Centre for Drug Research and Evaluation, National Institute of Health, Rome, Italy
e-mail: francesca.menniti@iss.it

F. Firenzuoli
CERFIT, Research and Innovation Center in Phytotherapy and Integrated Medicine, Careggi, University Hospital, Florence, Italy
e-mail: fabio.firenzuoli@unifi.it

J. Barnes (ed.), *Pharmacovigilance for Herbal and Traditional Medicines*,
https://doi.org/10.1007/978-3-031-07275-8_17

Dictionary for Regulatory Activities (MedDRA). Information on the content of food supplements is retrieved from the Ministry of Health archive, which contains all notified food supplements. To note, food supplements need to be notified before being marketed. For serious reactions, requiring hospitalization, clinical data and follow-up of the patients are retrieved. Causality assessment is performed using the World Health Organization standardized case causality assessment scale in use for drugs and adapted for food supplements and herbal containing preparations. A scientific committee is appointed for consultation, when necessary. The committee comprises experts in pharmacology, toxicology, pharmacognosy, phytotherapy and botany. When a case of suspected product contamination or adulteration arises, the implicated products are analysed in the laboratories of the Italian National Institute of Health. Specific safety surveillances, requested from the Ministry of Health, are performed on adverse reactions associated with magistral preparations of cannabis for medical use and on adverse reactions associated with magistral preparations for weight loss. Periodical reports of these surveillances are produced for the Ministry of Health.

Up to October 2020, 2315 reports were included in the database. Most (63%) related to women with a median age of 50 years (range: 1 month to 99 years). In 39% of the reports, concomitant drugs were indicated; 32% of the reports involved serious reactions (life-threatening events, hospitalization, death). Among 66% of the reports, one or more food supplements were present, magistral preparations were listed in 27% of the reports, and 7% related to other products, homemade decoctions of herbals, herbal "smart" drugs, such as *Argyreia nervosa* (Burm.f.) Bojer (Hawaiian baby woodrose), or *Salvia divinorum* Epling & Játiva (diving sage), and "energy drinks" containing caffeine, taurine, and other reputed performance-enhancing substances.

17.2 Safety Issues Identified Through the Phytovigilance System

Since the Phytovigilance system was established, many articles have been published from the data collected, documenting an intense evaluation activity of different safety issues that have emerged from the surveillance. Case reports have been published on rhabdomyolysis associated with *Commiphora mukul* (Hook. ex Stocks) Engl., a natural lipid-lowering agent [1], and allergic reactions to food supplements containing propolis, a resinous substance produced by honeybees from saliva, beeswax and botanical exudates, and which is reputed to have antibacterial, antiviral, antifungal, anti-inflammatory, antioxidant and chemopreventive activities [2]. Many reports in the Phytovigilance database related to hepatoxicity associated with ingredients from different plants, including: food supplements containing green tea extracts [3], greater celandine (*Chelidonium majus* L.) [4], *Actaea racemosa* L. (synonym: *Cimicifuga racemosa* (L.) Nutt.) [5], *Serenoa repens* (W.Bartram)

Small [6], self-prepared decoction of *Teucrium chamaedrys* L. [7], and *Garcinia gummi-gutta* (L.) Roxb. (synonym: *Garcinia cambogia* Desr.) [8]. Suspected interactions between food supplements containing plant ingredients and anticoagulant medicines were described in some reports [9]. Adverse reactions associated with food supplements or magistral preparations used for weight control [10, 11] and adverse reactions associated with herbal laxatives have also been documented [12]. Safety issues related to quality problems with food supplements were reported following chemical analyses of food supplements containing undeclared plants, including undeclared Rauvolfia species, or containing an excess dose of vitamin D [13, 14]. Adverse reactions to food supplements containing red yeast rice, products widely used for hyperlipidaemia, showed a risk profile comparable to that of lovastatin, explained by the chemical identical structure of monakolin K, contained in red yeast rice, and the drug lovastatin [15]. A specific analysis on adverse reactions occurring in children and adolescents was performed and the main risk factors associated with the occurrence of adverse reactions in this particular population were identified [16]. A recent analysis included 116 reports concerning 212 suspected adverse reactions to dietary supplements containing alfa-lipoic acid (ALA) collected within the Italian phytovigilance system [17]. The reports included mostly women (68.1%), aged between 14 and 89 years (mean age 57.1 years). Cases were mostly reported by physicians (57.7%) and pharmacists (22.4%). The reasons for the use of dietary supplements containing ALA were predominantly neuropathy (19.0%) and for treatment of lumbosciatalgia/cervical-brachialgia (17.2%), followed by carpal tunnel syndrome and osteoarticular disorders (both 7.8%). In some cases, food supplements were used by patients following the advice of a physician. With respect to adverse reactions, the most frequently reported system organ classifications (SOC) were "Skin and subcutaneous tissue disorders", "Gastrointestinal disorders" and "General disorders and administration site conditions". Ten cases of Hirata disease were collected within the Phytovigilance system, as compared with only five cases reported worldwide. Insulin autoimmune syndrome, also known as Hirata disease, is a life-threatening adverse reaction to ALA-containing dietary supplements, leading to severe hypoglycaemia. Although Hirata disease is well recognized in Japan, the diagnosis and imputability to ALA remains challenging in the Western world. Overall, in 45 (38.8%) cases the report was classified as serious [17].

Another recent signal emerged in 2019 from a cluster of reports of hepatitis, mostly cholestatic, associated with turmeric-containing supplements. Many actions were taken to manage the situation. To identify the substances potentially responsible for the adverse reactions observed, suspected products were collected and analysed. The analyses focused on identifying any intentionally added drugs, accidental contaminants, residues, and intentional synthetic adulterants. In particular, the products were checked for the presence of the following classes of substances: non-steroidal anti-inflammatory drugs (e.g. nimesulide); narcotic or psychotropic substances; heavy metals; aflatoxins; pesticides; pyrrolizidine alkaloids; and synthetic dyes [18]. Preliminary results relating to 7 of the 28 spontaneous reports of hepatitis associated to turmeric-containing food supplements have been published [19].

Finally, the Phytovigilance system in some cases has allowed health authorities to withdraw from the market products not compliant with the legislation, due to the presence of drugs or substances not permitted in dietary supplements. Other reports of adverse reactions have led to specific regulatory actions, such as the inclusion on the label of the warning "seek medical advice before use", or "do not use during pregnancy and lactation", or "do not use with other drugs" (such as with cholesterol-lowering drugs in the case of red rice supplements).

17.3 Conclusion

The Italian Phytovigilance system represents a rather unique tool in the European context to identify signals of potential harms associated with food supplements and compounded preparations, containing herbal ingredients. During the years of its operation to date, the system has been able to provide a scientific basis to the Ministry of Health to support regulatory actions to enhance consumer safety in relation to the use of these types of products and preparations.

References

1. Bianchi A, Cantù P, Firenzuoli F, Mazzanti G, Menniti-Ippolito F, Raschetti R (2004) Rhabdomyolysis caused by Commiphora mukul, a natural lipid-lowering agent. Ann Pharmacother 38(7–8):1222–1225. https://doi.org/10.1345/aph.1D486
2. Menniti-Ippolito F, Mazzanti G, Vitalone A, Firenzuoli F, Santuccio C (2008) Surveillance of suspected adverse reactions to natural health products: the case of propolis. Drug Saf 31(5):419–423. https://doi.org/10.2165/00002018-200831050-00007
3. Mazzanti G, Menniti-Ippolito F, Moro PA et al (2009) Hepatotoxicity from green tea: a review of the literature and two unpublished cases. Eur J Clin Pharmacol 65(4):331–341. https://doi.org/10.1007/s00228-008-0610-7
4. Moro PA, Cassetti F, Giugliano G et al (2009) Hepatitis from Greater Celandine (*Chelidonium majus* L.): review of literature and report of a new case. J Ethnopharmacol 124(2):328–332. https://doi.org/10.1016/j.jep.2009.04.036
5. Vannacci A, Lapi F, Gallo E et al (2009) A case of hepatitis associated with long-term use of *Cimicifuga racemosa*. Altern Ther Health Med 15(3):62–63
6. Lapi F, Gallo E, Giocaliere E et al (2010) Acute liver damage due to *Serenoa repens*: a case report. Br J Clin Pharmacol 69(5):558–560. https://doi.org/10.1111/j.1365-2125.2010.03618.x
7. Gori L, Galluzzi P, Mascherini V et al (2011) Two contemporary cases of hepatitis associated with Teucrium Chamaedrys L. decoction use: case reports and review of literature. Basic Clin Pharmacol Toxicol 109(6):521–526. https://doi.org/10.1111/j.1742-7843.2011.00781.x
8. Crescioli G, Lombardi N, Bettiol A et al (2018) Acute liver injury following Garcinia cambogia weight-loss supplementation: case-series and literature review. Intern Emerg Med 13(6):857–872. https://doi.org/10.1007/s11739-018-1880-4
9. Paoletti A, Gallo E, Benemei S et al (2011) Interactions between natural health products and oral anticoagulants: spontaneous reports in the Italian surveillance system of natural health products. Evid Based Complement Alternat Med 2011:612150. https://doi.org/10.1155/2011/612150

10. Vitalone A, Menniti-Ippolito F, Moro PA, Firenzuoli F, Raschetti R, Mazzanti G (2011) Suspected adverse reactions associated with herbal products used for weight loss: a case series reported to the Italian National Institute of Health. Eur J Clin Pharmacol 67(3):215–224. https://doi.org/10.1007/s00228-010-0981-4
11. Mazzanti G, Vitalone A, Da Cas R, Menniti-Ippolito F (2019) Suspected adverse reactions associated with herbal products used for weight loss: spontaneous reports from the Italian Phytovigilance System. Eur J Clin Pharmacol 75(11):1599–1615. https://doi.org/10.1007/s00228-019-02746-6
12. Vitalone A, Menniti-Ippolito F, Moro PA, Raschetti R, Renda F, Tartaglia L, Mazzanti G (2012) Surveillance of suspected adverse reactions to herbal products used as laxatives. Eur J Clin Pharmacol 68(3):231–238. https://doi.org/10.1007/s00228-011-1128-y
13. Gallo E, Giocaliere E, Benemei S et al (2012) Anything to declare? Possible risks for patients' health resulting from undeclared plants in herbal supplements. Br J Clin Pharmacol 73(3):482–483. https://doi.org/10.1111/j.1365-2125.2011.04115.x
14. Benemei S, Gallo E, Giocaliere E et al (2013) It's time for new rules on vitamin D food supplements. Br J Clin Pharmacol 76(5):825–826. https://doi.org/10.1111/bcp.12134
15. Mazzanti G, Moro PA, Raschi E, Da Cas R, Menniti-Ippolito F (2017) Adverse reactions to dietary supplements containing red yeast rice: assessment of cases from the Italian surveillance system. Br J Clin Pharmacol 83(4):894–908. https://doi.org/10.1111/bcp.13171
16. Lombardi N, Crescioli G, Bettiol A et al (2019) Safety of complementary and alternative medicine in children: a 16-years retrospective analysis of the Italian Phytovigilance system database. Phytomedicine 61:152856. https://doi.org/10.1016/j.phymed.2019.152856
17. Gatti M, Ippoliti I, Poluzzi E (2020) Assessment of adverse reactions to α-lipoic acid containing dietary supplements through spontaneous reporting systems. Clin Nutr S0261-5614(20):30393–30399. https://doi.org/10.1016/j.clnu.2020.07.028
18. Menniti-Ippolito F, Ippoliti I, Pastorelli AA et al (2020) Turmeric (Curcuma longa L.) food supplements and hepatotoxicity: an integrated evaluation approach. Ann Ist Super Sanita 56(4):462–469. https://doi.org/10.4415/ANN_20_04_08
19. Lombardi N, Crescioli G, Maggini V (2020) Acute liver injury following turmeric use in Tuscany: an analysis of the Italian Phytovigilance database and systematic review of case reports. Br J Clin Pharmacol 86(1):106–120. https://doi.org/10.1111/bcp.14460

Chapter 18
Pharmacovigilance for Herbal and Traditional Medicines in Bosnia and Herzegovina

Martin Kondža and Biljana Tubić

18.1 Pharmacovigilance in Bosnia and Herzegovina

Pharmacovigilance in Bosnia and Herzegovina (B&H) does not have a long history. Although B&H was part of the former Yugoslavia, which had a well-developed pharmacovigilance system, after the Yugoslav wars (from 1991 to 2001) and disintegration of Yugoslavia, B&H did not inherit the pharmacovigilance system. With the disintegration of Yugoslavia into Croatia, Slovenia, B&H, Serbia, Montenegro, and North Macedonia, a complicated political system was created in B&H, and the country was divided into the Federation of B&H (consisting of ten cantons each with its own administration), the Republic of Srpska, and the Brčko District. Each administrative unit regulated its own drug and medical policy. Only with the entry into force of the Act on Medicines and Medical Devices of Bosnia and Herzegovina (Official Gazette of Bosnia and Herzegovina 58/08 from 2008) [1] was a single national drug policy for the whole country of B&H created, and the Agency for Medicines and Medical Devices of Bosnia and Herzegovina (in Croatian: *Agencija za lijekove i medicinska sredstva Bosne i Hercegovine* (ALMBIH)) was founded in 2009. This Act explained the terms drugs and medical device, as well as the terms herbal and traditional medicines (HTMs). The Act on Medicines and Medical

M. Kondža (✉)
Faculty of Pharmacy, University of Mostar, Mostar, Bosnia and Herzegovina
e-mail: martin.kondza@farf.sum.ba

B. Tubić
Agency for Medicines and Medical Devices of Bosnia and Herzegovina, Banja Luka, Bosnia and Herzegovina

Faculty of Medicine, University of Banja Luka, Banja Luka, Bosnia and Herzegovina
e-mail: b.tubic@almbih.gov.ba

J. Barnes (ed.), *Pharmacovigilance for Herbal and Traditional Medicines*,
https://doi.org/10.1007/978-3-031-07275-8_18

Devices [1] was mostly based on similar regulations in the European Union (EU) [2], as B&H is developing a strong relationship with the EU through the Stabilization and Association Process, which came into force in 2015 [3]. The ALMBIH also has close relations with the European Medicines Agency (EMA) in terms of cooperation and stabilization of medicine regulations in accordance with EMA and EU regulations.

According to the Act on Medicines and Medical Devices, in B&H, an herbal medicine is "any medicine that, as active ingredients, contains exclusively one or more herbal substances, or one or more herbal preparations, or one or more herbal substances in combination with one or more herbal preparations. Herbal substances are whole, fragmented or cut plants, parts of plants, algae, fungi or lichens in unprocessed, dried or fresh form. Certain exudates that have not been processed are also considered herbal substances. Herbal substances are precisely defined by the part of the plant used and the botanical name according to the binomial system (genus, species, diversity, and author). Herbal preparations are those obtained by processing herbal substances by appropriate scientifically proven methods" [1, 2]. Traditional medicines are described as "medicines intended for self-medication in the manner specified in the package leaflet; they are used only for internal or external use and for which information is available relating to use as a medicine for at least 30 years, or at least 15 years in B&H or in EU Member States, and whose pharmacological effects, low risk of harm, and efficacy are assumed based on traditional use and experience" [1, 2].

The Act on Medicines and Medical Devices also defines the Rulebook on the manner of reporting, collecting, and monitoring adverse reactions to medicines [4]. These guidelines define how pharmacovigilance work is done for B&H. According to the provisions of these Rulebooks, health professionals (physicians, pharmacists, dentists, medical technicians, and others) and marketing authorization holders (MAH) in B&H are obliged to report suspected adverse reactions to medicines (and medical devices). The Mostar Pharmacovigilance Office (MPO) of ALMBIH operates as the National Pharmacovigilance Centre in B&H, which started working in 2009. The MPO is making progress in its work and has published two annual reports on adverse drug reactions (ADRs) for 2017 and 2018 [5, 6]. In addition, from 2019, B&H became a full member of the World Health Organization's (WHO) Programme for International Drug Monitoring operated by the WHO Uppsala Monitoring Centre (WHO-UMC). B&H is one of the last countries in Europe to join the WHO program. This is an important step and represents substantial progress in the improvement of ADR monitoring in B&H.

One issue, however, is that the existing legislation is outdated and in need of change. Among other things, there is a need to appropriately define how to collect, monitor, and analyze ADRs related to herbal and traditional medicines (HTMs). There are no such guidelines in the existing regulations. Therefore, processes for monitoring ADRs associated with HTMs in B&H are unclear. Under the existing Regulations [1, 4], only ADRs related to medicines officially registered and published in the B&H Medicines Registry are monitored. In addition, substances used in HTMs can be registered as dietary supplements. The regulation of dietary supplements is not the responsibility of the ALMBIH but is the responsibility of the entity ministries of health: the Federal Ministry of Health and the Ministry of Health and Social Welfare of the Republic of Srpska. There is no body in these systems that is tasked with responding to adverse reactions associated with dietary supplements. Thus, there is no infrastructure nor staff to systematically oversee adverse events of exclusively HTMs in B&H.

Nevertheless, there is a department in ALMBIH that is responsible for HTMs with respect to receiving and processing marketing authorization applications, applications for amendment, renewal, supplementation and maintaining licenses to place herbal and homeopathic remedies on the market. In doing so, the EU regulations and The International Council for Harmonization (ICH) guidelines [2, 7] are followed and applied when evaluating dossiers, working with experts to evaluate the safety and efficacy of the drug.

18.2 Herbal and Traditional Medicines in Bosnia and Herzegovina

Bosnia and Herzegovina has a long tradition of use of herbal remedies and preparations, and many people hold strong beliefs on the effectiveness of herbal and other traditional remedies. This is particularly true in rural areas of Bosnia and Herzegovina, where people still use so-called home remedies for the treatment of various diseases. Many of these home remedies are herbal and traditional remedies. Historically, much credit for the spread and promotion of HTMs in B&H is given to the Franciscan priests. Throughout B&H, they acted as *folk doctors* treating patients who did not have access to the formal healthcare system, copying old medical manuscripts into the vernacular and writing down their own prescriptions. Then, the so-called folk-medicine manuscripts were created, while the Franciscans were also engaged in the study of pharmacy and the cultivation of medicinal plants [8]. Since then, HTMs remain widely used throughout B&H, with each region cultivating its own recipes and variants.

In B&H, HTMs are most commonly used to treat gastrointestinal tract disorders, blood disorders, skin disorders, respiratory disorders, and urinary disorders [9]. Herbal remedies are mostly prepared in the form of infusions, ointments, balms, and decoctions. Some of the plant species commonly used in B&H for preparing HTMs [10] often include one or more of the following:

- Silver fir—*Abies alba* Mill. (Pinaceae)
- Noble yarrow—*Achillea nobilis* L. (Asteraceae)
- Sweet flag—*Acorus calamus* L. (Acoraceae)
- Horse-chestnut—*Aesculus hippocastanum* L. (Sapindaceae)
- Wild garlic—*Allium ursinum* L. (Alliaceae)
- Wormwood—*Artemisia absinthium* L. (Asteraceae)
- Common centaury—*Centaurium erythraea* Rafn (Gentianaceae)

- St John's wort—*Hypericum perforatum* L. (Hypericaceae)
- Winter savory—*Satureja montana* L. (Lamiaceae)
- Common comfrey—*Symphytum officinale* L. (Boraginaceae)
- Mountain germander—*Teucrium montanum* L. (Lamiaceae).

The use of similar botanical species is present in the folk medicine of other countries in this part of Central and South-Eastern Europe, such as Albania, Croatia, Serbia, and Slovenia [10–12].

All official medicinal products under the Act on Medicines and Medical Devices [1] are required to be registered with the ALMBIH, and patients obtain their medicines from public pharmacies. Pharmacies receive these medicinal products from a distributor or manufacturer, who is obliged to submit adequate extensive documentation for the registration of the drug/medicinal product to the ALMBIH.

There are several HTMs officially registered in B&H, most of which are registered for the treatment of respiratory diseases (Table 18.1).

Table 18.1 List of registered HTMs in Bosnia and Herzegovina

Area of use	HTM name and extract details (common name)	Number of registered products	Indications
Alimentary tract and metabolism	*Matricaria chamomilla* L. flower extract (German chamomile)	1	Mild inflammation of the gums and oral mucosa
	Salvia officinalis L. liquid leaf extract (sage)	1	Symptomatic treatment of inflammation of the mouth or throat associated with the common cold
	Senna alexandrina Mill. dry leaf extract (Alexandrian senna)	1	Constipation
	Thymus vulgaris L. liquid extract (thyme)	1	Symptomatic treatment of inflammation of the mouth or throat associated with the common cold
Nervous system	*Ginkgo biloba* L. leaf (ginkgo)	4	Circulatory and brain disorders (dementia), dizziness, tinnitus in the ears, circulatory disorders in the extremities
	Humulus lupulus L. dried flower extract (common hop)	1	Relieving mild symptoms of mental stress
	Hypericum perforatum L. dried extract (St John's wort)	1	Symptomatic treatment of mild depression
	Melissa officinalis L. dried extract (lemon balm)	2	Relieving mild mental symptoms, stress and as an aid in falling asleep
	Mentha x *piperita* L. dried leaf extract (peppermint)	1	To relieve mild mental symptoms, stress and as an aid in falling asleep
	Passiflora incarnata L. dried extract (maypop)	1	To relieve mild stress due to anxiety and tension
	Valeriana officinalis L. dried root extract (valerian)	2	Relieving mild stress due to anxiety and tension

Table 18.1 (continued)

Area of use	HTM name and extract details (common name)	Number of registered products	Indications
Respiratory system	*Althea officinalis* L. liquid root extract (marshmallow)	1	Dry, irritating coughs that occur in upper respiratory tract diseases
	Cetraria islandila L. dense extract (Iceland moss)	1	Dry, irritating cough, mild inflammation of the upper respiratory tract and irritation of the mucous membranes of the mouth and throat, including hoarseness and sore throat
	Citrus limon L. purified essential oil distillate (lemon)	1	Used for easier coughing in colds accompanied by cough
	Citrus x *sinensis* L. purified essential oil distillate (sweet orange)	1	Used for easier coughing in colds accompanied by cough
	Eucalyptus globulus Labill. purified essential oil distillate (southern blue gum)	1	Used for easier coughing in colds accompanied by cough
	Hedera helix L. dried leaf (common ivy)	1	Used as an expectorant in acute inflammation Respiratory system accompanied by cough, hypersecretion of mucus and difficulty breathing; for symptomatic treatment of acute exacerbations of chronic bronchitis
	Malva sylvestris L. flower (common mallow)	1	Dry, irritating coughs that occur in upper respiratory tract diseases
	Myrtus communis L. purified essential oil distillate (common myrtle)	1	Dry, irritating coughs that occur in upper respiratory tract diseases
	Pelagronium sidoides DC. whole plant (African geranium)	1	Symptomatic therapy for cold
	Plantago lanceolata L. liquid leaf extract (ribwort plantain)	1	Dry, irritating coughs that occur in upper respiratory tract diseases
	Primula vulgaris Huds. Liquid root extract (primrose)	1	Used for easier coughing in colds accompanied by cough
	Salvia officinalis L. dried extract (common sage)	1	Symptomatic treatment of inflammation of the mouth or throat (such as sore throat, hoarseness, and difficulty swallowing) associated with the common cold
	Thymus vulgaris L. dried extract (common thyme)	1	Relieves irritated mucous membranes of the throat and acts as expectorant for cold-related coughs
	Thymus vulgaris L. liquid extract (common thyme)	2	Symptomatic treatment of inflammation of the mouth or throat (such as sore throat, hoarseness, and difficulty swallowing) associated with the common cold
	Thymus vulgaris L. liquid plant extract (common thyme)	1	Alleviation of ailments in respiratory diseases caused by colds, with thick (viscous) mucus, relieving symptoms in acute bronchitis
	Thymus vulgaris L. standardized liquid plant extract (common thyme)	1	Symptomatic treatment of inflammation of the mouth or throat (such as sore throat, hoarseness, and difficulty swallowing) associated with the common cold

Due to traditional beliefs, price, poor infrastructure, and difficult access to healthcare, the B&H rural population is more likely to use HTMs than conventional medicines. Unfortunately, most HTMs are homemade, or sourced from local individuals, and are not registered with ALMBIH. This poses a substantial problem in monitoring the safety and effectiveness of HTMs. Specifically, it is very difficult to determine exactly which constituent of a plant is responsible for a particular effect, the amount and concentration of active and other constituents, and—because of the diversity of plant species—it is impossible to know the complete chemical composition of an individual HTM. One of the most popular HTM preparations is a unique balm in B&H called *melem* (Turkish *merhem*—ointment), which is usually prepared from various species of herbaceous plants that are chopped and pressed [9]. This sort of herbal treatment can be found in almost any household in B&H. It is applied locally on the skin, and is considered safe by users, although there is no high-quality evidence or studies to support this claim.

In addition to registered and unregistered products, preparations containing herbal ingredients are also available on the illicit market in B&H [13].

18.3 Pharmacovigilance for Herbal and Traditional Medicines in Bosnia and Herzegovina

As stated above, the existing legislation in B&H has not fully clarified and defined HTMs, or processes for monitoring adverse effects associated with these products and preparations. At present, B&H has a low frequency of adverse reaction reporting and has the lowest number of reported cases in the West-Balkan region and among the lowest in Europe [14]. In recent years, however, the situation has been improving and, for the first time, two annual reports on adverse reactions associated with medicines and medical devices in B&H have been issued [5, 6].

This is the first report on pharmacovigilance for Bosnia and Herzegovina since its independence, although there have been various departments and institutes for drug control at entity and county levels, and ALMBIH at the state level since 2009. The report systematically describes for the first time all ADRs for registered medicines that occurred in Bosnia and Herzegovina, which were received in ALMBIH on forms submitted by health professionals and MAHs. The report includes information about the reporter, their specialties, the sex and age of the patient, suspected drug(s) (international non-protected/proprietary name (INN) and protected name), anatomical-therapeutic-chemical classification (ATC), pharmaceutical form and dose, ADR(s) using the Medical Dictionary for Regulatory Activities (MedDRA), seriousness of adverse reaction, outcome, and outcome of causality assessment.

Since the start of pharmacovigilance monitoring in Bosnia and Herzegovina, from 2017 to 2019, there have been very few reported ADRs associated with the use of HTMs. In the short history of adverse reaction reporting in B&H, five adverse reactions related to the use of HTMs have been reported. All reports were received through the official ADR report form. According to the type of report, none of the applications received met the criteria for being classed as a serious ADR under the ALMBIH guidelines (i.e., resulted in death, life-threatening, hospital treatment or extension of hospital treatment, permanent or severe disability or incapacity, congenital anomaly, medically significant serious condition). Four of the reported ADRs related to use of HTMs described skin rashes. All products implicated in the ADR reports were registered products and all adverse reactions were non-serious.

The groups that reported these ADRs were health professionals (pharmacists and physicians). Generally, physicians and pharmacists in B&H are responsible for more than 95% of all reports [15]. All reported ADRs were described in the Summary of Product Characteristics (SmPC) for the products concerned. There were no unexpected ADRs, and no signal investigations were undertaken as a result of the reports of ADRs associated with HTMs. Generally, with conventional medicines, around 17% of unexpected ADRs received in B&H per year are investigated further as part of a signal detection process. The quality of the reports received is still not being evaluated by ALMBIH. This is an important part that needs to be added to the pharmacovigilance system in B&H. One other part of the pharmacovigilance system in B&H that needs a thorough review is patient reporting. Patient reporting is not allowed by the existing pharmacovigilance regulation; ADR reporting is only allowed for healthcare professionals and MAHs [1, 4]. If a patient wishes to report an ADR, it is not possible to do this by reporting it directly to ALMBIH: patients may only report through a healthcare professional. The healthcare professional then completes the necessary ADR form and sends it to ALMBIH via mail, fax, or post. This is not the case in the EU [16], so B&H is expected to go through changes regarding its existing regulation in order for future coordination with EU legislation.

The very low number of reports of suspected ADRs associated with HTMs in B&H should be considered in the context of numbers of all ADR reports submitted in B&H. The total numbers of ADR reports received for the years 2017 and 2018 were 227 and 292, respectively (data pending for 2019) with continuous data collection beginning in 2017. Also, B&H still has not met the criteria of the WHO regarding the development of a good pharmacovigilance system, as it receives fewer than 200 reports per one million population [17]. Thus, these numbers represent the only official data that B&H has regarding ADRs associated with HTMs. It is very important that the collection of these data continues in the future.

18.4 Future Perspectives

Although the number of reported ADRs associated with HTMs in B&H is very small (amounting for only about 0.4% of all reports received per year), this does not imply that natural health products, herbal or traditional medicines, are completely safe to use, or safer than conventional medicines. The administration of any medicinal product can pose a risk of harm for particular patients. As most HTMs are used in the rural population, prepared locally and mostly not registered with the authorities, the data discussed here cannot be considered representative. Much more research, annual analysis of ADR reports, and enhanced reporting of ADRs associated with HTMs are needed to develop a real picture of the harms profile associated with the administration of these products. In B&H, there is a general belief among users of HTMs that the so-called *natural products* are always completely safe to use and have no "side effects." As the components of HTMs are not necessarily known, and as individuals may have idiosyncratic reactions to medicinal products, there is a risk of allergic reactions occurring with HTMs, as with other medicines. Therefore, for HTMs, it is important to have standards and evidence-based information regarding their pharmaceutical quality, safety, and efficacy. Users of HTMs should be encouraged to consult a health professional if they are using HTMs in addition to, or instead of, conventional medicines, and if they have chronic health conditions; users should also be encouraged to report any information on new symptoms or adverse reactions experienced during or after use of HTMs to health professionals and the competent authority (where possible). It is very important to work on raising awareness of the necessity of reporting adverse reactions of not only conventional medicinal products, but also for HTMs. One of the necessary steps for raising awareness can be allowing patients to directly report adverse reactions to regulatory authorities, establishing a stable vigilance base for HTMs, encouraging healthcare and non-healthcare workers (for example, those in shops that offer HTMs) to report any known adverse events. By collecting more data, the regulatory bodies can have a representative picture and react to public health threats if necessary.

References

1. The Agency for Medicinal Products and Medical Devices Bosnia and Herzegovina. Act on Medicines and Medical Devices (Official Gazette of B&H, No. 58/08)
2. The European Parliament and the Council of the European Union (2004) Directive 2004/24/EC of the European Parliament and of the Council of 31 March 2004 amending, as regards traditional herbal medicinal products, Directive 2001/83/EC on the Community code relating to medicinal products for human use. Off J Eur Union L136:85–90
3. Stabilisation and Association Agreement between the European Communities and Bosnia and Herzegovina. Decision (EU, Euratom) 2015/998 concluding the Stabilisation and Association Agreement between the European Communities and Bosnia and Herzegovina
4. Expert Council of the Agency for Medicinal Products and Medical Devices (2012) Rulebook on the manner of reporting, collecting and monitoring of adverse drug reactions (Official Gazette of B&H, No. 58/12)
5. Kondža M, Tubić B (2018) Report on adverse reactions to drugs and medical devices in Bosnia and Herzegovina for the year 2017. Agency for Medicines and Medical Devices of Bosnia and Herzegovina, Mostar
6. Kondža M, Tubić B (2019) Report on adverse reactions to drugs and medical devices in Bosnia and Herzegovina for the year 2018. Agency for Medicines and Medical Devices of Bosnia and Herzegovina, Mostar
7. European Medicines Agency (1995) Validation of analytical procedures: methodology (CHMP/ICH/281/95)

8. Kujundžić N (2014) Great folk-medicine book of Sinj. Croatian Academy of Science and Arts, Faculty of Pharmacy and Biochemistry, Matica hrvatska, Zagreb, p 756
9. Šarić-Kundalić B, Fritz E, Dobeš C, Saukel J (2010) Traditional medicine in the pristine village of Prokoško lake on Vranica mountain, Bosnia and Herzegovina. Sci Pharm 78(2):275–290. https://doi.org/10.3797/scipharm.1003-06
10. Pieroni A, Giusti ME, Quave CL (2011) Cross-cultural ethnobiology in the western Balkans: medical ethnobotany and ethnozoology among Albanians and Serbs in the Pešter plateau, Sandžak, South-Western Serbia. Hum Ecol 39(3):333–349
11. Lumpert M, Kreft S (2017) Folk use of medicinal plants in Karst and Gorjanci, Slovenia. J Ethnobiol Ethnomed 13:16
12. Žuna Pfeiffer T, Krstin LJ, Špoljarić Manorić D et al (2019) An ethnobotanical survey of useful wild plants in the north-eastern part of Croatia (Pannonian region). Plant Biosyst 154(4):463–473. https://doi.org/10.1080/11263504.2019.1635222
13. Crno tržište:Rupu u zakonu popunili lažni preparati. https://startbih.ba/clanak/crno-trzisterupu-u-zakonu-popunili-lazni-preparati/116641. Accessed 11 May 2020
14. Glamočlija U, Tubić B, Kondža M, Zolak A, Grubiša N (2018) Adverse drug reaction reporting and development of pharmacovigilance systems in Bosnia and Herzegovina, Croatia, Serbia, and Montenegro: a retrospective pharmacoepidemiological study. Croat Med J 59(3):124–131
15. Kondža M, Tubić B (2018) The role of the pharmacist in reporting adverse drug reactions. XX simpozijum farmaceuta Republike Srpske s međunarodnim učešćem
16. Regulation (EC) No 726/2004 of the European Parliament and of the Council of 31 March 2004 laying down Community procedures for the authorization and supervision of medicinal products for human and veterinary use and establishing a European Medicines Agency. Off J Eur Union 136/1
17. Agency for Medicines and Medical Devices of Montenegro (2017) Izvještaj o rezultatima prijavljivanja neželjenih dejstava ljekova za 2016. godinu

Chapter 19
Pharmacovigilance for Herbal Products: A Canadian Perspective

Kevin Bernardo and Shahid Perwaiz

19.1 Introduction

Pharmacovigilance of health products is a multifaceted activity that usually begins when the products become available for sale on the market. While this post-market surveillance is typical for conventional medicines, which are required to undergo a rigorous pre-market approval process, it is also important for natural health products (NHPs), where, depending on the jurisdiction, market entry may occur with limited information or without pre-market review at all. Even when regulatory systems are in place (such as in Canada), and even for many NHPs with well-established historical use, post-market surveillance is crucial [1].

NHPs have been used to maintain health and to prevent and treat various medical conditions. In some jurisdictions, they may be advertised directly to consumers for chronic or serious illness, which may lead to misuse by consumers, thus increasing the risk of adverse effects [2]. The perception is that "natural" means "safe" and, therefore, not able to cause adverse reactions (ARs). Contrary to popular belief, HPs are not always "safe," particularly when used in combination with other medicines, and can result in negative health consequences. In the South-East Asia Region, 10–45% of Outpatient Department (OPD) visits in the public health sector were related to traditional herbal medicine use in 2015 [3–6].

Studies show that the use of NHPs is increasing in the general population, including among children, pregnant women, and seniors [7]. It has been reported that more than 70% of the population in Canada and the United States of America has used NHPs, such as vitamins and minerals, herbal products, and homeopathic medicines [7]. Furthermore, 80% of the developing world's population has reported using some form of traditional medicine [8, 9]. Around 12% of Canadians who use

K. Bernardo (✉) · S. Perwaiz
Marketed Health Products Directorate, Health Canada, Ottawa, ON, Canada
e-mail: kevin.bernardo@hc-sc.gc.ca; shahid.perwaiz@hc-sc.gc.ca

J. Barnes (ed.), *Pharmacovigilance for Herbal and Traditional Medicines*,
https://doi.org/10.1007/978-3-031-07275-8_19

NHPs report that they have experienced unwanted side effects (adverse reactions) [3, 7, 10–14]. A study reported that 49% of the population ($n = 1804$), sampled from a large Canadian tertiary care pediatric emergency room, used HPs [15]. According to the Natural Health Product Tracking Survey, around 72% of Canadians use NHPs for maintenance of health [16]. Given the widespread use of these products worldwide, and gaps in knowledge on safety and effectiveness, ongoing post-market monitoring is warranted.

19.2 Regulation of NHPs in Canada

The *Natural Health Products Regulations* [17], which came into force in Canada in 2004, cover pre-market product licensing requirements, site licensing, and post-market reporting requirements. Similar to other health product regulations in Canada, the NHP Regulations do not apply to the practice of compounding, nor to the practice of medicine, which fall under provincial/territorial jurisdiction. The main objective of the Regulations is to protect public health by assuring quality, effectiveness, and safety of NHPs marketed in Canada. The requirements and procedures to obtain marketing authorization for NHPs are identified in the *Food and Drugs Act* [18] and the *Natural Health Products Regulations* [17]. In Canada, all authorized NHPs undergo pre-market assessment for safety and efficacy; the degree of pre-market oversight varies depending on the risk suspected to be associated with the product.

In Canada, NHPs are described as naturally occurring substances that are used to restore or maintain good health. They include:

- Vitamins and minerals
- Herbal remedies
- Homeopathic medicines
- Traditional medicines, such as traditional Chinese and Ayurvedic medicines
- Probiotics
- Amino acids
- Essential fatty acids

Finished products come in a wide variety of forms, including tablets, capsules, tinctures, solutions, creams, ointments, and drops. Certain toothpastes, antiperspirants, shampoos, facial products, and mouthwashes are also classified as NHPs in Canada [19].

Products authorized for sale in Canada are issued a product license: a Natural Product Number (NPN) for an NHP or a Homeopathic Medicine Number (DIN-HM) for a homeopathic product. Health Canada has received over 100,000 license applications since 2004. Currently, over 90,000 products have been authorized for sale in Canada, including herbal medicines, vitamins and minerals, probiotics and homeopathic medicines; over 2500 companies hold product licenses for NHPs, and over

2600 sites have been licensed. In Canada, a site license is required to manufacture, package, label, or import an NHP.

The pre-market review of NHPs consists of the review of information to support the safety and efficacy of the product. In order to obtain approval for sale in Canada, NHPs must be safe, efficacious, of high quality, and carry detailed label information to allow people to make informed choices. Consumers can identify products that have been licensed for sale in Canada by looking for the eight-digit NPN or DIN-HM on the product label. A NPN or DIN-HM means that the product has been authorized for sale in Canada and is safe and effective when used according to the instructions on the label.

Health Canada has published various guidance documents that outline the evidence requirements needed to support the safety and efficacy of NHPs: the "Pathway for Licensing Natural Health Products Making Modern Health Claims" [19], the "Pathway for Licensing Natural Health Products used as Traditional Medicines" [20], and Evidence for Homeopathic Medicine [21]. Evidence from a range of sources, including clinical studies, pharmacopoeias, textbooks, peer-reviewed published articles, regulatory authority reports, and/or traditional references, is considered acceptable to support the safety and effectiveness of an NHP. In general, the level of evidence (type and amount) required varies depending on the proposed health claim(s) and the overall risk profile of the product or its ingredients [22]. The guidance document for NHPs making modern health claims includes evidence requirements to make basic general health claims as well as more specific health claims.

19.3 Pharmacovigilance for NHPs in Canada

Pharmacovigilance is the science and activity relating to the collection, detection, assessment, monitoring, and prevention of adverse effects associated with therapeutic products including herbal products [23, 24]. Surveillance, therefore, requires tools that can continually assess safety throughout the product life cycle, quickly generate hypotheses (i.e., through detection of potential safety signals), and methodically evaluate them to mitigate the identified health risk. Safety surveillance of NHPs in Canada comprises several elements. Safety signals are collected from domestic and international media and literature scans, foreign regulatory agencies, the Canada Vigilance database (reporting from industry, health professionals and consumers) as well as from the market authorization holders (industry safety reports such as periodic safety update reports/periodic benefit risk evaluation reports). The information received from these sources is combined to determine if a safety signal exists. The signals are assessed to determine whether a health risk to Canadians exists. Depending on the outcome of the signal assessment, the need for risk mitigation strategies (e.g., product recalls, labelling changes, risk communications) is determined.

19.4 Spontaneous Adverse Reaction Reports Associated with NHPs in Canada

Adverse reaction (AR) reporting is an important component of post-market surveillance for NHPs and is a collaborative effort among the various stakeholders. Canada Vigilance (CV) is a spontaneous adverse reaction monitoring program and database that has existed in Canada since 1965. It includes an online searchable database of AR case reports associated with NHPs, as well as pharmaceuticals, biologicals, biotechnology, and radiopharmaceuticals. It is a voluntary reporting system for use by health professionals and consumers; however, reporting by market authorization holders (MAH) of NHPs is mandatory. Every year, Health Canada receives tens of thousands of domestic AR reports associated with health products, including NHPs. For instance, in 2017, Health Canada received a total of 932 reports associated with the use of NHPs, 63,883 reports associated with pharmaceuticals, 41,743 reports associated with biologic products, and 569 reports associated with radiopharmaceutical products.

It is well recognized that under-reporting of ARs, especially those associated with NHPs, is a significant factor in the relatively low number of AR reports received by regulatory agencies worldwide. It should also be noted that under-reporting of suspected interactions between herbs and drugs is of increasing concern and arises from the same reasons as under-reporting of herbal adverse drug reactions in general [25–27]. Factors contributing to under-reporting of NHP ADRs include:

- Lack of association between herb and adverse effect.
- Patient stops using the herbal medicine when they feel unwell.
- Physician/patient unaware that herbal ADRs should be reported.
- Physician unaware of the use of herbal medicines as patient does not consider herbal and nutritional products to be "medicines" and does not disclose use.

As the use of NHPs increases, consumers may experience ARs, many of which go unreported, as a medical consultation is not always sought. Interactions between NHPs and drugs, other NHPs, or food, leading to clinically relevant events have been reported [23, 24]. These present some of the safety issues that need to be considered when reviewing the safety of these products. There is a perception that NHPs are safe, even if taken at the same time as prescription drugs [28]. NHPs may be used to treat the primary condition or to reduce the side effects of conventional treatment.

19.5 Challenges in Signal Evaluation for NHPs

Data gathered from multiple sources are evaluated together in order to assess any safety signals. A number of considerations (e.g., causality assessment of case reports, biological plausibility, dose-response curve, and the Canadian context for use of the particular product) are considered during an assessment of a safety signal. The assessment of safety signals associated with NHPs presents several unique challenges [26, 27].

An important challenge associated with the assessment of NHPs is their natural complexity and variability as NHPs can include complex mixtures containing hundreds of constituents, often with a large fraction of any botanical material consisting of unknown constituents. It is well known that different parts of plants have different medicinal properties as their profile is often not uniform and certain parts of the plant can be toxic. This makes it difficult to determine pharmacokinetics, pharmacodynamics, and toxicology and to establish which ingredient or combination of ingredients causes a safety concern [26, 27]. Other than the effects of the biologically active constituents of the plant, adverse effects may be due to herb-drug interactions or contaminants [29]. A study from Canada showed that most of the herbal products from a sample available for sale in the North American market were of poor quality, including considerable product substitution with cheap ingredients, contamination and use of fillers, or substances that are not listed in the label. In this study, 44 marketed herbal products, representing 12 companies, and 30 different species of herbs, and 50 leaf samples collected from 42 herbal species [30] were analyzed.

19.6 Risk Management Strategies

Recommendations related to risk management strategies are based on available evidence and the nature of the risk, which may vary significantly depending on the product and the ingredient involved in the safety assessment. Incomplete data can result in uncertainties in the true nature of the hazard and risk. Risk management decisions sometimes need to be made in a short timeframe (high-perceived risk) and on limited information. Without adequate information, such decisions may be conservative. However, these decisions can be revisited if/when new information becomes available. Some of the examples of risk management strategies that have been used to mitigate the risk associated with therapeutic products, including NHPs, in Canada are:

- Continuous surveillance, including monitoring of the Canada Vigilance database, scientific literature, and other sources.
- Requesting further data from industry (Issue-Related Summary Reports, periodic benefit risk evaluation reports), and from other regulatory agencies.
- Periodic review of signals (periodic safety update reports/periodic benefit risk evaluation reports; targeted monitoring of Canada Vigilance database).
- Updating product or ingredient information (monographs, labelling standards, labels, package inserts).
- Market suspension/withdrawal, product recall.

In 2017, Health Canada completed an assessment on the risk of liver toxicity associated with the use of green tea-containing NHPs, because of ongoing reports of serious liver injury worldwide, including Canadian reports of liver toxicity [31]. Health Canada's review concluded that there may indeed be an association between green tea extract-containing NHPs and the risk of liver injury. As a result, Health Canada decided to re-affirm/strengthen the existing cautionary risk statement in Health Canada's *Green Tea Extracts* monograph to include: "If you have a liver disorder, consult a health care practitioner prior to use. Stop use if you develop symptoms of liver trouble such as yellowing of the skin/eyes (jaundice), stomach pain, dark urine, sweating, nausea, unusual tiredness, and/or loss of appetite and consult a health care practitioner"; and "Rare, unpredictable cases of liver injury associated with green tea extract-containing products have been reported (in Canada and internationally)" [32].

The safety review also recommended that green tea extract products be used by adults only. Public communications were also issued to raise awareness among the Canadian public and healthcare professionals about this risk [31–35].

19.7 Risk Communication Strategies

Communicating risks associated with health products, including NHPs, is an important aspect of risk mitigation and risk management. In Canada, communicating health product-related risks is a shared responsibility—among the federal regulatory authorities, the market authorization holders/licensees, healthcare practitioners, and consumers. Each group plays a role in the way risks are communicated and received. While the overall objective is maintaining the health and safety of Canadians, risk communications are not intended as medical advice.

Health Canada uses several risk communication tools in disseminating new/emerging safety information on NHPs to Canadians. Many factors are considered in the evaluation of an emerging health product safety concern (e.g., availability and reliability of data, seriousness of the event) and the urgency of the communication. Health Canada has a number of risk communication tools that target consumers or health professionals, depending on the nature of the risk and the action to be taken. Some of Health Canada's risk communications strategies are Public Advisory, Dear Healthcare Practitioner Letter, Notice to Hospitals, and Health Product InfoWatch Publication [34, 35].

With regard to HPs and other NHPs, many safety-related communications involve product quality issues (i.e., adulteration of products with pharmaceutical drugs, microbial and heavy metal contamination). Health Canada has published several advisories, recalls and information updates for safety issues pertaining to NHPs licensed and sold in Canada.

19.8 Collaborative Efforts

A robust regulatory framework for NHPs includes collaboration with internal and external stakeholders. Domestically, Health Canada has collaborated with other organizations in order to obtain additional safety information and to communicate identified risks and possible risk mitigation strategies. For example, Health Canada has previously reached out to Canadian Poison Control Centres to help identify incidents associated with the use of specific NHPs. This type of information assists Health Canada to assess and communicate risks and risk mitigation measures to both consumers and healthcare practitioners. In addition, Health Canada has also collaborated with patient safety groups, such as the Institute for Safe Medications Practices Canada, in order to communicate risks associated with various NHPs to a wider audience.

Internationally, agreements such as memoranda of understanding, have been established between Health Canada and various regulatory agencies in order to support the mutual sharing of information. On a monthly basis, Health Canada communicates and collaborates with the United States of America's Food and Drug Administration, Australia's Therapeutic Goods Administration (TGA), Singapore's Health Sciences Authority (HSA), New Zealand's Medsafe, Swissmedic (Switzerland) and the United Kingdom's Medicines and Healthcare products Regulatory Agency (MHRA). This collaboration provides an international perspective on health-product-related safety concerns (including for NHPs) relevant to each jurisdiction. Such activities assist in expedited and efficient regulatory decision-making and action when safety concerns arise.

19.9 Future Perspectives and Challenges

The future of NHP vigilance is dependent upon the continued engagement of all stakeholders. Communication among regulatory agencies, industry, healthcare practitioners, and consumers is essential to the progression of NHP vigilance standards and practices. The environment surrounding this industry is ever-changing, with new combinations of ingredients constantly being introduced to the market. Innovative ingredient combinations, extraction methods, and dosages of herbal medicines, which have not been previously used by the general population, are emerging. As such, the real-world safety and effectiveness of these products is unknown and new information will continue to become available with continued use of these products. There is a demand for regulatory systems worldwide to modernize the approach to regulating NHPs, including herbal medicines. This includes creating awareness and understanding of the safe use of NHPs by consumers and the importance of communicating emergent safety issues within and among all impacted stakeholders [27].

References

1. Shaw D, Graeme L, Pierre D, Elizabeth W, Kelvin C (2012) Pharmacovigilance of herbal medicine. J Ethnopharmacol 140(3):513–518. https://doi.org/10.1016/j.jep.2012.01.051
2. Torok CB, Murray TH (2008) Wielding the sword of professional ethics against misleading dietary supplement claims. Glycobiology 18(9):660–663. https://doi.org/10.1093/glycob/cwn060
3. Brulotte J, Vohra S (2008) Epidemiology of NHP-drug interactions: identification and evaluation. Curr Drug Metab 9(10):1049–1054. https://doi.org/10.2174/138920008786927749
4. Giveon SM, Liberman N, Klang S, Kahan E (2004) Are people who use natural drugs aware of their potentially harmful side effects and reporting to family physician. Patient Educ Couns 53(1):5–11. https://doi.org/10.1016/S0738-3991(03)00241-6
5. Ipsos Mori on behalf of the Medicines and Healthcare products Regulatory Agency (2008) Public perception of herbal medicines: general public qualitative and quantitative research. https://www.ipsos.com/sites/default/files/migrations/en-uk/files/Assets/Docs/Polls/public-perceptions-of-herbal-medicines-report.pdf. Accessed 1 Mar 2020
6. Health Canada and the Public Health Agency of Canada (2016) Evaluation of the natural heath products Program. https://www.canada.ca/en/health-canada/services/drugs-health-products/natural-non-prescription/regulation/about-products.html. Accessed 17 Mar 2021
7. Goldstein LH, Elias M, Ron-Avraham G et al (2007) Consumption of herbal remedies and dietary supplements amongst patients hospitalized in medical wards. Br J Clin Pharmacol 64(3):373–380. https://doi.org/10.1111/j.1365-2125.2007.02878.x
8. Ekor M (2014) The growing use of herbal medicines: issues relating to adverse reactions and challenges in monitoring safety. Front Pharmacol 4:177. https://doi.org/10.3389/fphar.2013.00177
9. WHO guidelines on safety monitoring of herbal medicines in pharmacovigilance systems (2004). https://apps.who.int/iris/bitstream/handle/10665/43034/9241592214_eng.pdf. Accessed 1 Mar 2020
10. Chiu J, Yau T, Epstein RJ (2009) Complications of traditional Chinese/herbal medicines (TCM) - a guide for perplexed oncologists and other cancer caregivers. Support Care Cancer 17(3):231–240. https://doi.org/10.1007/s00520-008-0526-x
11. Ernst E (1999) Prevalence of complementary/alternative medicine for children: a systematic review. Eur J Pediatr 158(1):7–11. https://doi.org/10.1007/s004310051000
12. Kelly JP, Kaufman DW, Kelley K, Rosenberg L, Anderson TE, Mitchell AA (2005) Recent trends in use of herbal and other natural products. Arch Intern Med 165(3):281–286. https://doi.org/10.1001/archinte.165.3.281
13. Vohra S, Brulotte J, Le C, Charrois T, Laeeque H (2009) Adverse events associated with paediatric use of complementary and alternative medicine: results of a Canadian paediatric surveillance program survey. Paediatr Child Health 14(6):385–387. https://doi.org/10.1093/pch/14.6.385
14. Woodward KN (2005) The potential impact of the use of homeopathic and herbal remedies on monitoring the safety of prescription products. Hum Exp Toxicol 24(5):219–233. https://doi.org/10.1191/0960327105ht529oa
15. Goldman ED, Vohra S (2004) Complementary and alternative medicine use by children visiting a pediatric emergency department. Paper presented at the First Annual Complementary and Alternative Health Care & Paediatric Forum, University of Toronto, 2–3 Dec 2004
16. Health Canada (2011) Natural health product tracking survey - 2010. http://epe.lac-bac.gc.ca/100/200/301/pwgsc-tpsgc/por-ef/health/2011/135-09/report.pdf. Accessed 1 Mar 2020
17. About natural health product regulation in Canada. https://www.canada.ca/en/health-canada/services/drugs-health-products/natural-non-prescription/regulation.html. Accessed 1 Mar 2020
18. Government of Canada. Food and Drugs Act (last amended 2013-06-19). https://laws-lois.justice.gc.ca/eng/acts/f-27/. Accessed 1 Mar 2020

19. Health Canada (2012) Pathway for licensing natural health products making modern health claims. https://www.canada.ca/en/health-canada/services/drugs-health-products/natural-non-prescription/legislation-guidelines/guidance-documents/pathway-licensing-making-modern-health-claims.html. Accessed 1 Mar 2020
20. Health Canada. The approach to natural health products. https://www.canada.ca/content/dam/hc-sc/migration/hc-sc/dhp-mps/alt_formats/pdf/prodnatur/nhp-psn-eng.pdf. Accessed 1 Mar 2020
21. Health Canada (2015) Evidence for homeopathic medicines. https://www.canada.ca/en/health-canada/services/drugs-health-products/natural-non-prescription/legislation-guidelines/guidance-documents/evidence-homeopathic-medicines.html. Accessed 1 Mar 2020
22. World Health Organization (2017) Pharmacovigilance for traditional medicine products: why and how? https://apps.who.int/iris/bitstream/handle/10665/259854/Pharmacovigilane.pdf?sequence=1&isAllowed=y. Accessed 1 Mar 2020
23. Mann RD, Andrews EB (eds) (2002) Pharmacovigilance. Wiley, Chichester
24. De Smet PAGM (2007) Clinical risk management of herb–drug interactions. Br J Clin Pharmacol 63(3):258–267. https://doi.org/10.1111/j.1365-2125.2006.02797.x
25. Williamson EM, Driver S, Baxter K (eds) (2009) Stockley's herbal medicines interactions. Pharmaceutical Press, London
26. Jordan SA, Cunningham DG, Marles RJ (2010) Assessment of herbal medicinal products: challenges, and opportunities to increase the knowledge base for safety assessment. Toxicol Appl Pharmacol 243(2):198–216. https://doi.org/10.1016/j.taap.2009.12.005
27. Jordan SA, Jack SA, Pless RP (2018) Safety surveillance of dietary supplements: importance, challenges, and new horizons. Clin Pharmacol Ther 104(3):432–434. https://doi.org/10.1002/cpt.1101
28. MacDonald L, Murty M, Foster BC (2009) Antiviral drug disposition and natural health products: risk of therapeutic alteration and resistance. Expert Opin Drug Metab Toxicol 5(6):563–578. https://doi.org/10.1517/17425250902942302
29. Bent S (2008) Herbal medicine in the United States: review of efficacy, safety, and regulation: grand rounds at University of California, San Francisco Medical Center. J Gen Intern Med 23(6):854–859. https://doi.org/10.1007/s11606-008-0632-y
30. Newmaster SG, Grguric M, Shanmughanandhan D, Ramalingam S, Ragupathy S (2013) DNA barcoding detects contamination and substitution in North American herbal products. BMC Med 11:222. https://doi.org/10.1186/1741-7015-11-222
31. Health Canada (2017) Summary safety review - green tea extract - containing natural health products. https://hpr-rps.hres.ca/reg-content/summary-safety-review-detail.php?lang=en&linkID=SSR00173. Accessed 1 Mar 2020
32. Green tea extract monograph (2018). http://webprod.hc-sc.gc.ca/nhpid-bdipsn/atReq.do?atid=greentea_thevert&lang=eng. Accessed 1 Mar 2020
33. Health Canada (2017) Health Canada takes steps to help Canadians avoid rare risk of liver injury from use of green tea extract health products. https://healthycanadians.gc.ca/recall-alert-rappel-avis/hc-sc/2017/65114a-eng.php. Accessed 1 Mar 2020
34. Health Canada (2017) Green tea extract-containing natural health products - rare risk of liver injury. https://healthycanadians.gc.ca/recall-alert-rappel-avis/hc-sc/2017/65100a-eng.php. Accessed 1 Mar 2020
35. Health Canada (2011) Risk communication - protecting Canadians through information. https://www.canada.ca/en/health-canada/services/drugs-health-products/reports-publications/medeffect-canada/risk-communication-protecting-canadians-through-information-health-canada-2011.html. Accessed 1 Mar 2020

Chapter 20
Pharmacovigilance of Cannabis Products for Medical and Non-medical Purposes

Stephanie Jack

20.1 Legal Status of Cannabis

On October 17, 2018, Canada became the second country, after Uruguay, to legalize and regulate cannabis (*Cannabis sativa* L.) for medical and non-medical purposes under the *Cannabis Act* and its Regulations [1]. Previously, personal possession and production of cannabis was only legal in Canada for medical patients with an identified need under legal exemption, and subsequently under medical access regulations.[1]

Uruguay was the first country to legalize cannabis for medical and non-medical purposes through a legal access model of sale through pharmacies, cannabis social clubs, and home cultivation [2]. In other countries worldwide, cannabis is legal for medical purposes only [3, 4], with the exception of certain jurisdictions that have taken steps towards legalization of cannabis for non-medical purposes including some U.S. states and territories, Australian Capital Territory, Georgia, Luxembourg, and South Africa [5–9].

At the international level, there are ongoing discussions on the legal status of cannabis. In 2018, the World Health Organization's Expert Committee on Drug Dependence (WHO ECDD) recommended to the United Nations (UN) Secretary General that preparations considered to be pure cannabidiol (CBD) not be placed under international control, and recommended rescheduling of cannabis and several cannabis-related substances (cannabis plant and cannabis resin, extracts and tinctures of cannabis, delta-9-tetrahydrocannabinol (THC), and isomers of THC) [10,

[1] Marihuana Medical Access Regulations (2001); Marihuana for Medical Purposes Regulations (2014); Access to Cannabis for Medical Purposes Regulations (2016).

S. Jack (✉)
Office of Cannabis Science and Surveillance, Controlled Substances and Cannabis Branch, Health Canada, Ottawa, Canada
e-mail: stephanie.jack@hc-sc.gc.ca.ca

J. Barnes (ed.), *Pharmacovigilance for Herbal and Traditional Medicines*,
https://doi.org/10.1007/978-3-031-07275-8_20

11]. A decision by the United Nations Commission on Narcotic Drugs (CND) was pending at the time of writing [12].

20.2 Canadian Legislative and Regulatory Framework for Cannabis Products

20.2.1 A Public Health Approach to Cannabis Legalization

The purpose of the *Cannabis Act* is to protect public health and public safety by strictly regulating and controlling the production, distribution, sale, and possession of cannabis across Canada with three aims: keep cannabis out of the hands of youth; keep profits out of the pockets of the illicit market; and protect public health and safety by allowing adults access to legal cannabis [1]. This public health approach includes evidence-based education and awareness along with measures to protect the health and safety of vulnerable subpopulations (i.e., children, youth, pregnant and breastfeeding women, individuals with certain pre-existing health conditions) including health warning messages, child-resistant packaging, labelling, and promotion and access restrictions.

In 2017, the Task Force on Cannabis Legalization and Regulation completed its final report, *A Framework for Legalization and Regulation of Cannabis in Canada* [13], which identified the need for ongoing research and surveillance in order to monitor trends including population demographics and patterns of cannabis consumption (e.g. frequency of cannabis use, changes in behavior); to monitor harms, including new or incompletely documented health effects; and to identify areas that require additional research and surveillance.

According to the National Cannabis Survey (Third Quarter 2019)[2] [14], about 17% of Canadians aged 15 and older reported using cannabis in the previous 3 months (mid-August to mid-September). Younger adults were more likely to consume cannabis as compared to older adults (26% at ages 15–24 and 25% at ages 25–44, relative to 10% at ages 45–64, and 7% at ages 65 and older). However, seniors were the age group showing the most growth in cannabis usage, including new consumers. Seniors were also more likely to use cannabis for medical reasons, whereas middle aged adults reported both medical and non-medical use of cannabis and younger adults and youth reported mostly non-medical use. These data highlight that cannabis consumers are heterogeneous in nature and may involve aging adults with complex health histories such as chronic disease(s), comorbidities(s), and use of multiple health products or drug(s) (i.e., polypharmacy) that pose additional challenges in safety monitoring of cannabis products [15].

[2] Selected as relevant survey data at the time of writing this book chapter.

20.2.2 Cannabis Regulations

The *Cannabis Regulations* [16] outline several different regulatory provisions, including requirements for licensing, good production practices, use of pest control products, classes of cannabis products (including restriction on ingredients and THC concentration), promotions, packaging and labelling, and reporting and disclosure including adverse reaction reporting. A licence issued by Health Canada is required to conduct activities with cannabis including cultivation, processing (including sale for non-medical purposes), sale for medical purposes, import and export of cannabis, analytical testing, and research [17].

Classes of cannabis products that may be sold under this regulatory framework include dried and fresh cannabis, cannabis topicals, edible cannabis, and cannabis extracts (as well as cannabis plants and seeds). Although they do not undergo pre-market product approval, all new cannabis products must be notified to Health Canada at least 60 days before making the new cannabis product available for sale (other than cannabis plants or seeds) [18].

Individuals may access cannabis products through two pathways in Canada:

- Patients (including pediatric patients) who have a medical need for cannabis may obtain a medical authorization document from a healthcare practitioner (physician or nurse practitioner) that grants the patient or their caregiver the ability to purchase cannabis for medical purposes directly from federal licence holder(s); to produce a limited amount of cannabis for their own medical purposes; or to designate someone to produce it for them. Patients are authorized to possess the lesser of 150 g or a 30-day supply of dried cannabis (or the equivalent in cannabis product), in addition to the 30 g allowed for non-medical purposes for adults [19].
- Adult consumers may purchase cannabis for non-medical purposes from provincial or territorial authorized retailers (online or in retail stores) or may cultivate their own cannabis for personal use (four plants per residence), subject to the restrictions on age limit and growing in each province and territory.

Health products containing cannabis, or for use with cannabis, are regulated under a different legislative framework in Canada, the *Food and Drugs Act* and its Regulations. Two prescription drugs containing cannabis, namely Sativex® (THC and CBD) and Marinol® (THC; voluntarily withdrawn from market by the manufacturer) as well as the synthetic cannabinoid nabilone (Cessamet®), have been approved for sale. Natural health products containing permitted parts of the cannabis plant (e.g., hemp seed, hemp seed oil, hemp seed protein) containing no more than 10 ppm of THC) are also licensed by Health Canada [20].

20.2.3 Adverse Reaction Reporting Requirements

Under the *Cannabis Regulations* [16], companies who produce and/or sell a cannabis product for medical or non-medical purposes must comply with mandatory adverse reaction reporting requirements. Specifically, a holder of a licence that sells or distributes a cannabis product must:

- Within 15 days after becoming aware of a serious adverse reaction to the cannabis product, provide a detailed report containing all information in their possession that is associated with the use of the cannabis product by the individual who experienced the reaction.
- Prepare an annual summary report that contains a concise and critical analysis of all adverse reactions to the cannabis product that the holder became aware of during the previous 12 months.

The holder must retain the reports for at least 25 years after the day on which they are prepared.

An adverse reaction is defined in the *Cannabis Regulations* as a noxious and unintended response to a cannabis product, while a serious adverse reaction is defined as a noxious and unintended response to a cannabis product that requires inpatient hospitalization or a prolongation of existing hospitalization, causes congenital malformation, results in persistent or significant disability or incapacity, is life-threatening, or results in death. Other medically important events may not be immediately life-threatening, or result in death or hospitalization, but may be considered serious if they require significant medical intervention to prevent an outcome that is serious [21] and should also be reported.

Both domestic adverse reactions occurring inside of Canada, as well as foreign adverse reactions occurring outside of Canada with cannabis products that have been exported by a licence holder for medical use in other jurisdictions, are in scope of the reporting requirements outlined in Section 248 of the *Cannabis Regulations*. However, these rules do not apply to cannabis that is sold to be used for the purpose of a clinical trial, whether inside Canada or outside of Canada (i.e., cannabis exported for scientific research or experimental study). As outlined in the *Cannabis Exemption Regulations* [22], cannabis is considered an investigational drug if used under conditions of a clinical trial and is subject to mandatory reporting requirements under the applicable clinical trials regulatory framework(s) of the jurisdiction(s) in which the trial is being conducted, such as the *Food and Drugs Act* and the *Food and Drug Regulations* in Canada. As well, marketed drugs or health products containing cannabis would also be under adverse reaction reporting provisions under the applicable regulatory framework(s) (e.g., the *Food and Drug Regulations*).

20.3 Vigilance Framework for Cannabis Products

Under the *Cannabis Regulations*, Health Canada has implemented a vigilance framework for cannabis products based on the principles of pharmacovigilance, and adapted to meet the unique characteristics of cannabis products. Although pharmacovigilance was initially developed for pharmaceutical drugs, it has evolved to include other types of products, such as herbal and traditional medicines (i.e., natural health products or dietary supplements), vaccines, cosmetic products [23, 24]; and cannabis products for medical and non-medical purposes.

The aim of Health Canada's vigilance framework for cannabis products is the timely collection, monitoring, detection, and assessment of adverse reactions with cannabis products to support evidence-based decision-making, knowledge translation, public communication and compliance and enforcement activities. It involves a set of tools that work collaboratively, including regulations, guidance, systems, and procedures. Health Canada also monitors emerging adverse reactions with cannabis (unfinished or from undetermined or illicit sources), particularly for issues of public health importance (e.g., vaping-associated lung illness, accidental ingestion of edibles in children).

Many factors may contribute to the occurrence of adverse reactions with cannabis products, including the patient characteristics (e.g., sex, age), condition(s) of use of cannabis product(s) (product form, dose, route of administration, cannabinoid concentrations, other ingredients), alone and/or in combination with health products or other substances (e.g., prescription and non-prescription drugs, natural health products, food, alcohol, tobacco, illicit drugs), reason for use (for medical or non-medical purposes) as well as unintentional use(s) (e.g., medication errors, accidental ingestion), misuse/abuse, product quality issues (e.g., contamination, adulteration) or packaging or labelling issues. Consumers may also obtain products from the informal (e.g., friends or family) or illicit market, without quality control measures. Additionally, consumers may be healthy individuals consuming cannabis for non-medical purposes; patients with medical condition(s) consuming cannabis for medical purposes in relation to particular health condition(s); or consumers with pre-existing health condition(s) but consuming cannabis for non-medical purposes. As such, consumers or patients may have varying risk profiles as well varying risk tolerance(s) to adverse effects with cannabis.

20.4 Pharmacovigilance Databases for Cannabis Products

In Canada, adverse reaction reports with cannabis products are collected and housed in Health Canada's Canada Vigilance database (pharmacovigilance database) [25], which was originally developed to collect adverse reaction reports associated with health products including prescription and non-prescription drugs, vaccines, biologics, and natural health products. Internationally, other systems that collect spontaneous reports of adverse reactions associated with cannabis include Italy's Phytovigilance database [26], the United States Food and Drugs Administration's FAERS Database [27], and VigiBase, the World Health Organization's (WHO) global database of individual case safety reports, developed and maintained by the Uppsala Monitoring Centre [28]. As a member of the WHO program for International Drug Monitoring, Health Canada submits adverse reaction reports from the Canada Vigilance database to the UMC. The use of an established pharmacovigilance system to collect adverse reaction reports with cannabis products is advantageous for several reasons:

- It leverages existing systems and tools, such as standardized reporting form(s) both for industry as well as for voluntary reporters [29]. The use of a standardized reporting format helps to capture important data elements for cannabis reports, including:
  - Reporter
  - Patient
  - Suspect product(s) details: brand name(s), product form(s), route of administration, batch number(s), license holder (i.e., manufacturer or producer); concentrations of cannabinoids (THC/CBD), other labelled ingredients, duration and frequency of use, dosage and posology
  - Reason for use (medical or non-medical purpose)
  - Description of the reaction: adverse events (AEs), time to onset, temporal relationship, positive dechallenge or rechallenge, treatment or intervention(s), laboratory test(s)
  - Other suspect product(s)
  - Additional case details, including concomitant health product(s) or other substance(s) used, and relevant medical history, including previous use of cannabis product(s); whether the patient is cannabis naïve
- Adverse reactions reports are collected into a single centralized database irrespective of the source (e.g., cannabis industry, health product industry, consumers, patients, healthcare practitioners, retailers, or other sources). This is important as multiple suspect products and concomitant products may be involved, including cannabis products, prescription and non-prescription drugs, biologics, vaccines, natural health products, as well as accessories or medical devices. By centralizing the data, this enables signal detection across the various program areas, in particular, for suspected cases of interaction (pharmacokinetic or pharmacodynamic), thus enabling a more comprehensive understanding of factors that may be contributing to adverse reactions with cannabis products.
- The use of an established pharmacovigilance system also enables the use of internationally accepted standards, such as the Medical Dictionary for Regulatory Activities (MedDRA) [30], and International Council for Harmonization of Technical Requirements (ICH guidelines on pharmacovigilance) [31]. The use of MedDRA enables identification of reports with specific signs or symptoms according to standardized adverse event terms (i.e., preferred terms [PTs]), and also enables identification of case reports under clinically relevant groupings (e.g., HLT, HGLT, and SOC[3]) or defined medical condition(s) (e.g., SMQ[3]). Similarly to herbal medicines, coding of adverse reaction reports with cannabis products according to international standards facilitates data aggregation, searching, and signal identification at the national level, as well as monitoring and searching of data at the international level, i.e., in VigiBase [24, 32]. Additional activities supported by a centralized pharmacovigilance system include data collection, verification and management, conducting and documenting follow-up and identifying duplicates or linked reports for data accuracy [32].

[3] High Level Term; High Level Grouping Term; System Organ Class; Standardized MedDRA Query.

20.5 Sources of Adverse Reactions with Cannabis Products

Adverse reaction reports associated with cannabis products may originate from a number of different sources in Canada. Spontaneous adverse reactions must be reported to Health Canada by licence holders as part of their mandatory reporting obligations for serious adverse reactions. In addition, spontaneous reports may be reported directly to Health Canada on a voluntary basis by healthcare practitioners including primary health care practitioners as well as healthcare practitioners who are specifically managing patients using cannabis for medical purposes (e.g., part of a medical cannabis clinic); these cases are considered medically confirmed. Other types of voluntary reporters include consumers, patients, and provincial or territorial authorized retailers. Spontaneous adverse reactions may also originate from the literature (published case reports), as well from other Canadian surveillance programs (i.e., incident reporting programs for pesticides, consumer products, cosmetics, or food). Although cannabis as a sole suspect product is not within scope of Mandatory Reporting of Serious Adverse Reactions by Hospitals in Canada [33], cases involving both co-suspect drugs and cannabis products would still be required to be reported, which is important for detecting potential cases of interaction(s). Hospitals, as well as regional and provincial health authorities, may also report adverse reactions with cannabis products (as a sole suspect product) on a voluntary basis.

Adverse reaction reports may also originate from studies or other organized data collection systems, including observational studies conducted by licence holders, which are also within scope of the adverse reaction reporting requirements under the *Cannabis Regulations*. In addition, research conducted in humans involving cannabis or cannabis products that are not clinical trials (e.g., studies undertaken to evaluate taste or sensory attributes of cannabis product(s) for consumer preference) must also meet conditions for serious adverse reaction reporting outlined in their Cannabis Research Licence.

Active surveillance of cannabis-related adverse health outcomes may be derived from observational studies that collect new data on the use of cannabis or cannabis products from primary data sources (e.g., surveys) or from secondary data sources examining existing data collected for other purposes such as administrative healthcare data (e.g., administrative claims or electronic health records) [34, 35]. Health Canada monitors aggregate-level administrative healthcare data in Canada through the Canadian Institute for Health Information (CIHI), including emergency department visits and inpatient hospitalization data associated with cannabis and other substances (using ICD diagnostic codes) [36]. Health Canada also monitors other sources of data, including from Canadian Poison Centers and from the Canadian Hospitals Injury Reporting and Prevention Program (CHIRPP), which is an emergency department-based injury and poisoning sentinel surveillance system that collects data on substances including cannabis from 11 pediatric and 6 general emergency departments across Canada [37]. These data sources are important in the ongoing surveillance of hospitalizations and injuries, but these data are limited in their ability to distinguish cases according to the type and form of cannabis product used (i.e., legal status, brand name, licence holder, product form, concentration(s) of cannabinoids [THC/CBD], or other ingredients). An additional challenge in using administrative health data, such as insurance claims data, is that although access to cannabis for medical purposes requires a medical authorization document, it is not systematically covered under Canadian provincial or private drug insurance plans (with some exceptions). Additionally, healthcare is administered by provinces and territories, meaning that electronic health records are collected within each jurisdiction and are subject to privacy restrictions. This limits the ability to use administrative health data for cannabis products to aggregate and compare across provincial and territorial datasets, although efforts have been made to develop a common data model [38].

In contrast, certain organized data systems have been developed specifically to collect data on patients using cannabis. For example, the Quebec Cannabis Registry[4] is a research database with aim to collect data on indication(s), dosage(s), benefits, and adverse events associated with the use of cannabis for medical purposes [39]. DATACANN (DATAbase for CANNabinoid Consumption and Study) is a provincial pain registry in Ontario that will serve to collect longitudinal data on patients' use of cannabis over the course of their treatment period [40]. Further, the Canadian Pediatric Surveillance Program (CPSP) is also conducting studies examining serious and life-threatening events reported by pediatricians associated with exposure to cannabis for medical or non-medical purposes by children and youth [41, 42].

There are other Canadian surveillance tools that collect survey-based data on cannabis utilisation, knowledge and behaviors including the Canadian Cannabis Survey [43]; the National Cannabis Survey [44], the Canadian Alcohol and Drugs Surveys (adult and student surveys) [44]; COMPASS, a longitudinal study exploring youth health behaviors of high school students over time, including use of cannabis [45], among others.

In jurisdictions outside Canada, the use of medical cannabis is being monitored through patient cohort registries on a national (e.g., Israel, United Kingdom) [46, 47], state level (e.g., Florida, Minnesota, Hawaii, Illinois) [48] or institution-level basis (e.g., hospitals) [49].

20.6 Signal Management Process with Cannabis Products

Adverse reaction reports with cannabis products received in the Canada Vigilance database are monitored on a continuous basis, and serious cases are assessed for causality using the WHO-UMC causality classification system [50]. A signal may be generated from two or more cases involving clinically related events originating from spontaneous adverse reaction data or other sources (literature, studies, or other organized data collection systems) suggesting a new potentially causal association,

[4]Launched in May 2015 and recruitment ending in October 2018 with 3400 patients from 71 physician-collaborators across 11 regions in Quebec.

or new aspect of a known, or partially documented, association between a cannabis product and an adverse health effect. These case reports may be further prioritized (signal prioritization) and evaluated as a case series (signal evaluation) in order to validate the signal. A case definition is formulated in order to develop search parameters to identify and retrieve other relevant cases. Additional sources of data are also integrated into the signal evaluation, including foreign adverse reaction data (e.g., from WHO VigiBase or published case reports), and scientific literature. These data together are analyzed for clinical relevance, temporal association, biological plausibility, and strength of association (positive dechallenge and rechallenge), in order to confirm or rule out an association [51]. If a case report includes a suspected product quality issue, this may prompt the regulator to take steps towards verifying compliance with the regulations, including request for additional information, audit of documents, on-site inspection of the manufacturing facility, or, possible recall if the product is found to be out of compliance [52].

Signal evaluations are also important for determining whether additional educational and awareness resources are required to inform consumers, patients, and healthcare practitioners about cannabis and its health effects. Adverse reaction data are also analyzed on a cumulative basis annually for trends according to seriousness, intended use (medical or non-medical purposes), product form, route of administration, cannabinoid concentrations, and patient demographics. By comparing these data on a cumulative basis, trends may be evaluated over time for safety signals [53, 54].

20.7 Challenges in the Vigilance of Cannabis Products

Unique challenges exist in the safety monitoring for cannabis products; however, by identifying issues early in the development of a vigilance framework, it is hoped that these may be overcome in order to establish international best practices and standards for the vigilance of cannabis products for medical and non-medical purposes.

20.7.1 Limited Evidence on Cannabis Products

The evidence on the safety and efficacy of cannabis (*Cannabis sativa* L.) has been restricted by a long history of prohibition worldwide, which has limited research and surveillance. Knowledge has largely been derived from evidence on THC-dominant dried cannabis that is smoked, or derived from clinical evidence on the safety and efficacy of cannabinoid-based prescription drugs authorized for specific medical indications [53]. These data, although informative, are not necessarily reflective of the broad array of cannabis products legally available for medical and/or non-medical purposes. Through ongoing efforts to improve monitoring and collection of adverse reaction data with cannabis products, greater specificity in the data will be achieved, enabling a better understanding of health effects according to cannabis product form, route of administration, cannabinoid concentration(s) and the potential contribution of other constituents or ingredients, which permits more precision in signal detection and evaluation.

20.7.2 *Variability of Cannabis Products*

Cannabis products for medical and non-medical purposes include fresh and dried herbal material (whole or milled flowers and leaves, pre-rolled); cannabis extracts including ingestible cannabis extracts (e.g., oils, sprays, capsules, strips) inhaled chemically concentrated extracts (highly concentrated extracts, such as vaping liquids or cartridges, butane hash oil, shatter, budder, and wax[5]) and physically concentrated extracts (loose trichomes or pressed resin from the plant, such as hash, kief, resin or rosin[6]); cannabis topicals (cannabis extracts in topical preparations, e.g., lotions, creams, gels, salves, patches, bath products); and edible cannabis products (extracts of cannabis contained in beverages, baked goods, confectionary, baking ingredients, or other shelf-stable products) [54, 55]. These can vary in their routes of administration, including oral use (including ingestion as well as sublingual or buccal administration); inhalation by smoking or vaping; topical administration, and others.

Cannabis also contains over a 100 different cannabinoids, of which delta-9-tetrahydrocannabinol (THC) and cannabidiol (CBD) are the most well-known [53]. Certain products may be higher in THC relative to CBD ("THC dominant" or "THC leaning"), be relatively equivalent in THC and CBD ("THC:CBD balanced" or "equilibrated"), or be higher in CBD relative to THC ("CBD dominant" or "CBD leaning"). Other cannabis constituents may be present in variable quantities or concentrations (e.g., minor cannabinoids and terpenes), and other ingredients may also be present (e.g., food ingredients in edibles; cosmetic ingredients in topicals; carrier oils or solvents in cannabis extracts).

These factors add to the variability and complexity of cannabis products and may influence their pharmacokinetics and pharmacodynamics. For example, inhaled cannabis products have a more rapid onset of action, achieve higher plasma concentrations of cannabinoids and shorter duration of pharmacodynamic effects; in contrast, oral cannabis products have a slower onset of action, lower peak concentrations of cannabinoids, and a longer duration of pharmacodynamic effects [53].

[5]Chemically concentrated extracts are made by using solvents such as butane or carbon dioxide resulting in extracts of varying consistency from liquid (e.g., vaping liquid, butane hash oil) to cream or wax-like (e.g., wax, honey, or budder) or solid form (e.g., shatter).

[6]Physically concentrated extracts are made by using physical aids (e.g., screens, ice bath, pressure) to cannabis plant material resulting in extracts of varying consistency from sticky (e.g., rosin and resin; hash) to dry or crumbly (sift, crumble).

Additionally, oral cannabinoids are subject to first-pass metabolism in the liver through several cytochrome P450 enzymes that are involved in metabolism of other drugs, increasing the potential risk of cannabis-drug interactions [53, 56].

Cannabis products may also involve the use of accessories (e.g., dried cannabis used with vaporizers, bongs, pipes; cannabis extracts used with vape pens, pipes/bongs or dabbing rig/nail[7]), meaning that additional factors should be considered, such as consumer behavior (i.e., depth and duration of inhalation and breath hold; whether used according to manufacturer's instructions or modified; use with non-cannabis substances such as flavorings or nicotine), and device function or performance (i.e., functioning properly; malfunctioning, such as clogging; battery issue(s), such as overheating; whether new, borrowed or second-hand). Additionally, all legal cannabis products are required to meet safety and quality specifications under their respective legal frameworks; however, product quality or compliance issues may still arise (e.g., packaging or labelling issues; potential contamination; and accessories issues). The use of products or accessories obtained from the illicit market must also be considered; therefore the source or place of purchase is important to clarify.

In summary, although cannabis was previously regarded as a homogeneous substance, cannabis products vary significantly and additional factors may also play a role, all of which are important to consider when collecting, coding, and assessing adverse reaction data with cannabis products.

20.7.3 International Nomenclature

Cannabis sativa L. is formally recognized in the scientific community as a single and highly polymorphic species with subspecies indica and sativa (*C. sativa* L. subsp. *sativa*; *C. sativa* L. subsp. *indica*) [57, 58]. However, the terms "sativa," "indica," and "hybrid" (based on morphology) have been used erroneously in the common vernacular to classify cannabis according to phenotype(s) and "expected" physiological effect(s) based on anecdotal evidence. The use of the aforementioned terms in this manner is not reflective of formal nomenclature, nor of the concentration of cannabinoids or other compounds (e.g., terpenes) in cannabis products that may be associated with clinical effects [57, 59–61]. While the inaccurate use of botanical terms has been exacerbated by the illicit market in cannabis for many years, recent developments in the regulation and standardization of cannabis for medical purposes are expected to improve the appropriate use of terminology for cannabis and cannabis products [62–66].

It remains to be seen whether cannabis products will be integrated into existing international nomenclature and therapeutic classification for herbal substances that

[7] Dab or dabbing refers to a method of applying a cannabis extract to a hot surface (such as a nail or pen attached to a dab rig, pipe or bong) heated by a torch or electronically and inhaling the vapor (high heat vaporization).

are used for pharmacovigilance purposes, such as the Herbal Anatomical Therapeutic Classification (HATC) [67]. The HATC is a framework for the nomenclature and therapeutic classification of herbal substances and their combinations that classifies active substances according to the organ or system on which they act and their therapeutic, pharmacological, and chemical properties. In contrast, the ATC classification is used for prescription drugs containing cannabinoids, such Epidiolex® (CBD oral solution) [68] and Sativex® (THC and CBD buccal spray) [69]. To date, the HATC/ATC classification has not been systematically applied to cannabis products for medical or non-medical purposes. This is important as WHODrug Global builds on the HATC/ATC system as the international reference for medicinal product information maintained by the Uppsala Monitoring Centre [70] that is used as the drug dictionary for coding suspect and concomitant drugs and herbal remedies in adverse reaction data in the WHO's Vigilyze database; this resource is also relevant for cannabis products.

In Canada, cannabis products are coded according to an internal standard for capturing cannabis product data in the Canada Vigilance database, including brand name, licence holder, THC and CBD concentrations, dosage form, and route of administration. Dosage forms are developed from international standards such as ICH E2B; those that do not exist in E2B are also manually created. Case reports are further classified according to reported reason or indication for use of the cannabis product(s) (i.e., medical purposes or non-medical purposes). All adverse events are coded according to MedDRA terminology.

One challenge is that coding of cannabis products at the local level in the national or regional database does not necessarily get translated when data extracts are transmitted as individual case safety reports (ICSRs) to the WHO VigiBase, resulting in challenges in coding and characterization of case reports at the international level. For example, certain ICSRs associated with "CBD-dominant" cannabis products (e.g., "CBD drops oil") may be inadvertently coded as "cannabidiol (CBD)" under WHODrug active ingredient; however, this should in fact be coded as *Cannabis sativa* L. (whole extracts) and not as a pure cannabinoid, as this product would contain a greater concentration of CBD but also some THC. Furthermore, ICSRs coded as *Cannabis sativa* as the active ingredient often do not include tradename description(s), despite additional information existing in the local or national pharmacovigilance database (e.g., Canada Vigilance database). This limits the ability to distinguish between pharmaceutical or herbal products containing cannabis from cannabis products used for medical or non-medical purposes that may contain varying concentrations or quantities THC and CBD and other cannabinoids. Additionally, although MedDRA is used to code adverse event terms overall, there are limited codes specific to cannabis (cannabis abuse, cannabis dependence, cannabis withdrawal). Terms relating to the broader physiological effects of cannabis remain to be developed. For example, at the time of the writing of this book chapter there was no specific MedDRA code for "cannabis hyperemesis syndrome," a known but often under-recognized adverse effect from the use of cannabis [71].

As such, development of international nomenclature for cannabis used for medical and non-medical purposes in Canada and in other jurisdictions would help greatly in the safety monitoring and clinical data management for these products at an international level.

20.8 Conclusion

The vigilance of cannabis products is relatively new and has yet to be implemented in a standardized manner at the international level given that legalization of cannabis is still evolving globally. However, in applying some key foundational principles of pharmacovigilance developed for pharmaceuticals and herbal medicines, the development of a vigilance system for cannabis products is possible, and is important in the ongoing monitoring, detection, assessment, and management of adverse reactions with cannabis products and furthering real-world evidence under legal frameworks moving forwards.

References

1. Government of Canada (2019) Cannabis Act. https://laws-lois.justice.gc.ca/PDF/C-24.5.pdf. Accessed 31 Oct 2019
2. European Monitoring Centre for Drugs and Drug Addiction. Cannabis policy: status and recent developments. http://www.emcdda.europa.eu/publications/topic-overviews/cannabis-policy/html_en. Accessed 11 Nov 2019
3. Aguilar S, Gutierrez V, Sanchez L et al (2018) Briefing paper: Medicinal cannabis policies and practices around the world. International Drug Policy Consortium. http://fileserver.idpc.net/library/Medicinal%20cannabis%20briefing_ENG_FINAL.PDF. Accessed 11 Nov 2019
4. European Monitoring Centre for Drugs and Drug Addiction (2018) Rapid communication. Medical use of cannabis and cannabinoids: questions and answers for policymaking. http://www.emcdda.europa.eu/publications/rapid-communications/medical-use-of-cannabis-and-cannabinoids-questions-and-answers-for-policymaking_en. Accessed 11 Nov 2019
5. National Conference of State Legislatures (2019) State medical marijuana laws. http://www.ncsl.org/research/health/state-medical-marijuana-laws.aspx. Accessed 11 Nov 2019
6. Wayne S (2018) Smoking marijuana legalised in Georgia. http://georgiatoday.ge/news/11592/Smoking-Marijuana-Legalized-in-Georgia. Accessed 11 Nov 2019
7. European Monitoring Centre for Drugs and Drug Addiction (2019) Australian capital territory legalised personal cannabis growing and consumption. http://www.emcdda.europa.eu/news/2019/australian-capital-territory-legalises-personal-cannabis-growing-and-consumption_en. Accessed 12 Nov 2019

8. BBC News (2018) South Africa's highest court legalizes cannabis use. https://www.bbc.com/news/world-africa-45559954. Accessed 12 Nov 2019
9. Health Europa (2019) Meet the first European country to legalise cannabis: Luxembourg. https://www.healtheuropa.eu/first-european-country-legalise-cannabis-luxembourg/92793/. Accessed 30 Nov 2019
10. World Health Organization (2018) WHO expert committee on drug dependence. https://apps.who.int/iris/bitstream/handle/10665/279948/9789241210225-eng.pdf?ua=1. Accessed 12 Nov 2019
11. World Health Organization (2018) Q & A: WHO expert committee on drug dependence review of cannabis. https://www.who.int/medicines/access/controlled-substances/CannabisQAv2.pdf?ua=1. Accessed 12 Nov 2019
12. United Nations Office on Drugs and Crime (2019) WHO's Recommendation on cannabis and cannabis-related substances. UN Commission on narcotic drugs (UNCND). https://www.unodc.org/unodc/en/commissions/CND/Mandate_Functions/current-scheduling-recommendations.html. Accessed 12 Nov 2019
13. Government of Canada (2016) A framework for the legalization and regulation of cannabis in Canada. The final report of the task force on cannabis legalization and regulation. https://www.canada.ca/en/health-canada/services/drugs-medication/cannabis/laws-regulations/task-force-cannabis-legalization-regulation/framework-legalization-regulation-cannabis-in-canada.html. Accessed 2 Dec 2019
14. Statistics Canada (2019) National cannabis survey third quarter results. https://www150.statcan.gc.ca/n1/pub/13-610-x/cannabis-eng.htm. Accessed 2 Dec 2019
15. Nobili A, Garattini S, Mannucci PM (2011) Multiple diseases and polypharmacy in the elderly: challenges for the internist of the third millennium. J Comorb 1:28–44
16. Government of Canada (2019) Cannabis Regulations SOR/2018-144. https://laws-lois.justice.gc.ca/PDF/SOR-2018-144.pdf. Accessed 1 Nov 2019
17. Government of Canada (2019) Applying for a licence under the Cannabis Act and Regulations. https://www.canada.ca/en/health-canada/services/drugs-medication/cannabis/industry-licensees-applicants/applying-licence.html. Accessed 13 Nov 2019
18. Government of Canada (2019) Notice of new cannabis product guide. https://www.canada.ca/en/health-canada/services/publications/drugs-health-products/notice-new-cannabis-product/guide.html. Accessed 22 Nov 2019
19. Government of Canada (2018) Cannabis for medical purposes under the Cannabis Act. Information and improvements. https://www.canada.ca/en/health-canada/services/drugs-medication/cannabis/medical-use-cannabis.html. Accessed 8 Nov 2019
20. Government of Canada (2019) Health products containing cannabis or for use with cannabis: guidance for the Cannabis Act, the Food and Drugs Act, and related regulations. https://www.canada.ca/en/health-canada/services/drugs-health-products/drug-products/applications-submissions/guidance-documents/guidance-cannabis-act-food-and-drugs-act-related-regulations/document.html. Accessed 5 Nov 2019
21. International Conference on Harmonization of Technical Requirements for Registration of Pharmaceuticals for Human Use (1994) ICH Topic E2A clinical safety data management: definitions and standards for expedited reporting. https://www.ema.europa.eu/en/documents/scientific-guideline/international-conference-harmonisation-technical-requirements-registration-pharmaceuticals-human-use_en-15.pdf. Accessed 3 Dec 2019
22. Government of Canada (2020) Cannabis Exemption (Food and Drugs Act) Regulations. https://laws-lois.justice.gc.ca/eng/Regulations/SOR-2016-231/index.html. Accessed 27 Jan 2020
23. Tissier MH, Lepagnol F (2002) Cosmetovigilance: a French pharmacovigilance system for cosmetics developed by the French health products safety agency. A proposal for the future. Thérapie 57(3):273–282
24. Shetti S, Kumar CD, Sriwastava NK et al (2011) Pharmacovigilance of herbal medicines: current state and future directions. Pharmacogn Mag 7(25):69–73

25. Health Canada (2018) Canada vigilance program. https://www.canada.ca/en/health-canada/services/drugs-health-products/medeffect-canada/canada-vigilance-program.html. Accessed 31 Oct 2019
26. Crescioli G, Lombardi N, Bettiol A et al (2020) Adverse events following cannabis for medical use in Tuscany: an analysis of the Italian phytovigilance database. Br J Clin Pharmacol 86(1):106–120
27. United States Food and Drugs Administration. FDA adverse event reporting system (FAERS) public dashboard. https://www.fda.gov/drugs/questions-and-answers-fdas-adverse-event-reporting-system-faers/fda-adverse-event-reporting-system-faers-public-dashboard. Accessed 28 Jan 2020
28. World Health Organization Uppsala Monitoring Centre WHO Vigilyze database. https://www.who-umc.org/vigibase/vigilyze/. Accessed 2 Dec 2019
29. Health Canada (2019) Report a side effect. https://www.canada.ca/en/health-canada/services/drugs-health-products/medeffect-canada/adverse-reaction-reporting.html. Accessed 2 Dec 2019
30. Medical Dictionary for Regulatory Activities (MedDRA). https://www.meddra.org/. Accessed 9 Dec 2019
31. ICH. Efficacy guidelines: E2A-E2F pharmacovigilance. https://www.ich.org/page/efficacy-guidelines. Accessed 9 Dec 2019
32. World Health Organization (2004) WHO Guidelines on safety monitoring of herbal medicines, Geneva. https://apps.who.int/medicinedocs/documents/s7148e/s7148e.pdf. Accessed 9 Dec 2019
33. Health Canada (2018) Mandatory reporting of serious adverse drug reactions and medical device incidents by hospitals. Draft guidance document. https://www.canada.ca/en/health-canada/services/drugs-health-products/public-involvement-consultations/medeffect-canada/consultation-draft-guidance-hospital-mandatory-reporting-regulations/guidance-document.html. Accessed 12 Dec 2019
34. European Network of Centres for Pharmacoepidemiology and Pharmacovigilance (ENCePP) (2010) Guide on methodological standards in pharmacoepidemiology (Revision 7). EMA/95098/2010. http://www.encepp.eu/standards_and_guidances/documents/ENCePPGuideonMethStandardsinPE_Rev7.pdf. Accessed 12 Dec 2019
35. Huang YL, Moon J, Segal JB (2014) A comparison of active adverse event surveillance systems worldwide. Drug Saf 37(8):581–596
36. Canadian Institute for Health Information (2019) Hospital stays for harm caused by substance use among youth aged 10 to 24. https://www.cihi.ca/sites/default/files/document/hsu-youth-report-2019-en-web.pdf. Accessed 5 Jan 2020
37. Rao DP, Abramovici H, Crain J et al (2018) The lows of getting high: sentinel surveillance of injuries associated with cannabis and other substance use. Can J Public Health 109(2):155–163
38. Platt R, Dormuth CR, Chateau D et al (2016) Observational studies of drug Safety in multi-database studies: methodological challenges and opportunities. EGEMS (Wash DC) 4(4):1221
39. Ware MA, Vigano A, Beaulieu P et al (2019) The Quebec cannabis registry, a database on the use of cannabis for medical purposes: final results. In: Abstracts of the 29th annual symposium of the International Cannabinoid Research Society (ICSR), Bethesda, Maryland
40. Michael G DeGroote Centre for Medicinal Cannabis Research (2020) DATACANN. Database for cannabis consumption and study. McMaster University. https://datacann.mcmaster.ca/. Accessed 3 Jan 2020
41. Canadian Pediatric Society Program (2018) 2018 results: serious and life-threatening events associated with non-medical (recreational) cannabis use in Canadian children and youth. https://www.cpsp.cps.ca/uploads/publications/CPSPResults2018.pdf. Accessed 5 Dec 2019
42. Canadian Pediatric Society Program (2019) Serious adverse events related to cannabis used for medical purposes questionnaire. https://www.cpsp.cps.ca/surveillance/study-etude/serious-adverse-events-related-to-cannabis-used-for-medical-purposes. Accessed 30 Jan 2020

43. Government of Canada (2019) Canadian cannabis survey 2019 - summary. https://www.canada.ca/en/health-canada/services/publications/drugs-health-products/canadian-cannabis-survey-2019-summary.html. Accessed 3 Jan 2020
44. Government of Canada (2019) Cannabis stats hub, health. https://www150.statcan.gc.ca/n1/pub/13-610-x/cannabis-eng.htm. Accessed 3 Jan 2020
45. Zuckermann AME, Battista K, de Groh M et al (2019) Pre-legalisation patterns and trends of cannabis use among Canadian youth: results from the COMPASS prospective cohort study. BMJ Open 9:e026515
46. Aviram J, Pud D, Schiff-Keren B et al (2018) The Israeli multi-center registry of medical cannabis (MC) for chronic pain: current findings. Rambam Maimonides Med J 91:13–14
47. Wise J (2019) Medical cannabis registry launched to build up evidence base. BMJ 367:16360. https://www.bmj.com/content/367/bmj.l6360. Accessed 3 Jan 2020
48. Minnesota Department of Health. Medical cannabis registry. https://www.health.state.mn.us/people/cannabis/registry.html. Accessed 3 Jan 2020
49. Children's Hospital of Philadelphia (2019) Medical cannabis registry, Philadelphia. https://www.chop.edu/clinical-trial/medical-cannabis-registry. Accessed 3 Jan 2020
50. World Health Organization. The use of the WHO-UMC system for standardized case causality assessment. https://www.who.int/medicines/areas/quality_safety/safety_efficacy/WHOcausality_assessment.pdf. Accessed 31 Oct 2019
51. Potts J, Genov G, Segec A et al (2019) Improving the safety of medicines in the European union: from signals to action. Clin Pharmacol Therap 107:521–529. https://doi.org/10.1002/cpt.1678
52. Health Canada (2019) Cannabis voluntary recall guide. https://www.canada.ca/en/health-canada/services/publications/drugs-health-products/cannabis-voluntary-recall/guide.html. Accessed 7 Nov 2019
53. Health Canada (2018) Information for health care professionals: cannabis (marihuana, marijuana) and the cannabinoids. https://www.canada.ca/en/health-canada/services/drugs-medication/cannabis/information-medical-practitioners/information-health-care-professionals-cannabis-cannabinoids.html. Accessed 31 Oct 2019
54. Health Canada (2019) About cannabis. https://www.canada.ca/en/health-canada/services/drugs-medication/cannabis/about.html. Accessed 31 Oct 2019
55. Health Canada (2019) What you need to know if you choose to consume cannabis. https://www.canada.ca/en/health-canada/services/drugs-medication/cannabis/resources/what-you-need-to-know-if-you-choose-to-consume-cannabis.html. Accessed 31 Oct 2019
56. Antoniou T, Bodkin J, Ho J (2020) Drug interactions with cannabinoids. CMAJ 192(9):E206. https://doi.org/10.1503/cmaj.191097
57. McPartland JM (2018) Cannabis systematics at the levels of family, genus, and species. Cannabis Cannabinoid Res 3(1):203–212
58. Pollio A (2016) The name of cannabis: a short guide for nonbotanists. Cannabis Cannabinoid Res 1(1):234–238
59. Mudge EM, Murch SJ, Brown PN (2018) Chemometric analysis of cannabinoids: chemotaxonomy and domestication syndrome. Sci Rep 8:13090
60. Rahn B (2018) What's the difference between cannabis types? Leafly. https://www.leafly.ca/news/cannabis-101/sativa-indica-and-hybrid-differences-between-cannabis-types. Accessed 6 Feb 2020
61. Sawler J, Stout JM, Gardner KM et al (2015) The genetic structure of marijuana and hemp. PLoS One 10(8):e0133292
62. ASTM (2019) ASTM international technical committee D37 on cannabis. astm.org/COMMIT/D37_Fact_Sheet_2019.pdf. Accessed 3 Feb 2020
63. Smalls E (2017) Classification of Cannabis sativa L. in relation to agricultural, biotechnological, medical and recreational utilization. In: Chandra S, Lata H, Elshohly M (eds) Cannabis sativa L.- botany and biotechnology. Springer, Cham

64. Manns D (2019) Cannabis for medicinal use: development of pharmacopoeia as a quality standard. Bundesgesundheitsblatt, Gesundheitsforschung, Desundheitsschutz 62(7):806–810
65. World Health Organization (2018) Cannabis plant and cannabis resin. Section 1: Chemistry. WHO expert committee on drug dependence pre-review. https://www.who.int/medicines/access/controlled-substances/Section1.CannabisPlant.Chemistry.pdf?ua=1
66. Sarma ND, Waye A, ElSohly MA et al (2020) Cannabis inflorescence for medical purposes: USP considerations for quality attributes. J Nat Prod 83(4):1334–1351
67. World Health Organization (2019) The herbal anatomical therapeutic chemical classification system. WHODrug. https://www.who-umc.org/whodrug/whodrug-portfolio/whodrug-global/herbal-atc/
68. GW Research Ltd. (2018) Epidiolex oral solution. Highlights of the prescribing information, Carlsbad, California. https://www.accessdata.fda.gov/drugsatfda_docs/label/2018/210365lbl.pdf. Accessed 3 Jan 2020
69. GW Pharma Ltd. (2020) Sativex product monograph, Histon, Cambridge, UK. https://pdf.hres.ca/dpd_pm/00054388.PDF
70. World Health Organization Uppsala Monitoring Centre (2020) WHODrug Global. https://www.who-umc.org/whodrug/whodrug-portfolio/whodrug-global/. Accessed 3 Jan 2020
71. Galli JA, Sawaya RA, Friedenberg FK (2011) Cannabis hyperemesis syndrome. Curr Drug Abuse Rev 4(4):214–224

Chapter 21
Pharmacovigilance for Herbal Medicines in Brazil

Marcelo Vogler de Moraes and Juliana de Castro Zoratto

21.1 Introduction

Brazil covers 8.5 million km^2 and occupies almost half of South America, encompassing several climatic zones that provide great ecological variations and forming distinct biogeographic zones or biomes. The variety of its biomes (Amazon Rainforest, Pantanal, Cerrado, Caatinga, Pampas, and Atlantic Forest) reflects the enormous wealth of Brazilian flora and fauna, leading Brazil to harbor the greatest biodiversity on the planet. This abundant variety of life—translated into more than 20% of the total number of species on Earth—elevates Brazil to the position of the main nation among the 17 countries with the greatest biodiversity [1].

In addition to this genetic collection, Brazil has a rich cultural and ethnic diversity that has resulted in a considerable accumulation of traditional knowledge and technologies, transmitted from generation to generation, among which stands out the vast knowledge collection on the management and use of medicinal plants [2].

Since 2006, the guarantee of safe access and the rational use of medicinal plants and herbal medicinal products (HMPs) in Brazil have been subject of public policies (National Policy for Medicinal Plants and Herbal Medicinal Products and the National Policy for Integrative and Complementary Practices in *Sistema Único de Saúde* (SUS), which guiding principles are the expansion of therapeutic options and improvement of health care for the Brazilian public health system (*Sistema Único de Saúde—SUS*) users [2, 3]. According to 2017 data, Herbal Medicine Services

M. Vogler de Moraes (✉) · J. de Castro Zoratto
Regulation Specialist, Brazilian Health Regulatory Agency - Anvisa,
Brasília, Brazil
e-mail: marcelo.moraes@anvisa.gov.br; juliana.zoratto@anvisa.gov.br

J. Barnes (ed.), *Pharmacovigilance for Herbal and Traditional Medicines*,
https://doi.org/10.1007/978-3-031-07275-8_21

were offered in 1108 municipalities and, in the same year, 12 types of herbal medicines that are listed on the National List of Essential Medicines (*Relação Nacional de Medicamentos Essenciais—RENAME*) were freely distributed to the population, representing 2,183,098 pharmaceutical units (where one unit is one bottle, or one blister pack of tablets) of herbal medicines dispensed [4, 5].

The minimum quality requirements for medicinal plants, among other specifications, are described along 83 monographs in the sixth edition of the Brazilian Pharmacopeia, including 22 monographs for tinctures (herbal preparations), 19 monographs for fluid extracts (herbal preparations), and 25 monographs for oils, fats, and waxes.

The Brazilian Pharmacopeia also includes particular documents related to herbal medicinal products and medicinal plants, such as the *Memento de Fitoterápicos* (MFFB) and the *Formulário de Fitoterápicos* (FFFB). The *Memento de Fitoterápicos* consists of a document for a quick consultation, by prescribing professionals, to guide the prescription of medicinal plants and herbal medicinal products. Monograph contents are based on scientific evidence that may assist the prescriber's therapeutic conduct. The second compendium focuses mainly on compounding practices and dispensing of herbal medicinal products, contributing to Herbal Medicine Services and pharmacies across the country [6, 7].

21.2 Herbal Medicinal Products Regulatory Framework

21.2.1 Marketing Authorization

In Brazil, the medicinal plant is defined as the plant species itself, cultivated or not, used for therapeutic purposes, either in its fresh state after harvest/collection or after a drying process. A fresh medicinal plant (or its parts) can be subjected to stabilization processes, when applicable, and drying, taking the whole, torn, comminuted, or powdered forms, constituting what is called a "herbal drug." The product of an extraction from the fresh medicinal plant or herbal drug, which contains the substances responsible for the therapeutic action, is called a "vegetal derivative" (extract, fixed and volatile oil, wax, exudate, and others) [7].

The herbal drug, being the active ingredient in the formulation, can be marketed in this way (without further processing), as a medicinal tea for use in extemporaneous preparations, or it can be marketed in other pharmaceutical dose forms, such as capsules, for example, which may also contain excipients. When the formulation consists of herbal derivatives, whether associated with excipients or not, it can be administered under different pharmaceutical dose forms [8].

An herbal medicinal product is defined as the product obtained from active vegetal raw material (medicinal plant, herbal drug, or herbal derivative), except those that include isolated or highly purified active substances (synthetic, semisynthetic, or natural) or the associations of these with other extracts (herbal or other sources such as animal), with prophylactic, curative, or palliative purpose. This product category includes the herbal medicines and traditional herbal products (THPs), which can be simple (when the active ingredient comes from a single medicinal plant species) or compound (when the active ingredient comes from more than one medicinal plant species) [9].

The difference between an herbal medicine and a THP is related to the proofs of efficacy and safety when applying for a marketing authorization. The first is based on clinical evidence and is characterized by consistent quality. The second is based on data on safe and effective use (traditional use), demonstrated in technical-scientific documentation, with no known or informed evidence of risk to the user's health. In addition, THPs are designed to be used without a physician's supervision for diagnostic, prescription, or monitoring purposes; cannot refer to diseases, disorders, conditions, or actions considered as serious; cannot contain known hazardous chemical groups in concentrations above safe limits; and should not be administered by injectable and ophthalmic routes [8]. Evidence by traditional use is a form of proof recommended by the World Health Organization (WHO) and seen in major international legislation frameworks for herbal medicinal products, such as those from the European Community, Canada, Australia, Mexico, and Brazil [8, 9].

In Brazil, herbal medicines are subject to "simplified marketing authorization/regular marketing authorization" processes, and THPs are subject to both "simplified marketing authorization/regular marketing authorization" processes and "notification." THPs constitute a new class of medicines created by Brazilian Health Regulatory Agency (Anvisa) in order to make it clear to consumers whether the product they are using has gone through all clinical tests for safety and efficacy proof, or if it has been approved for its effective and safe use on the basis of long-term (traditional use) [8].

Marketing authorization and post-approval changes of herbal medicinal products, as well as notification of THPs, currently follow specific regulations that have been published by Anvisa in 2014 and in line with international regulations. These comprise:

- Resolution-RDC n° 26/2014, which provides for the marketing authorization of herbal medicines and the marketing authorization and notification of THPs [9].
- Resolution-RDC n° 38/2014, which provides for post-approval changes applications for herbal medicines and THPs and other measures [10].

Additional regulatory frameworks of a complementary nature were also published to detail the rules and procedures of external scope with additional guidance to the Resolutions, containing, for example, the "List of herbal medicines for simplified marketing authorization" and "List of THP for simplified marketing authorization"; the Guidance for marketing authorization of Herbal Medicine and marketing authorization and notification of THP and procedures related to Product Change History protocol and the deadline for analyzing post-approval changes applications for herbal medicines and THPs, based on the provisions of Resolution-RDC n° 38/2014 [8, 11, 12].

Beyond specific standards for herbal medicinal products, other general and transversal regulations applicable to different categories of medicines are also applicable here, such as Good Manufacturing Practices; Clinical research for the purpose of proving the safety and efficacy of medicines; Request procedures at Anvisa; List of therapeutic indications exempt from medical prescription; Drug importing rules; Pharmacovigilance guidelines for drug registration holders; Requirements for conducting stability studies; Validation guidelines for analytical methodologies, leaflet and labeling requirements; and advertising rules [8].

Despite the potential for growth in the Brazilian market for herbal medicinal products, and the public policies aimed at expanding its use, the number of licensed herbal medicinal products is considered small when compared to that in other countries, and non-native plant species prevail in the composition of herbal medicinal products in Brazil. There has been a decline in the number of licensed herbal medicinal products over the years. In 2008, 512 herbal medicines were licensed, of which 432 were simple and 80 were combination (multiple ingredient) products. In 2011, there were 382 licensed herbal medicines (357 simple, 25 combination products). In the last survey carried out in 2018, 359 herbal medicines were licensed, 332 as simple products and 27 as combination products. Several factors contributed to this scenario, such as the presence of more restrictive regulatory frameworks, the need for a medical prescription for many herbal medicinal products that do not have this restriction in other countries, and the delay in the analysis of those marketing authorization applications which had to be adapted to the new regulations, among others [13].

21.2.2 Manufacturing and Compounding

The quality of an industrialized herbal medicinal product must be ensured by controlling all stages of its manufacturing, that is, from applying the principles of Good Agricultural Practices (GAP) through Good Manufacturing Practices (GMP) for raw materials, to Good Manufacturing Practices for medicines. Regarding production of plant species for use in herbal medicinal products, GAP must be observed; this provides guidance on the correct cultivation, collection/harvesting, processing, drying, and storage of the medicinal plant. However, the regulation of this activity comes under the Ministry of Agriculture, Livestock and Supply (MAPA) in Brazil, and is not in Anvisa's scope of control, which begins with GMP for Vegetal Active Pharmaceutical Ingredient.

GMP compliance by the manufacturing companies for herbal medicines and THP production is required and a Certificate of Good Manufacturing Practices is the document issued by the Brazilian Health Regulatory Agency stating that the licensed facility complies with the requirements of this regulatory framework [14–16].

Medicine compounding, in general, has its own regulation such as Resolution-RDC n° 67/2007 (Good Practices of Compounding for Magistral and Official

Preparations for Human use in Pharmacies) and Resolution-RDC n° 87/2008 (Amends the Technical Regulation on Good Practices of Compounding in Pharmacies). However, to support the already mentioned public policies outlined for herbal medicinal products, in 2010, the *Farmácia Viva* program was instituted in the SUS through Ordinance n°886/GM/MS. The program aims to produce accessible herbal medicine products to the population and carry out all stages from cultivation, collection, processing, storage of medicinal plants to compounding and the dispensing of Magistral and Official Preparations of medicinal plants and herbal medicinal products. In order to guide the execution of these activities, Resolution-RDC n°18/2013 was published, which provides for the good practices of processing and storage of medicinal plants, preparation and dispensing of Magistral and Official products of medicinal plants and herbal medicinal products in *Farmácias Vivas* within the scope of the SUS [17–20].

21.3 Pharmacovigilance

Pharmacovigilance actions in Brazil are executed by several institutions within the SUS, but are coordinated by the National Monitoring Centre, which is located in the Anvisa's—Brazilian Health Regulatory Agency—Pharmacovigilance Office (*Gerência de Farmacovigilância—GFARM*), according to Ordinance Anvisa's GM/MS n° 696/2001 [21]. Both the Regional Monitoring Centres as well as the Local Health Regulatory Authorities of the 27 Federation Units and the Ministry of Health make up the Brazilian Pharmacovigilance System, together with Anvisa. Each of these institutions has a defined role, with Anvisa coordinating this system and carrying out pharmacovigilance actions at the national level. The Ministry of Health, more specifically, is responsible for pharmacovigilance of medicines distributed by the SUS, with an emphasis on vaccines.

In 2001, Brazil became the 62nd member of the World Health Organization (WHO) Programme for International Drug Monitoring and, since then, pharmacovigilance and other post-market actions for medicinal products have stood out as important tools for drug sanitary control in the country.

In Brazil, ICSRs can be provided/reported by citizens, all health professionals and marketing authorization holders (MAH). It is only necessary to obtain a login and password to access the electronic system for MAHs and health units. Other reporters can provide ICSRs through the Anvisa website. Although numbers of ICSRs are still far from those received in developed countries, Brazil has seen an increasing number of ICSRs submitted per million inhabitants in recent years. From the implementation of a new management system for spontaneous reporting in December 2018, and following a more effective coordination of the National Pharmacovigilance System by Anvisa, Brazil received 13,461 ICSRs in 2018 and 21,896 ICSRs in 2019.

21.3.1 Work Processes

Data collection for detection of signals of adverse events in Brazil is carried out by passive and active pharmacovigilance methods, such as stimulated reporting and observational studies, among other methods. However, it is in passive pharmacovigilance that the country has advanced significantly in a short period of time. In 2018, the management system for spontaneous reporting—Notivisa® (operative since 2006)—was replaced by VigiFlow®, a system made available by the Uppsala Monitoring Centre (UMC). Locally named VigiMed®, this system supports the collection, processing, and sharing of ICSRs, in order to facilitate effective data analysis. In addition, it uses the Medical Dictionary for Regulatory Activities (MedDRA) Terminology, as well as meets the requirements of the International Council for Harmonization of Technical Requirements for Pharmaceuticals for Human Use (ICH) E2B guideline, regarding the internationally harmonized format and standards for data transmission [22].

ICSRs are received, analyzed, and processed daily by GFARM, with the aim of detecting signals. When a signal is detected, an investigative process is opened to search for more evidence in order to strengthen the signal (or to refute it if other evidence does not support it). Restricted (specialized) consultations are directed to Sentinel Hospitals to verify whether that event has also emerged in any other Sentinel Network unit (*Rede Sentinela*).[1] In addition, bibliographic searches are performed on research platforms, websites of international regulatory authorities are consulted, and searches are conducted using VigiLyze®, a powerful research and analysis tool that provides access to more than 20 million ICSRs held in VigiBase®, the WHO global database of ICSRs, contributed by more than 130 countries. VigiLyze® includes data on allopathic drugs, herbal medicinal products, and biological products, including vaccines [23].

Investigations have an undefined term in which to be completed (days, months, or years), depending on their complexity. Sometimes, it is necessary for the marketing authorization holder to provide documents (e.g., updated Periodic Benefit-Risk Evaluation Report) or even the execution of Good Pharmacovigilance Practices inspections at their facilities. When the investigation is completed, some actions may be triggered, depending on the case in question. These are:

- Safety alert publication on the Anvisa website.
- Publication of a letter to health professionals on the Anvisa website.
- Package leaflet amendment demand issued by Anvisa.
- Adoption of precautionary measures (suspension of import, manufacture, distribution, trade, use, etc.).
- Marketing authorization cancellation [24].

---

[1] *Rede Sentinela* is a strategy, launched in the middle of 2001, with the purpose of being an active observatory for safety and performance of regularly used health products: medicines, medical devices, sanitizers, cosmetics, blood and its components and so on.

21.3.2 Regulatory Frameworks

Basically, there are three regulatory frameworks that cover most of the legal requirements related to pharmacovigilance in Brazil.

21.3.2.1 Resolution-RDC No. 36/2013

This regulatory instrument institutes actions for patient safety in health services, making it mandatory for each health service to create a Patient Safety Centre that must prepare a Patient Safety Plan. This document should identify risk situations and describe the strategies and actions defined by the health service for risk management aimed at the prevention and mitigation of incidents, from admission to transfer, discharge, or death of the patient in the health service [25].

The Patient Safety Centre must also monitor incidents and adverse events that have occurred, reporting them using the electronic tools provided by Anvisa. Adverse events that result in death must be reported within 72 h of the event (death) occurring.

21.3.2.2 Resolution-RDC No. 406/2020 and Normative Instruction N° 63/2020

Resolution-RDC n° 406/2020 and Normative Instruction n° 63/2020 provide for pharmacovigilance rules applied to drug marketing authorization holders. This regulation defines the scope of pharmacovigilance, establishes the obligation to report adverse events and their respective deadlines to do it, as well as the request to submit the Risk Management Plan/Risk Minimization Plan and the Periodic Benefit-Risk Evaluation Reports to Anvisa [26, 27].

These regulations are the result of updating the previous regulation to meet the requirements of the International Council for Harmonization of Technical Requirements for Pharmaceuticals for Human Use (ICH) Guidelines E2B, E2D and M1. Brazil, through Anvisa, was accepted as an ICH observer in December 2015 and became a regular member in the following year. To remain in this condition, Brazil needs to implement five ICH level II guidelines by November 2021, three of which are related to pharmacovigilance and have already been mentioned.

21.3.2.3 Ordinance GM/MS N° 1660/2009

Through this regulatory framework, the Adverse Event Investigation and Report System (*Sistema de Investigação e Notificação em Vigilância Sanitária—VIGIPOS*) has been established as an integral part of the SUS, with the aim of monitoring, analyzing, and investigating adverse events and technical complaints related to services and products under health regulation in the post-approval phase. This type of work was previously performed by Anvisa and the Ministry of Health in a very specific way, but in 2009 the opportunity arose to formalize and give greater importance to the post-approval monitoring of these products. Ordinance GM/MS n° 1660/2009 clearly establishes the responsibilities and attributions of each of the institutions belonging to the SUS and defines that Anvisa's electronic system for adverse events reporting will be the reference system [28].

21.3.3 Pharmacovigilance for Herbal Medicinal Products

There are no specific regulatory frameworks nor special attention placed on pharmacovigilance for herbal medicinal products in Brazil. Due to the technological limitations of the previous management system for spontaneous reporting (Notivisa), searching for data and reports related to herbal medicinal products is a difficult and laborious process. It is also important to note that, in the previous system (Notivisa), notification by citizens was difficult, as it was necessary to obtain a password and login, unlike the simplified process today with VigiMed®. Nevertheless, 31 ICSRs involving 18 herbal medicines were found in Notivisa and, more recently, in VigiMed, for the period 2006–2020. The largest number of ICSRs is related to the use of *Senna alexandrina*—senna (1 serious ICSR and 6 non-serious ICSRs), *Ginkgo biloba*—ginkgo (2 serious ICSRs and 2 non-serious ICSRs), *Aesculus hippocastanum*—horse chestnut (1 serious ICSR and 2 non-serious ICSRs), and *Piper methysticum*—*kava kava* (2 serious ICSRs and 1 non-serious ICSR). All ICSRs came from hospitals.

It is expected that from the use of the new management system for spontaneous reporting—VigiMed®—together with other technological alternatives that Anvisa is developing, it will be possible to increase the number of ICSRs and facilitate data searching. However, if the need for more specific monitoring for a given herbal medicinal product is identified, active pharmacovigilance can be triggered in a partnership with the Ministry of Health or another institution belonging to the Brazilian pharmacovigilance system, as it is the case today with antimalarial and multiresistant tuberculosis drugs.

21.4 Final Considerations

With the emergence of a greater number of technologies every day and the need to expand the population's access to new products more and more quickly, the role played by post-market monitoring becomes essential to ensure that patients have access to and use effective, safe, and good quality medicinal products. In this scenario, special attention is given to pharmacovigilance, which has evolved substantially in recent decades, especially in relation to the tools that are available for its practice. However, when it comes to herbal medicinal products, Brazil still faces difficulties regarding under-reporting. As in many other countries, the likely under-reporting related to herbal medicines may also be a consequence, at least in part, of the way these products are named, distributed, purchased, and perceived by the user.

Certainly, there is much to be done to make pharmacovigilance of herbal medicinal products more effective in Brazil. Nevertheless, advances such as the increased number of ICSR observed in recent years, the adoption of a more efficient management system for spontaneous reporting, the updating of pharmacovigilance regulatory framework directed to marketing authorization holders and incorporation of ICH Guidelines will serve as the basis for pharmacovigilance actions for certain types of drugs, including herbal medicinal products.

References

1. Ministério do Meio Ambiente (BR). Biodiversidade Brasileira. https://www.mma.gov.br/biodiversidade/biodiversidade-brasileira. Accessed 13 Mar 2020
2. Ministério da Saúde (BR), Secretaria de Ciência, Tecnologia e Insumos Estratégicos. Departamento de Assistência Farmacêutica (2016) Política e Programa Nacional de Plantas Medicinais e Fitoterápicos. http://bvsms.saude.gov.br/bvs/publicacoes/politica_programa_nacional_plantas_medicinais_fitoterapicos.pdf. Accessed 13 Mar 2020
3. Ministério da Saúde (BR), Secretaria de Atenção à Saúde, Departamento de Atenção Básica (2015) Política nacional de práticas integrativas e complementares no SUS: atitude de ampliação de acesso. http://bvsms.saude.gov.br/bvs/publicacoes/politica_nacional_praticas_integrativas_complementares_2ed.pdf. Accessed 13 Mar 2020
4. Ministério da Saúde, Secretaria de Ciência, Tecnologia e Insumos Estratégicos, Departamento de Assistência Farmacêutica e Insumos Estratégicos (2017) Relação Nacional de Medicamentos Essenciais: RENAME 2017. https://bvsms.saude.gov.br/bvs/publicacoes/relacao_nacional_medicamentos_rename_2017.pdf. Accessed 15 Apr 2020
5. Ministério da Saúde (BR). Plantas Medicinais e Fitoterápicos no SUS. http://saude.gov.br/acoes-e-programas/programa-nacional-de-plantas-medicinais-e-fitoterapicos-ppnpmf/plantas-medicinais-e-fitoterapicos-no-sus. Accessed 15 Apr 2020
6. Agência Nacional de Vigilância Sanitária (2016) Farmacopeia Brasileira. Memento Fitoterápico. http://portal.anvisa.gov.br/documents/33832/2909630/Memento+Fitoterapico/a80ec477-bb36-4ae0-b1d2-e2461217e06b. Accessed 15 Apr 2020
7. Agência Nacional de Vigilância Sanitária (2011) Formulário de Fitoterápicos da Farmacopéia Brasileira. http://portal.anvisa.gov.br/documents/33832/259456/Formulario_de_Fitoterapicos_da_Farmacopeia_Brasileira.pdf/c76283eb-29f6-4b15-8755-2073e5b4c5bf. Accessed 20 Apr 2020
8. Agência Nacional de Vigilância Sanitária—Anvisa. Instrução Normativa—IN n° 04, 18 Jun 2014. Diário Oficial da União. 20 Jun 2014; Section 1.86. https://bvsms.saude.gov.br/bvs/saudelegis/anvisa/2014/int0004_18_06_2014.pdf. Accessed 8 May 2020
9. Agência Nacional de Vigilância Sanitária—Anvisa. Resolução—RDC n° 26, 13 May 2014. Diário Oficial da União. 14 May 2014; Section 1:52. https://bvsms.saude.gov.br/bvs/saudelegis/anvisa/2014/rdc0026_13_05_2014.pdf. Accessed 8 May 2020

10. Agência Nacional de Vigilância Sanitária—Anvisa. Resolução—RDC n° 38, 18 Jun 2014. Diário Oficial da União. 20 Jun 2014; Section 1:82. https://bvsms.saude.gov.br/bvs/saudelegis/anvisa/2014/rdc0038_18_06_2014.pdf. Accessed 8 May 2020
11. Agência Nacional de Vigilância Sanitária—Anvisa. Instrução Normativa—IN n°02, 13 May 2014. Diário Oficial da União. 14 May 2014; Section 1:58. https://bvsms.saude.gov.br/bvs/saudelegis/anvisa/2014/int0002_13_05_2014.pdf. Accessed 8 May 2020
12. Agência Nacional de Vigilância Sanitária—Anvisa. Instrução Normativa—IN n° 05, 18 Jun 2014. Diário Oficial da União. 20 Jun 2014; Section 1:86. https://bvsms.saude.gov.br/bvs/saudelegis/anvisa/2014/int0005_18_06_2014.pdf. Accessed 8 May 2020
13. Carvalho ACB, Lana TN, Perfeito JPS, Silveira D (2018) The Brazilian market of herbal medicinal products and the impacts of the new legislation on traditional medicines. J Ethnopharmacol 212:29–35. https://doi.org/10.1016/j.jep.2017.09.040
14. Agência Nacional de Vigilância Sanitária—Anvisa. Resolução—RDC n° 301, 21 Aug 2019. Diário Oficial da União. 22 Aug 2019; Section 1:64. http://bvsms.saude.gov.br/bvs/saudelegis/anvisa/2019/rdc0299_12_08_2019.pdf. Accessed 8 May 2020
15. Agência Nacional de Vigilância Sanitária—Anvisa. Instrução Normativa—IN n° 39, 21 Aug 2019. Diário Oficial da União. 22 Aug 2019; Section 1:87. http://portal.anvisa.gov.br/documents/10181/5389382/IN_39_2019_.pdf/f190e75b-e1d6-43ec-b67a-067a0a9c83ab. Accessed 8 May 2020
16. Agência Nacional de Vigilância Sanitária—Anvisa. Resolução—RDC n° 69, 8 Dec 2014. Diário Oficial da União. 9 Dec 2014; Section 1:43. http://bvsms.saude.gov.br/bvs/saudelegis/anvisa/2014/rdc0069_08_12_2014.pdf. Accessed 8 May 2020
17. Agência Nacional de Vigilância Sanitária—Anvisa. Resolução—RDC n° 67, 8 Oct 2007. Diário Oficial da União. 9 Oct 2007; Section 1:29. https://bvsms.saude.gov.br/bvs/saudelegis/anvisa/2007/rdc0067_08_10_2007.html. Accessed 8 May 2020
18. Agência Nacional de Vigilância Sanitária—Anvisa. Resolução—RDC n° 87, 21 Nov 2008. Diário Oficial da União. 24 Nov 2008; Section 1:58. https://bvsms.saude.gov.br/bvs/saudelegis/anvisa/2008/res0087_21_11_2008.html. Accessed 8 May 2020
19. Ministério da Saúde (BR). Minister's Office. Portaria n° 886, 20 Apr 2010. Diário Oficial da União. 22 Apr 2010; Section 1:75. http://bvsms.saude.gov.br/bvs/saudelegis/gm/2010/prt0886_20_04_2010.html. Accessed 8 May 2020 and revoked by: Ministério da Saúde (BR). Minister's Office. Portaria n° 5, 28 Sept 2017. Diário Oficial da União. 3 Oct 2017; 12 Mar 2020; Section 1:185. http://bvsms.saude.gov.br/bvs/saudelegis/gm/2017/prc0005_03_10_2017_comp.html. Accessed 8 May 2020
20. Agência Nacional de Vigilância Sanitária—Anvisa. Resolução RDC n° 18, 3 Apr 3rd 2013. Diário Oficial da União. 5 Apr 2013; Section 1:67. https://bvsms.saude.gov.br/bvs/saudelegis/anvisa/2013/rdc0018_03_04_2013.html. Accessed 8 May 2020
21. Ministério da Saúde (BR). Minister's Office. Portaria N° 696, 7 May 2001. Diário Oficial da União. 8 May 2001; Section 1:14. https://bvsms.saude.gov.br/bvs/saudelegis/gm/2001/prt0696_07_05_2001.html. Accessed 15 Apr 2020
22. International Council for Harmonisation of Technical Requirements for Pharmaceuticals for Human Use. ICH Guidelines. https://www.ich.org/page/ich-guidelines. Accessed 17 Apr 2020
23. Uppsala Monitoring Centre. Get the answers you need. https://www.who-umc.org/vigibase/vigilyze/. Accessed 17 Apr 2020
24. Vogler M (2019) Por que notificar eventos adversos a medicamentos? Boletim Farmacoterapêutico 3:5–10
25. Agência Nacional de Vigilância Sanitária—Anvisa. Resolução-RDC n° 36, 25 Jul 2013. Diário Oficial da União. 26 Jul 2013; Section 1:32. https://bvsms.saude.gov.br/bvs/saudelegis/anvisa/2013/rdc0036_25_07_2013.html. Accessed 17 Apr 2020
26. Agência Nacional de Vigilância Sanitária—Anvisa. Resolução-RDC n° 406, 22 Jul 2020. Diário Oficial da União. 29 Jul 2020; Section 1:64. https://www.in.gov.br/en/web/dou/-/resolucao-de-diretoria-colegiada-rdc-n-406-de-22-de-julho-de-2020-269155491. Accessed 22 Sept 2020

27. Agência Nacional de Vigilância Sanitária—Anvisa. Instrução Normativa n° 63, 22 Jul 2020. Diário Oficial da União. 29 Jul 2020; Section 1:69. https://www.in.gov.br/en/web/dou/-/instrucao-normativa-in-n-63-de-22-de-julho-de-2020-269155503. Accessed 22 Sept 2020
28. Ministério da Saúde (BR). Minister's Office. Portaria n° 1.660, 22 Jul 2009. Diário Oficial da União. 24 Jul 2009; Section 1:45. https://bvsms.saude.gov.br/bvs/saudelegis/gm/2009/prt1660_22_07_2009.html. Accessed 15 Apr 2020

Chapter 22
Pharmacovigilance for Indian Traditional Medicines

Vivekanandan Kalaiselvan and Galib Ruknuddin

22.1 Introduction

As stated by the World Health Organization (WHO), over 100 types of traditional systems of medicines are being used in different countries. In India, the popular traditional systems of medicines are familiar as Ayurveda, Yoga and Naturopathy, Unani, Siddha and Homoeopathy (AYUSH). Very recently the Sowa-Rigpa system, a traditional medical system popular in Himalayan societies and other parts of India, has also been recognized in India as an official medical system [1]. Ayurveda is an holistic, age-old science and practice which has elaborated sets of guidelines for healthy, peaceful and happy living and maintenance and protection of physical and psychological health to achieve longevity. This system is associated with the Vedic civilization in India and the literature for the treatment is mentioned in oldest scriptures known as Vedas. The philosophy of Ayurveda is based on the theory of five elements of which all the objects and living bodies are composed. This medical system aims to keep the state of equilibrium of the bodily elements, in order to maintain good health. Siddha and Unani systems of medicine also consider ailments in the human body as a result of imbalance of bodily humours. All these AYUSH systems utilize ingredients from natural resources, such as plant, animal, mineral, and marine organisms to maintain health and treat diseases.

Of a population of 1.3 billion people in India, still 70–80% uses traditional medicines for the treatment of various diseases or disorders [2]. The Indian herbal drugs

V. Kalaiselvan (✉)
Indian Pharmacopoeia Commission, Ministry of Health & Family Welfare, Government of India, Ghaziabad, India
e-mail: kalaiselvan.ipc@gov.in

G. Ruknuddin
All India Institute of Ayurveda, An Autonomous Organization Under the Ministry of AYUSH, Government of India, New Delhi, India

J. Barnes (ed.), *Pharmacovigilance for Herbal and Traditional Medicines*,
https://doi.org/10.1007/978-3-031-07275-8_22

industries utilize around 960 medicinal plants for making different formulations, and their cumulative turnover is more than INR 80 billion. The AYUSH medicines contribute to 3% of the total pharmaceutical exports of India [2]. As AYUSH medicines are becoming increasingly used by the global community, it becomes essential to establish safety data, through a structured pharmacovigilance approach and by inculcating contemporary scientific and technological advancement.

As defined by the WHO, pharmacovigilance is the science and activities relating to the detection, assessment, understanding and prevention of adverse events or any other possible drug-related problems [3]. The WHO extends its scope of pharmacovigilance to traditional system of medicines too, as it encourages the national pharmacovigilance centre of the respective member states to also focus on the safety of these products and preparations. Although AYUSH systems of medicines have been used for centuries in India, still there is a lack of systematic documentation on safety and occurrence of adverse events. The strong perception prevails among the general public and practitioners that AYUSH medicines are free from adverse events as they are derived from natural sources. Factors such as environmental factors, industrial pollutants, pesticides, adulteration, poor compliance with pharmacopoeial standards, heavy metal content, concomitant use of allopathic medicines, over the counter sale, etc. are challenging in evaluating safety aspects of traditional medicines [4]. Thus, to establish and understand the safety profile of AYUSH medicines has become an essential part of pharmacovigilance in India. This chapter outlines the regulation and system and procedure for reporting of adverse events associated with AYUSH medicines in India.

22.2 AYUSH Medicines Regulation

There are adequate provisions under the Drugs and Cosmetics Act, 1940, and Drugs and Cosmetics Rules, 1945, for the regulatory framework and for monitoring the quality, safety and efficacy of drugs belonging to AYUSH systems. In India, Schedule T of this Act prescribes specifications with respect to the good manufacturing practices for Ayurveda, Siddha and Unani medicines, whereas Schedule M-1 of the Act provides specifications for Homoeopathy medicines. Licensing Authorities are appointed by the State Governments to oversee the enforcement of legal provisions for the manufacture and sale of AYUSH medicines. Good Manufacturing Practices and adherence to standards for drugs as prescribed in the Ayurvedic Pharmacopoeia of India are mandatory for the manufacturing of licensed products to ensure quality, safety and efficacy of AYUSH medicines. The Ayurveda, Siddha, Unani Drugs Technical Advisory Board (ASUDTAB) and the Ayurveda, Siddha, Unani Drugs Consultative Committee (ASUDCC) are statutory bodies under the Drugs and Cosmetics Act tasked with advising the Central and State Governments on technical matters The National AYUSH Mission, Ministry of AYUSH, also has a vision to promote quality standards for AYUSH drugs [1].

22.3 Pharmacovigilance System for Ayurveda, Siddha, Unani and Homoeopathy (ASU&H) Medicines

The quality issues relating to ASU&H medicines are raised from various sources and it is felt necessary in the interest of public health to oversee the impact of ASU&H medicines taken by patients from the perspective of their safety profile. Dissemination and advertisement of improper drug information is also a matter of concern that needs to be addressed with systematic surveillance and regulatory action. Also, clinical validation and documentation of adverse events associated with use of ASU&H medicines is one of the tools to promote their acceptance globally. In view of the above, towards detection, assessment, understanding, prevention and regulatory action of adverse events of ASU medicines, the then department of AYUSH, Ministry of Health & Family Welfare, Government of India, has launched a National Pharmacovigilance Programme for ASU (NPP-ASU) Drugs in 2007. In addition to NPP-ASU, there is also a vertical nationwide Pharmacovigilance Programme of India (PvPI) to monitor the safety of drugs and pharmaceuticals, which is managed by the Indian Pharmacopoeia Commission as a National Coordination Centre. The functioning of both NPP-ASU and PvPI is given in Fig. 22.1.

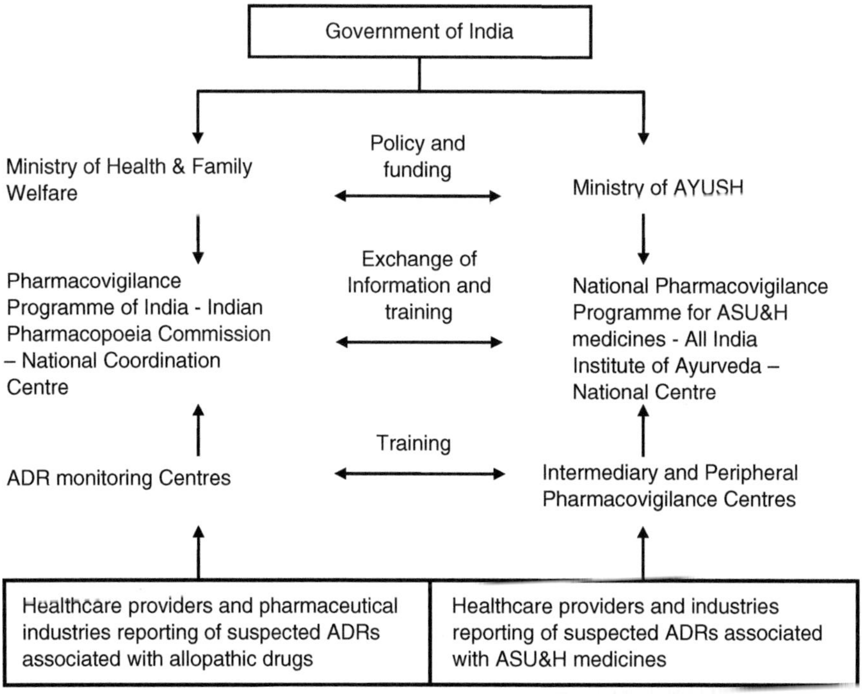

Fig. 22.1 Pharmacovigilance system for allopathic and Ayurveda, Yoga and Naturopathy, Unani, Siddha and Homoeopathy (AYUSH) systems of medicines in India

In order to ensure effective implementation of the NPP, the programme has been restructured by the Ministry of AYUSH, Government of India, under the Central Sector Scheme to include the Homoeopathy component (i.e. ASU&H drugs) during 2017 in support and guidance of the Indian Pharmacopoeia Commission and concerned programme officers of the WHO Country Office, India. The All India Institute of Ayurveda, New Delhi, has been recognized as a National Pharmacovigilance Coordination Centre (NPvCC) for this programme. The purpose of the initiative is to collect, collate and analyse data related to suspected adverse drug reactions (ADRs) and undertake surveillance of advertisements related to ASU&H drugs thus to establish evidence clinical safety of these drugs in a scientific manner. The scheme mainly aims at (1) inculcating a reporting culture among consumers as well as ASU&H practitioners to facilitate documentation of suspected ADRs and instances of misleading advertisements for Ayurveda, Siddha, Unani and Homoeopathy drugs; (2) developing a system-wide database of ADRs associated with ASU&H drugs and (3) evolving evidence-based recommendations regarding the clinical safety and improper advertisements of ASU&H drugs for regulatory actions.

22.4 Structural Framework of NPP-ASU&H

There is a three-tier structure comprising a National Pharmacovigilance Coordination Centre (NPvCC), Intermediary Pharmacovigilance Centres (IPvCs) and Peripheral Pharmacovigilance Centres (PPvCs). The All India Institute of Ayurveda, as an NPvCC, is responsible for the implementation of the NPP-ASU&H medicines. The NPvCC receives inputs in terms of suspected ADRs from the IPvCs, and the IPvCs provide support to the PPvCs. The functions of NPvCC, IPvCs and PPvCs are given in Table 22.1. The NPvCC, in consultation with the concerned IPvC, identifies the PPvCs by applying certain criteria. Such recognized PPvCs are primarily responsible for collecting reports of suspected adverse events associated with ASU&H medicines from doctors, pharmacists, industry, consumers and others through their physicians.

22.5 ADR Reporting Criteria

The pharmacovigilance programme is encouraging reporting of all suspected ADRs associated with use of ASU&H drugs alone or along with any other drugs. Reactions to any other drugs which are suspected of significantly affecting patient's management, including serious suspected reactions (death, life threatening, hospitalization, disability and so forth) are also encouraged to be reported. All healthcare

Table 22.1 Functions and responsibilities of pharmacovigilance centres under National Pharmacovigilance Program for Ayurveda, Siddha, Unani and Homoeopathy Drugs

National Pharmacovigilance Coordination Centre (NPvCC)	• Document and monitor suspected ADRs associated with ASU&H medicines
	• Management of periodic safety update reports (PSURs) from manufacturing companies for their ASU&H medicines for all patent and proprietary products
	• Organize awareness and capacity building workshops for stakeholders
	• Undertake causality assessment for suspected ADRs associated with ASU&H medicines and recommend necessary regulatory action to the Ministry of AYUSH
	• Provide information to end users through seminars, drug alerts and other means
	• Report objectionable advertisements
	• Other related activities
Intermediary Pharmacovigilance Centres (IPvCs)	• Document and monitor suspected ADRs associated with ASU&H medicines
	• Collect information regarding suspected ADRs associated with ASU&H medicines from the respective PPvCs
	• Report suspected ADRs associated with ASU&H medicines to NPvCC at regular intervals for causality assessment
	• Organize awareness building and capacity building workshops for stakeholders
	• Scrutinize project proposals received from the PPvCs and forward the same to the NPvCC along with recommendations
	• Report objectionable advertisements
Periphery Pharmacovigilance Centres (PPvCs)	• Document and monitor suspected ADRs associated with ASU&H medicines
	• Report suspected ADRs associated with ASU&H drugs to the concerned IPvC at regular intervals
	• Report objectionable advertisements

professionals can report suspected ADRs. On observation of a suspected ADR, the healthcare professional concerned will report this, along with their assessment using the Naranjo probability scale [5], to the nearby PPvC. These reports are then reported to respective IPvCs and then to the NPvCC.

Awareness, training of AYUSH physicians in pharmacovigilance aspects, proper screening methods, and ADR reporting culture all play a pivotal role in the successful functioning of the programme that ultimately helps the systems to join the mainstream. Considering its importance, the basic concepts of pharmacovigilance have been introduced into the curriculum of Ayurveda postgraduate studies. Although the programme was implemented 2 years ago, a lack of expertise in performing causality analysis with traditional medicines appears to be one of the most challenging aspects.

To develop a culture of suspected ADR notification (reporting), various stakeholders, including healthcare professionals, physicians and pharmacists of ASU&H systems, are being informed on the concept of pharmacovigilance through various awareness programmes across the country. A web portal (https://www.ayushsuraksha.com/) has been launched for online reporting of suspected ADRs associated with ASU&H medicines.

22.6 Implementation and Monitoring of NPP-ASU&H

The implementation of the NPP-ASU&H is under the supervision of a monitoring committee that comprises experts from the Ministry of AYUSH, the Indian Pharmacopoeia Commission (IPC), the Central Drugs Standard Control Organisation (CDSCO) and the Pharmacopoeial Commission for Indian Medicine and Homoeopathy (PCIM&H). A Technical Advisory Committee is also constituted involving representatives from the Ministry of AYUSH, the IPC, the PCIM&H, the CDSCO, the Chairmen of the Ayurveda Pharmacopoeia Committee, the Unani Pharmacopoeia Committee, the Siddha Pharmacopoeia Committee, the Homoeopathy Pharmacopoeia Committee, pharmacology experts from the All India Institute of Medical Sciences, New Delhi, or the Indian Council of Medical Research, New Delhi, and the WHO representative of the pharmacovigilance programme, WHO country office, India. A Central Signal Detection & Causality Assessment Committee has also been constituted that comprises AYUSH clinical experts, pharmacy experts for AYUSH systems and pharmacology/toxicology experts. These committees are mainly concerned with reviewing and analysing the ADRs reported at different levels and suggesting proper remedial measures.

In order to monitor the implementation of the programme and measure its efficacy, the NPvCC established several key indicators, such as (1) process, (2) outcome and (3) impact of the programme. The process indicators assess the number of ADR monitoring centres participating in the programme and the surveillance of advertisements of ASU&H medicines, the number of personnel trained in ADR monitoring and surveillance of advertisements of ASU&H medicines and practices, and utilization of financial resources for effective implementation of the programme. The outcome indicators evaluate the number of ADR reports received and processed each year, and the number of misleading advertisements relating to ASU&H medicines and practices reported and processed each year. The impact indicators assess the number of signals generated and confirmed, the number of safety-related alerts issued by the Ministry of AYUSH, and the number of misleading advertisements relating to ASU&H drugs and practices rectified or withdrawn.

The Pharmacovigilance Programme of India (PvPI) at Indian Pharmacopoeia Commission receives spontaneous ADR data associated with the use of drugs or other medical products (such as medical devices, blood products, vaccines) in the form of individual case safety reports (ICSRs) from its recognized ADR monitoring centres, doctors, pharmacists, the pharmaceutical industry and others. ICSRs reported with ASU&H as suspected/concomitant medicine(s) along with allopathic drugs are transferred to NPP-ASU&H.

22.7 Assessment of Adverse Events Associated with ASU&H Medicines

During the last 7 years, from assessment of adverse events data associated with the use of ASU&H medicines, it is observed that most reports are attributable either to poor-quality products or to improper use, and can be categorized under two headings, i.e. drug-related and clinic-related. Drug-related factors mainly concern the quality of the ASU&H medicine, which may be hampered by poor-quality raw material, poor-quality plant preparations, and ASU&H medicines not being prepared following specific procedures. Good dispensing practices ensure that an effective form of the correct medicine is delivered to the right patient, in the correct dosage and quantity, with clear instructions, and in a package that maintains the potency of the medicine. It is observed that in some cases medicines are dispensed in loose packets with an instruction to the patient to take the drug with an approximate weight basis (i.e. an approximate quantity of the material, which has clear implications for efficacy and toxicity). The efficacy of the drug also depends on its correct administration as described in the form of prescribing information/package inserts or leaflets.

22.8 Overview of Spontaneous Reports with Respect to ASU&H Medicines

The PvPI and NPP-ASU&H are heavily depending on spontaneous ADRs associated with ASU&H medicines; the PvPI receives data from its recognized regional ADR monitoring centres. These centres have been trained on reporting of ADRs associated with the use of medicines, vaccines, medical devices, ASU&H medicines and others. ADR reporting associated with the use of ASU&H medicines to PvPI is minimal; moreover, most of the reports are found to have allopathic medicines as concomitant medicines. Therefore, causality assessment for these cases becomes more challenging.

Disease management in ASU&H systems involves various practices, including medication with natural substances, with varied methodologies and philosophies, and in a personalized (individualized) manner after thorough examination of various factors. Unlike synthetic medicines, herbal medicines are complex products and not isolated single active molecules. Thus, evaluating ADR reports associated with use of medicines in ASU&H systems becomes a big challenge. The suspected ADR reports are often related to the gastro-intestinal system; the reported reactions include nausea, loss of appetite, hyperacidity, diarrhoea, constipation, abdominal bloating, are usually mild and non-serious in nature.

Despite global concerns around medication safety, there is limited ADR reporting among healthcare professionals. This is observed with medicines used in ASU&H systems too. The awareness sessions being conducted through the pharmacovigilance programme described above have inculcated reporting culture among healthcare professionals, and the numbers of suspected ADR reports being received into the programme is observed to be increasing over the past 2 years.

Any healthcare professional (including registered medical practitioners of ASU&H systems, and other paramedical personnel who are involved in providing healthcare services, including nurses, pharmacists, primary healthcare workers, etc.) may report suspected ADRs. For the time being, the programme shall not accept reports from lay members of the public or anyone else who is not a healthcare professional. Others can report through their physicians under whom he/she had undergone treatment.

Herbal drugs used in ASU&H medicines are coded by their biological (scientific) name, or common name, or vernacular name.

ADRs associated with the use of a supercritical carbon dioxide extract of *Artemisia annua*, including abnormal liver function and QT prolongation, reported from New Zealand have raised concerns [6–10]. As *Artemisia annua* and its derivatives are widely used in traditional and modern medicines in India, the above safety signal has triggered the PvPI and NPP-ASU&H to be vigilant about the possibility of ADRs associated with *Artemisia annua* in India.

22.9 Challenges in Pharmacovigilance for ASU&H Medicines in India

The current PvPI and NPP-ASU&H encounter challenges in causality assessment of adverse events associated with ASU&H drugs as well as under-reporting of suspected ADRs associated with these products/preparations. Unlike allopathic drugs, most of the ASU&H formulations are composed of multiple herbal ingredients; therefore, it may not always be possible to establish a causal/temporal relationship between the responsible ASU&H drug and the adverse event. For example, some traditionally used herbal medicines, such as garlic, senna, and digitalis, have been suspected to cause eruptions, diarrhoea, and anorexia respectively. However, it was not possible to perform causality assessment in these cases as there is a lack of clinical evidence, or because the herbal substances were administered with other herbal medicines/allopathic medicines [4]. In addition, there is a high prevalence among the population of taking ASU&H medicines concurrently with allopathic drugs [11]. The availability of a few ASU&H drugs over the counter as nutraceuticals or dietary supplements, incidences of concomitant use of these with conventional drugs, and meagre information on herb-drug interactions also pose challenges. There are subtleties in the legal differentiation between food supplements and ASU&H medicines. Thus, ADRs that might develop in such cases may not be entered into ADR databases. This becomes more challenging in causality assessment and signal detection as well.

Maintenance of health and disease management in ASU&H systems deals with empirical practices, including medication with preparations of natural substances, and focusing on overall wellness with varied methodologies and philosophies. Understanding human individuality through assessment of each individual constitution type forms the fundamental basis of such practices, explained by the belief that every individual is unique in his/her constitution. Considering such a unique characteristic approach in relation to drug administration in causality assessment is difficult.

It is observed that most ASU&H medicines available in the market do not contain required mandatory prescribing information in the package insert. Further, for proprietary ASU&H medicines, in addition to pre-marketing safety evaluation, post-marketing safety evaluation should also be given importance. Therefore, in the years ahead, there is a need to establish a comprehensive ADR database for ASU&H medicines.

To conclude, ASU&H medicines are widely used in healthcare on the Indian subcontinent. However, considering the safety concerns that may have a negative impact on traditional medicine practices, and to increase recognition of ASU&H practices, safety monitoring mechanisms through pharmacovigilance systems are very much essential. To overcome the challenges in pharmacovigilance for these products and preparations, enabling methodologies, such as frequent training programmes, should be conducted.

References

1. Ministry of AYUSH. https://www.ayush.gov.in. Accessed 30 Oct 2020
2. Nihariha S, Padmavati M (2013) Herbal drug regulation and commercialization: an Indian industry perspective. J Altern Complement Med 19(12):957–963. https://doi.org/10.1089/acm.2012.0275
3. World Health Organization. Pharmacovigilance. https://www.who.int/medicines/areas/quality_safety/safety_efficacy/pharmvigi/en/. Accessed 30 Oct 2020
4. Kalaiselvan V, Saurabh A, Kumar R, Singh GN (2015) Adverse reactions to herbal products: an analysis of spontaneous reports in the database of the pharmacovigilance programme of India. J Herb Med 5(1):48–54. https://doi.org/10.1016/j.hermed.2015.01.006
5. Naranjo CA, Busto U, Sellers EM et al (1981) A method for estimating the probability of adverse drug reactions. Clin Pharmacol Ther 30(2):239–245. https://doi.org/10.1038/clpt.1981.154
6. Tay C (2018) Liver fears: NZ health ministry issues further warning over supplement ingredient. https://www.nutraingredients-asia.com/Article/2018/12/03/Liver-fears-NZ-health-ministry-issues-further-warning-over-supplement-ingredient. Accessed 30 Oct 2020

7. Therapeutic Goods Administration (2019) Arthrem capsules. https://www.tga.gov.au/alert/arthrem-capsules. Accessed 30 Oct 2020
8. Heinrich M, Barnes J, Prieto-Garcia JM, Gibbons S, Williamson EM (2018) Methods in natural product analytical chemistry. In: Fundamentals of pharmacognosy and phytotherapy, 3rd edn. Elsevier, Edinburgh, p 106
9. Savage RL, Hill GR, Barnes J, Kenyon SH, Tatley MV (2019) Suspected hepatotoxicity with a supercritical carbon dioxide extract of Artemisia annua in grapeseed oil used in New Zealand. Front Pharmacol 10:1448. https://doi.org/10.3389/fphar.2019.01448
10. WHO Pharmaceuticals Newsletter No. 5 (2019). https://www.who.int/medicines/publications/PharmaNewsletter5-19/en/. Accessed 30 Oct 2020
11. Jayanti R, Debadatta C, Rudrajit P, Kunal S (2018) Prevalence of the use of complementary and alternative medicine in an eastern Indian population with emphasis on tribal/ethnic minority group. J Taibah Univ Med Sci 13(4):384–389. https://doi.org/10.1016/j.jtumed.2018.04.001

Chapter 23
Pharmacovigilance for Traditional Chinese Medicinal Drugs in China

Li Zhang and Tianyi Yang

23.1 Introduction

In its thousands of years of history, Traditional Chinese Medicine (TCM), with its strong theoretical systems, has played a key role in health protection and disease in China [1]. TCM is as important as Western Medicine (WM) in the Chinese healthcare system. However, with the increasing acceptance of TCM, and its extensive use worldwide, safety issues concerning Traditional Chinese Medicine drugs (TCMDs) have attracted attention globally [2, 3]. Whether TCMDs are used alone, or in combination with western medicinal drugs (WMDs), evaluation methods, and the quality of clinical research on their safety and efficacy, need to be strengthened. In fact, ever since TCMDs have been applied in clinical practice, Chinese physicians and pharmacists have paid attention to the assessment of benefits and harms associated with TCMDs and, in TCM theory, there are many classical principles processing methods and "compatibility" of TCMDs in "prescription" formula to ensure rational and safe use. Nowadays, many new pharmaceutical technologies have been used during research and development (R&D) of TCMDs, alongside progress in scientific technology, and many new dosage forms for TCMDs have emerged. The introduction of these new TCMD products into the market, with extensive exposure among the population, brings new potential harms.

Improvements in monitoring systems for the whole life cycle of drugs in China led to the National Medical Products Administration's (NMPA) revision of regulations concerning pharmacovigilance, which, previously, were weak during the R&D

L. Zhang (✉)
Dongfang Hospital Beijing University of Chinese Medicine, Beijing, China

T. Yang
Rollins School of Public Health Emory University, GA, Atlanta, USA

J. Barnes (ed.), *Pharmacovigilance for Herbal and Traditional Medicines*,
https://doi.org/10.1007/978-3-031-07275-8_23

phase of drug development. Now, since China joined the International Council for Harmonisation of Technical Requirements for Pharmaceuticals for Human Use (ICH), and the revision of the *Drug Administration Law of the People's Republic of China* (DAL) [4], NMPA has established an individual case safety reports (ICSRs) online submission system database for pre-market drugs. The new pre-marketing submission system replaces the old system, which received reports via email and fax. Together with the existing post-market submission system, this forms an important way for collecting safety data for the whole life cycle for drugs. Suspected, unexpected, and serious ADR ICSRs for drugs in clinical trials and post-marketing phases are collected and reported to the pre-marketing pharmacovigilance system and the National Center for Adverse Drug Reaction Monitoring (NCADRM) Spontaneous Reporting System (SRS), namely the National Adverse Dug Reaction Monitoring System (NADRMS) [5–8]. In order to improve the efficiency and quality of ICSRs, and to promote active surveillance on post-marketing drugs, in 2016, NMPA began to use real-world electronic medical records (EMR) and to explore the establishment of a China Hospital Pharmacovigilance System (CHPS) and the Chinese ADR Sentinel Surveillance Alliance (CASSA) program. CASSA is organized and coordinated by NCADRM and is a program that connects NCADRM with provincial ADR monitoring centers and competent sentinel hospitals, and all member sentinel hospitals are assigned by NCADTM. CHPS developed by NCADRM provide an effective working platform for CASSA hospitals, connects to healthcare information systems (HIS) and makes it possible to extract necessary structured data from the HIS system of CASSA hospitals. The data processed by CHPS could simultaneously be uploaded and applied by both SRS and active monitoring [9, 10]. Although NADRMS has played an important role in risk management for TCMDs over the past 30 years, pharmacovigilance for TCMSs is still facing substantial challenges. Against this background, this chapter reviews the current situation and challenges in TCMDs' pharmacovigilance and proposes measures to improve pharmacovigilance and risk management for TCMDs in China.

23.2 Regulatory Oversight and Management for TCMDs in China

In China, TCM and WM exist simultaneously and are practiced alongside each other at every level of the healthcare system. They are equally important, and both are covered by health insurance. About 90% of general hospitals include a TCM department and provide TCM services for patients [11]. NMPA, the Chinese agency for regulating drugs and medical devices (formerly the China Food and Drug Administration (CFDA)), focuses equally on TCMDs and Western medicines/drugs (WMDs) during supervision and administration of their whole product life cycles. To assure the quality, safety, and efficacy of drugs post-marketing, NMPA has promulgated a series of regulations and provisions. Viewed from NMPA administration, the Chinese Pharmacopoeia 2015 (ChP 2015) [12] of TCMDs includes Chinese Materia Medica (CMM)/prepared slices of Chinese crude drugs (PSCCDs), herbal oils and extracts, Chinese "patent" medicines (CPMs), and "simple preparations." PSCCDs, also known as "TCM decoction pieces" (or Yin Pian) are CMM processed according to TCM practices and principles. PSCCDs are used as prescription drugs for decoctions by physicians, and as raw materials for production of CPMs. CPMs and "simple preparations" refer to products and preparations that use PSCCDs as the material and have a certain type of formulation, specification(s), function(s) and/or caution(s), and which can be used directly for the treatment of diseases under the guidance of TCM practices and principles. As for WMDs, all CPMs and some PSCCDs need to be strictly evaluated and approved by NMPA: almost all policies and regulations promulgated by NMPA are applicable to TCMDs. As with chemical and biological patent products, the quality, safety, and efficacy of all CPMs should be proved and assured under the guidance of Good Laboratory Practice (GLP), Good Clinical Practice (GCP), Good Manufacturing Practice (GMP), and Good Supply Practice (GSP). It is illegal for pharmaceutical companies to produce and sell unlicensed CPMs in China.

23.2.1 Introduction of Updated Regulations on Pharmacovigilance Under the New Regulatory Situation

Since *Opinions on the Reform of Review and Approval Process for Drugs and Medical Devices* [13] was issued in August 2015, NMPA has developed a series of supportive guidelines to encourage new drug R&D, including reforming clinical trial management, accelerating review and approval processes, prioritizing reviewing procedures, and expansion of the Marketing Authorization Holder (MAH) Program. Since NMPA joined the International Conference on Harmonization of Technical Requirements for Registration of Pharmaceuticals for Human Use (ICH) in June 2017 [14], NMPA regulation has further aligned with international practice. The *Drug Administration Law of the People's Republic of China* (DAL) [4] released recently promotes the implementation of the MAH system, which clearly stipulates the legal liability of MAHs for the whole life cycle of drugs. Concerning pharmacovigilance, as well as continuing to implement regulations regarding PSURs from MAHs, and spontaneous reports from healthcare professionals, MAHs, and others [15], new regulations and guidance, covering pharmacovigilance from R&D to post-marketing, have been developed. NMPA has officially implemented the MAH direct-reporting regulation in 2018, and formulated corresponding guidelines and standards [16]. A new pharmacovigilance system comprising E2B (R3) electronic data transmission was launched on January 1st, 2020, comprising high standards and requirements for submission of ICSRs [5–8, 17]. The newly formed R&D ADR reporting system [6] provides a convenient way for reporting and detecting signals and harms identified during R&D. Moreover, new regulations require MAHs to submit annual reports of their pharmacovigilance activities to NMPA, including the production and sale of drugs, post-marketing research, risk management and so forth [18].

23.2.2 Regulations on Safety Surveillance Throughout the Whole Product Life Cycle for TCMDs

Because of the different characteristics and uses of CPMs and PSCCDs, in order to ensure safety of public medication and product efficacy, NMPA takes corresponding and specific regulatory measures.

23.2.2.1 Chinese Patent Medicines (CPMs)

During the registration period for CPMs, all documents and research relating to registration should be prepared carefully, and all pharmaceutical research, including preclinical studies and clinical trials, should strictly comply with the respective laws and regulations [19]. It is important for NMPA to support tradition as well as innovation for TCMDs through establishing and optimizing registration management and technologic evaluation systems combining TCM characteristics with the scientific requirements of drug R&D. For example, TCM compound prescriptions (e.g., Kaixin Powder and Xuefuzhuyu Capsules, which contain 4 and 11 different PSCCDs, respectively) selected from classical ancient TCM texts could be considered in developing new drugs according to TCM syndrome differentiation. In 2018, to ensure that scientific standards are met for R&D with this kind of CPM, NMPA introduced guidelines that aim to provide basic guidance on the development of clinical trials and the evaluation of efficacy and safety for new CPMs in development for treatment of TCM syndromes [20]. In addition, NMPA recently strengthened post-marketing safety surveillance and evaluation for CPMs [21]. For instance, to effectively capture and identify signals of drug-induced liver injury (DILI), and accurately assess causality between TCMDs and DILI, NMPA issued a guideline for researchers and physicians [22]. Moreover, in order to standardize the planning and revision of CPMs' specifications, and ensure rational use, NMPA issued "*Requirements and guidelines on the writing of the package leaflet (PL) of CPMs and Natural Medicines*" in 2006 [23]. Revising package leaflets/inserts (PLs) is another risk management measure used by the NMPA (see Sect. 23.2.1): PLs may be revised by NMPA, or by MAHs at the requirement of the NMPA. By March 2020, 70% of PLs for CPMs described in the ADR Information Bulletin (ADRIB) had been revised in accordance with drug safety issues.

23.2.2.2 Prepared Slices of Chinese Crude Drugs (PSCCDs)

Use of PSCCDs as ingredients affects the clinical efficacy of traditional decoctions and the quality of CPMs. PSCCDs are required to be prepared under the guidance of GMP provisions and to meet the Chinese Pharmacopoeia and Health Ministry standards. NMPA has implemented a series of regulations relating to these aspects [24]. Since 2018, NMPA has begun specific enforcement of the quality of PSCCDs [25]. This project aims to investigate and punish illegal activities, circulation, and sales of PSCCDs, through exposing and publishing names of offending companies. NMPA achieves social participation in co-governance through enhancing the eligibility criteria for PSCCDs industries' access, revision and filing of provincial-level and national PSCCDs processing standards for public inquiry [26]. For promoting the standardization, specialization, and large-scale production of PSCCDs and improving technical management systems, NMPA takes several measures. For example, promptly improving pharmacopeial standards, revising Good Agriculture Practice (GAP) regulations encourages industries to engage in the production of PSCCDs from authentic locally grown herbs.

In addition, NMPA has strengthened the supervision of imported medicinal materials (IMMs) for herbal medicines, requiring strict standards for these materials [19]. For IMMs with different sources, the order of the implementation of the standards is: ChP 2015 [12] (current edition); standards for IMMs; Health Ministry standards [19]. The national minority medicinal materials may comply with the corresponding provincial standards, or the materials standard of autonomous regions [18]. At the same time, an improved and unified IMMs-informatic platform was established with the authorization, recording-filling, and port inspection to track the source of medical information in favor of the supervision and management of IMMs.

In short, to ensure the safety and efficacy of TCMDs, NMPA formed a whole-chain supervision system for TCMDs from R&D inspection to post-marketing surveillance, and established supervision procedures combining routine inspections, inspections without prior notification to the manufacturers, and random inspections. Pharmacovigilance is an important part in the whole chain of supervision.

23.3 Current Status of Pharmacovigilance for TCMDs

Pharmacovigilance for TCMDs is important in the whole life cycle supervision process. However, despite the long history of use of TCMDs in China, post-marketing safety surveillance for TCMDs began much later than for WMDs. There are several reasons for this, including differences in pharmacovigilance concepts in TCM, extent of public interest, social response, and lack of skilled personnel. Further, there is a lack of innovation in methods for post-marketing safety surveillance for TCMDs. TCMDs share the same pharmacovigilance systems, such as SRS [24] and CHPS, with WMDs, so it is difficult to collect information on risk factors solely associated with TCMDs from these systems [10]. ADR monitoring only focuses on CPMs at present. The number of ICSRs associated with PSCCDs collected by NCADRMD and CHPS is limited and these reports have not yet been analyzed.

23.3.1 Signal Detection for TCMDs Safety Concerns from NADRMS

Signals of safety concerns associated with drugs post-marketing arise mainly from NADRMS, scientific literature and information from overseas medicines regulatory agencies. Compared with scientific information and resources relating to safety of WMDs, safety information for TCMs is very limited. The lack of scientific literature on preclinical research, ICSRs, and post-marketing clinical evidence relating to TCMDs makes it difficult to understand potential harms associated with these products and preparations [24]. The NADRMS is important in identifying and disseminating early warnings about risks associated with TCMDs over the past 30 years, including discovering potential risks of harms arising from poor-quality products. Through data mining and manual clinical assessment of reports, signals of safety concerns that warrant attention can be identified. Some of these signals specify serious ADRs, a high frequency of reports, and implicate products/preparations used extensively; some of these are for TCMDs and further investigation is needed. From 1999 to 2019, the NADRMS received 15.19 million ICSRs of suspected ADRs in total; in 2019, 1.5 million ICSRs were received. As one of the important indicators to measure the level of national ADR monitoring, the average (mean) number of reports per million people in China was 1130 in 2019. According to category-specific statistics for suspected drugs, reports relating to chemical products, TCMDs, bio-products, and unspecified products accounted for 84.9%, 12.7%, 1.6%, and 0.8% of submitted reports, respectively [27]. From 2009 to 2019, the proportion of ICSRs in the NADRMS involving TCMs ranged from 13.3% to 14.6%. The most common types of oral CMPs are Chinese and WM compound products (CWMCPs) ($n = 11$; 64.7%), dermatologic drugs (3; 17.6%), and orthopedic drugs (2; 11.8%). Compared to pure TCMDs, the risks of CWMCPs stand out the most. There are two specific warnings relating to PSCCDs published in ADRIB: one concerned TCMDs containing toxic aristolochic acids associated with causing renal injury; the other related to *Polygonum multiflorum* Thunb. (Polygoni Multiflora Radix, He shou wu; accepted name: *Reynoutria multiflora* (Thunb.) Moldenke) and reports of liver injury. Safety monitoring for TCMDs should focus on injections, oral preparations containing toxic PSCCDs, CWMCPs [1], and dermatologic and orthopedic oral preparations. Attention should be given to the potential for allergic reactions with TCMIs, and the risks of liver injury with certain oral preparations of TCMDs.

In accordance with signals detected, NMPA implements risk management through several approaches, including arranging manufacturer communication meetings, modification of PLs, and restriction, suspension, or withdrawal of products [1, 21, 24]. For instance, given the known risks associated with TCMIs [28, 29], NMPA issued seven guidelines for post-marketing re-evaluation of TCMIs [30, 31], in addition to revising PLs and promoting rational use of TCMIs [32]. Further, NMPA promotes post-marketing studies (PMS) in China, particularly clinical research. Academic institutions and MAHs in China have tried to use different databases as sources to develop clinical research, such as HIS and health insurance databases. These studies of TCM post-marketing safety surveillance have provided useful information and led to active monitoring and PMS in China [24, 33, 34]. NMPA has issued guidance on key monitoring of post-marketing drugs in 2013 [35].

23.3.2 China ADR Sentinel Surveillance Alliance (CASSA) and CHPS

After years of use, the active monitoring mode of post-marketing safety surveillance has been gradually replaced by CHPS, which is an active monitoring model based on hospital-based electronic health records (EHR) of CASSA [10, 36]. It is an active surveillance approach developed by the NADRM in 2016 with 150 CASSA hospitals participating by the end of 2019. Through the data interface with CHPS, its facilities capture of routine data from HIS (including EMR system, pharmacy information management system, laboratory information system, etc.) of sentinel hospitals. CHPS is an information system for the detection, reporting, and evaluation of ADRs/adverse events, for undertaking key monitoring, PMS, and for obtaining alert information for medical devices. Now, an improved system, CHPS V2.0, can penetrate the independent information system barriers in different hospitals, and has led to improvements in the quality and efficiency of ADR reporting.

23.4 Challenges and Strategies in Safety Monitoring for TCMDs

The development of safety monitoring for TCMDs has gone through a difficult process. Since only about 15% of ICSRs relate to TCMDs, it is difficult to identify TCMDs' signals from the total database of ICSRs [28]. Signal detection mainly relies on daily monitoring, weekly summaries, and manual quarterly analysis. ICSRs were subject to automated classification statistics and summaries for TCMDs, chemical products and biological products until the launch of the third version of NADRMS in 2011; this substantially improved the efficiency of identifying signals associated with TCMDs. However, the functions and principles of automated signal detection for TCMDs from NADRMS still need to be established. Guidelines on safety monitoring and evaluation for TCMDs have not yet been developed. The following measures are proposed to help progress pharmacovigilance for TCMDs.

23.4.1 Establishing Post-marketing Surveillance Models for TCMDs Led by Detecting Important Signals from NADRMS, and Widening the Channels for Collecting ICSRs

Due to the lack of information and the large number and variety of TCMDs available on the market, it is difficult to select targets and points for PMS of high-risk CPMs for benefit-harm assessment [24]. NADRMS has always been the main way to obtain information on safety concerns associated with TCMDs in China. Signals detected for TCMDs have all come from the NADRMS, and related TCMIs, CPMs containing toxic PSCCDs, CWCMPs, drugs used in dermatology and orthopedics, and other species/substances identified for monitoring. Selecting significant signals related to CMPs for monitoring could be improved in the following ways:

- Further improving CHPS to widen the channels for collecting ICSRs.
- Obtaining higher quality ICSRs through promoting active monitoring.
- Establishing a specific ADR reporting system for patients to strengthen the monitoring of CPMs' oral preparations and non-prescription ("over-the-counter") drugs.
- Further analysis of ICSRs received by NADRMS to summarize the characteristics and trends observed in SRS data.
- Using SRS data to identify signals of safety concerns associated with TCMs.
- Indicating further research topics for post-authorization safety studies (PASS) and post-authorization efficacy studies (PAES) for CPMs.

23.4.2 Establishing Post-marketing Surveillance Approaches in Accordance with the Characteristics of TCMDs

It is necessary to establish an appropriate monitoring and evaluation model that is aligned with the characteristics of TCMDs and their use. Compared with WMs, the components of TCMDs are complex and most CPMs are compound preparations. Thus, there are many ingredients, as well as other factors, that could influence their safety profile. Thus, early warnings might indicate quality defects or inappropriate use. With respect to the drugs themselves, due to the different sources of raw material, agricultural practices, and processing of PSCCDs, it is difficult to define what are the toxic substances of CPMs: CPMs typically contain three or more PSCCDs, giving rise to the potential for additive, opposing and/or synergistic interactions among these ingredients. Concerning clinical applications for TCMDs: it is difficult to promote rational use of TCMDs, as more than half of TCMDs are used by Western physicians, and TCMDs are frequently used in combination with WMDs. As far as monitoring and signal detection are concerned: so far, SRS and CHPS, which are used for both TCMDs and WMs, are technical platforms designed according to the characteristics of WMDs [22, 28]. It is difficult to capture the special risk factors for harms associated with TCMDs [37, 38] using the current capabilities of SRS and CHPS. As far as signal detection is concerned: the underlying database is incomplete: there are no standard coding and classification databases for drug names and TCM medical terminology in the spontaneous and active pharmacovigilance systems. Moreover, although some epidemiological methods are used in PASS for TCMIs, few characteristics of TCMDs are indicated and the research quality needs to be improved. Based on the above problems, it is necessary to modify the system of information collection, data mining technology and evaluation systems to accommodate the characteristics of TCMDs.

Specific recommendations include:

- Including TCM diagnosis and treatment information into the ICSR collection system.
- Improving the ability of detecting signals and risk factors by strengthening the construction of a basic database, such as developing an internationally recognized classification and coding system and rules for TCMDs and developing processes for signal detection for drug-drug interactions with TCMDs.
- Establishing a model for comprehensive safety evaluation for TCMDs, and improving the extent and quality of evidence-based drug evaluation for TCMDs by using multiple data sources, such as data from SRS, basic pharmaceutical, pharmacological and toxicological research, real-world evidence, etc.
- Learning from long-standing experience with WMDs, using the appropriate epidemiological approaches to develop post-marketing studies, improving the quality of PMS of TCM and providing evidence for scientific supervision.
- Paying more attention to safety issues caused by the modern techniques applied in production processes for TCMDs, and encouraging MAHs to undertake the necessary basic research [28, 29].

For example, regarding the last point, according to an analysis of ADRIB, the CPMs Zhuanggu Guanjie Pill and Xianling Gubao Capsule, which contains *Psoralea corylifolia* L. (Psoraleae Fructus, Buguzhi) and *Epimedium brevicornu* Maxim. (Epimedii Folium, Yinyanghuo) may potentially cause DILI. However, this toxicity was not previously recorded in ancient TCM texts. Toxicology research has found that using ethanol as the extraction solvent may increase the quantity of toxic components of Buguzhi and Yinyanghuo. Moreover, pairs of TCMDs, such as *Curculigo orchiodes* Gaertn. (Curculiginis Rhizoma, Xianmao) and Yinyanghuo, are often used in clinical practice. Toxicology studies have shown that toxic components of the paired TCMDs are mainly curculigoside and enriched in ethanol extracts. The toxicity is positively related to the content of curculigoside; however, water extracts show no obvious toxicity [39]. This illustrates the importance of using proper extraction methods in the production process for CPMs.

23.4.3 *Strengthening Surveillance of PSCCDs, and Healthcare Products Containing PSCCDs Approved by NMPA*

The World Health Organization (WHO) emphasizes that the quality, safety, and efficacy of medicinal plants should be scientifically evaluated, to ensure rational use of plant-based products in an integrated approach [40].

Although PSCCDs and TCM decoctions are widely used in China, it is difficult to determine their risks due to the complexity of their components, interactions between components, and inappropriate use of PSCCDs and decoctions. There remains a lack of legal requirements for these types of preparations and basic and clinical research concerning their safety [24, 28]. Potential harms from PSCCDs from real-world use have not been fully recognized. Therefore, there is still a long way to go to optimize the relevant regulations and technical systems. In addition, healthcare products containing PCSSDs approved by NMPA are commonly used by consumers for self-care. Factors affecting their safety, such as adulteration with chemical drugs, are complex, but corresponding pharmacovigilance systems to monitor these have not yet been set up in China. It is necessary to strengthen pharmacovigilance for these preparations to ensure consumer safety.

23.4.4 *Implementing Responsibilities for MAHs in the Whole Product Life Cycle for Benefit-Harm Assessment of TCMDs*

As the primary responsible stakeholder for ensuring the optimal benefit-harm balance during the whole product life cycle of TCMDs, MAHs should undertake the following work in accordance with the requirements of the new *Drug Administration Law of the People's Republic of China* [4]:

- Establish a PV system, actively perform ADR monitoring, PASS and basic research, improve capabilities for signal detection and risk management.
- Carry out essential effectiveness evaluations for TCMDs in accordance with the new registration regulations and technical requirements to reevaluate and highlight their precise indication and clinical value: benefit-harm evaluation is throughout the whole product life cycle [41, 42]. As some CPMs that have been marketed for long periods of time have not been investigated in clinical trials, indications for these CPMs are often very general and not for specific medical conditions. The corresponding MAHs should initiate PAES to explore the clinical benefits of these CPMs, and provide an evidence base for the government's distribution of medical resources, such as medical selections of National Essential Drug List and National Essential Insurance Medicines list.
- Introduce the concept of quality risk management to contribute to the whole life cycle of CPMs: MAHs should strengthen quality control for PSCCDs to prevent

risks from incorrectly sourced material and comply with regulations relating to drug production and distribution. While clarifying the key points of quality control in each link throughout the whole supply-chain for CPMs, MAH should establish their own CPMs' traceability system [28] and an early-warning safety mechanism to obtain prompt responses, and effective prevention and control.

In summary, although the Chinese pharmacovigilance system plays an important role in harm prevention and risk management for TCMDs, the establishment of a complete pharmacovigilance system for TCMDs still has a long way to go. Promoting rational TCMD use, establishing and perfecting scientific supervision and technical evaluation systems aligned with the characteristics of TCMDs, fully implementing responsibilities for MAHs, and introducing the concept of quality risk management are effective strategies to ensure TCMDs' safety and efficacy, sustainable development, and internationalization and are critical measures to improve the core competitiveness of MAHs in China.

References

1. Zhang L, Yan J, Liu X et al (2012) Pharmacovigilance practice and risk control of traditional Chinese medicine drugs in China: current status and future perspective. J Ethnopharmacol 140(3):519–525. https://doi.org/10.1016/j.jep.2012.01.058
2. Zhang L, Yang X, Deng Y (2009) Liver damage related to Polygonum multiflorumand its preparations: domestic literature review and analysis. China J Chin Mater Med 34(8):2414–2418
3. Medicines and Healthcare products Regulatory Agency (MHRA) (2014) Banned and restricted herbal ingredients. https://www.gov.uk/government/publications/list-of-banned-or-restricted-herbal-ingredients-for-medicinal-use. Accessed 5 Feb 2020
4. National Medical Products Administration (2019) Drug Administration Law of the People's Republic of China (DAL). http://www.nmpa.gov.cn/WS04/CL2076/357712.html. Accessed 20 January 2020
5. National Medical Products Administration, National Center for ADR Monitoring (2019) Adverse drug reaction reporting system. http://www.cdr-adr.org.cn/drug_1/adrReport_1/. Accessed 5 Feb 2020
6. National Medical Products Administration, National Center for ADR Monitoring (2019) Direct reporting system for MAH. http://daers.adrs.org.cn/#/login?_k=z0gf96. Accessed 5 Feb 2020
7. National Medical Products Administration, National Center for ADR Monitoring (2011) National Center for ADR Monitoring spontaneous reporting system for healthcare professionals and marketing authorization holder. http://111.202.232.186/sso/login?service=http%3A%2F%2F111.202.232.186%2FPPF%2FcasAuthUser. Accessed 5 Feb 2020
8. Hou Y, Li X, Wu G et al (2016) National ADR monitoring system in China. Drug Saf 39(11):1043–1051. https://doi.org/10.1007/s40264-016-0446-5
9. Li H, Tan N, Zhuang H et al (2020) Effect of Chinese hospital pharmacovigilance system on the reporting of adverse drug reactions in a hospital. Herald Med 39(2):265–267
10. Hou Y, Song H, Liu H et al (2019) Practice and discussion on active surveillance by Chinese hospital pharmacovigilance system. Chin J Pharmacovigil 16(4):212–214
11. The State Administration of Traditional Chinese Medicine (2011) China statistical yearbook of Chinese medicine. http://www.satcm.gov.cn/1987-2010/start.htm. Accessed 5 Feb 2020
12. Pharmacopoeial Commission of PRC (2017) Pharmacopoeia of the People's Republic of China. China Medical Science Press, Beijing

13. The State Council of the PRC (2015) Opinions of the state council on reforming the drug and medical device evaluation and approval system. http://english.www.gov.cn/archive/state_council_gazette/2015/09/10/content_281475187020362.htm. Accessed 25 February 2020
14. FDA News (2017) China joins ICH as full regulatory member, pledges to implement guidelines. https://www.fdanews.com/articles/182330-china-joins-ich-as-full-regulatory-member-pledges-to-implement-guidelines. Accessed 10 Mar 2020
15. MOH and China Food and Drug Administration (2011) Adverse drug reaction reporting and monitoring provision. http://www.nmpa.gov.cn/WS04/CL2174/300642.html. Accessed 10 Mar 2020
16. National Medical Products Administration (2018) Announcement on direct report of ADR by marketing authorization holder. Accessed 25 Feb 2020
17. National Medical Products Administration, National Center for ADR Monitoring (2019) Announcement on on-line trial operation of electronic E2B (R3) transmission system. http://www.cdr-adr.org.cn/drug_1/zcfg_1/zcfg_zdyz/201912/t20191231_47006.html. Accessed 25 Feb 2020
18. National Medical Products Administration, National Center for ADR Monitoring (2019) Announcement on issuing guideline on the preparation of annual pharmacovigilance report for MAH (Interim). http://www.cdr-adr.org.cn/tzgg_home/201911/t20191129_46857.html. Accessed 25 Feb 2020
19. National Medical Products Administration (2019) Administrative provision for imported medicinal materials. http://www.nmpa.gov.cn/WS04/CL2138/373553.html. Accessed 25 Feb 2020
20. National Medical Products Administration (2018) Technical guidance for clinical trial of TCM new drug based on TCM syndrome theories. http://www.nmpa.gov.cn/WS04/CL2138/331783.html. Accessed 25 Feb 2020
21. Zhang L, Wong LY, He Y et al (2014) Pharmacovigilance in China: current situation, successes and challenges. Drug Saf 37(10):765–770. https://doi.org/10.1007/s40264-014-0222-3
22. National Medical Products Administration (2018) Guidance for clinical evaluation of drug-induced liver injury (DILI). http://www.nmpa.gov.cn/WS04/CL2138/229510.html. Accessed 25 Feb 2020
23. National Medical Products Administration (2006) Notification on the requirements and guidance for the printing and distribution of the formats and contents of the for the prescription of traditional Chinese medicine and natural medicine. http://www.nmpa.gov.cn/WS04/CL2196/323564.html. Accessed 25 Feb 2020
24. Zhang L (2018) Pharmacovigilance of herbal and traditional medicines. In: Bate A (ed) Evidence-based pharmacovigilance: clinical and quantitative aspects. Springer, New York, pp 37–65
25. National Medical Products Administration (2018) Notification on publishing the work programme for strengthening the quality of prepared slices of Chinese crude drugs. http://www.nmpa.gov.cn/WS04/CL2196/330076.html. Accessed 25 Feb 2020
26. National Medical Products Administration (2020) Notification on procedures and requirements for keeping records of provincial norm for the PSCCMs' process. http://www.nmpa.gov.cn/WS04/CL2196/374149.html. Accessed 25 Feb 2020
27. National Medical Products Administration (2020) Annual report for national ADR monitoring. http://www.nmpa.gov.cn/WS04/CL2155/376451.html. Accessed 16 May 2020
28. Zhang L, Yang X, Cao L (2005) Pondering on current status and development of monitoring on adverse reaction of traditional Chinese medicine in China. Chin J Integr Trad West Med 25(7):581–584
29. Zhang L, Ye Z, Ji S (2012) Safety monitoring of traditional Chinese medicine injections: the current status and risk management. World Sci Technol Modern Tradit Chin Med Mater Medica 12(6):845–850

30. China Food and Drug Administration (2009) Notification on doing well in the reevaluation of TCM injections. http://www.nmpa.gov.cn/WS04/CL2196/323740.html. Accessed 20 Mar 2020
31. China Food and Drug Administration (2010) Notice on publication the 7 principals for technical guideline on TCM injection safety re-evaluation concerning production process, quality control clinical study and so on. http://www.ccpie.org/cn/yjxx/yphzp/webinfo/2010/10/1481297440376509.htm. Accessed 20 Mar 2020
32. National Medical Products Administration (2008) Notification on further strengthening the production, management and clinical rational use of TCMI. http://www.nmpa.gov.cn/WS04/CL2079/333373.html. Accessed 20 Mar 2020
33. Yan YY, Yang YH, Wang WW et al (2017) Post-marketing safety surveillance of the Salvia miltiorrhiza depside salt for infusion: a real world study. PLoS One 12(1):e0170182. https://doi.org/10.1371/journal.pone.0170182
34. Li XX, Zhuo L, Zhang Y et al (2019) The incidence and risk factors for adverse drug reactions related to Tanreqing injection: a large population-based study in China. Front Pharmacol 10:1523. https://doi.org/10.3389/fphar.2019.01523
35. China Food and Drug Administration (2013) Notification on promoting key monitoring of pharmaceuticals manufacturing industries (draft for comments). http://www.nmpa.gov.cn/WS04/CL2101/228805.html. Accessed 20 Mar 2020
36. Li X, Li H, Deng J et al (2018) Active pharmacovigilance in China: recent development and future perspectives. Eur J Clin Pharmacol 74(7):863–871. https://doi.org/10.1007/s00228-018-2455-z
37. Zhang L, Yang X (2004) Cause analysis and countermeasure of ADR caused by traditional Chinese medicine. China J Tradit Chin Med Pharm 19(8):496–498
38. Zhang L, Yang X (2009) Pharmacovigilance idea should be introduced sufficiently into the safety monitoring and evaluation process of Chinese drugs. Chin J Integr Trad West Med 29(9):843–846
39. Zhu F, Yang M, Chen W et al (2015) Correlation between toxicity and content of Curculiginis Rhizoma and Epimedii Folium with different compatible proportion. Chin J Exp Tradit Med Formulae 21(5):175–177. https://doi.org/10.13422/j.cnki.syfjx.2015050175
40. WHO Regional Office for South-East Asia (2009) The use of herbal medicines in primary health care. WHO Regional Office for South-East Asia. https://apps.who.int/iris/handle/10665/206476
41. Zhang L, Wang G, Lin R (2009) Analysis and discussion on the pharmacovigilance information and measures of risk control in China (1). Chin Pharm Affairs 23(8):735–739
42. Zhang L, Wang G, Lin R (2009) Analysis and discussion on the pharmacovigilance information and measures of risk control in China (2). Chin Pharm Affairs 23(9):853–856

Chapter 24
Pharmacovigilance for Herbal and Traditional Medicines in Japan

Lida Teng

24.1 Types of Herbal Products in Japan

24.1.1 Traditional Medicines Commonly Known as Kampo

In Japan, traditional folk medicine was influenced by ancient Chinese medical theory around 1500 years ago. Traditional medicine used to be a mainstream healthcare approach until the end of the Edo Period (1603–1868), before the implementation of the Western medical system in the Meiji Periods (from 1860s). Since then, to distinguish it from conventional Western medicine, traditional medicine was commonly named Kampo (漢方) (Kan:漢 means Chinese and Po:方 means prescription).

Kampo shares some of the traditional Chinese holistic medical theories, such as *Shokanron* (*Shanghanglun* in Chinese, namely, treatise on cold damage, before 220 AD) and *Kinkiyoryaku* (*Jinguiyaolue* in Chinese, namely, essential prescriptions from the golden cabinet, approximately 1000 years ago). However, the practice of Kampo has been revised and developed in a Japanese way to suit the local use by Kampo practitioners over hundreds of years. The number and types of crude drugs and Kampo products in Japan are considerably fewer than those used in Traditional Chinese medicine in China. The selection of plant and animal species and the composition of formulae used in Kampo are preferentially based on the local available plant species and local preferable uses, even under the same medicinal name as used in traditional Chinese medicine.

Over the past decades, typical practice of Kampo medicine has changed from individualized pattern diagnosis (a strictly traditional holistic way) into generalized

L. Teng (✉)
Department of Health Economics and Outcomes Research, Graduate School of Pharmaceutical Sciences, The University of Tokyo, Tokyo, Japan

J. Barnes (ed.), *Pharmacovigilance for Herbal and Traditional Medicines*,
https://doi.org/10.1007/978-3-031-07275-8_24

diagnosis based on the patient's symptoms (similar to conventional Western medical practice).

Kampo medicines are provided in two major types: crude drugs (shoyaku) and herbal preparations. Crude drugs are processed from plant or animal parts, and they are commonly used for making decoctions. The popular types of herbal preparation are dry extract powder, tablets, liquid, and pills. The most common dosage form for prescription use is dry extract powder.

24.1.2 Non-Kampo Herbal Medicines

Western herbal medicines have become more popular in Japan within recent years. For example, a proprietary product containing red vine (*Vitis vinifera* L.) leaf extract has been registered as an "over-the-counter" (OTC) product in Japan since 2011, with the indication to improve blood circulation and reduce leg swelling. Another recent approval is for a proprietary product containing dried *Vitex agnus-castus* L. extract for premenstrual syndrome (PMS), which was registered as an OTC product in 2014. These two OTC products will have an 8-year re-evaluation status and are sold as OTC products under the category of "drug requiring guidance" from a pharmacist (pharmacist-only) [1].

However, the definition of actual "Western herbal medicine" is still unclear in Japan. In most cases, this term is used to differentiate the herbal formulae that are not traditionally practiced under Kampo medicine principles. This type of herbal medicine is officially classified as "non-Kampo crude drug products" [2]. Non-Kampo crude drug products may include pharmaceutical compounds, such as vitamins.

24.1.3 Herbal Products as Health Foods or Unlicensed Medicines

Several herbal products with ingredients beyond the range of medicines (i.e., not on the official *List of Raw Materials Exclusively Used as Pharmaceuticals* [3]) are sold as health foods or unlicensed medicines.

24.2 Current Regulatory Framework for Herbal Products

Unlike some other Asian countries, there is no specific statutory regulation issued specifically for herbal medicine practitioners. Herbal medicines are sold either as OTC products, including pharmacist-only, or are prescribed by statutory-registered conventional Western medical doctors. Other herbal products are sold as health foods and unlicensed herbal medicines.

The latest version of the Japanese Pharmacopoeia (JP) 17th edition (Japanese version; JP17), published in April 2016, includes 349 monographs for single crude herbs, or herbal preparations, around 9% of which are Kampo formulae [4].

In addition to the JP, other official regulations for herbal medicines are *The Japanese standards for non-Pharmacopoeial crude drugs (Non-JP crude drug standards or Non-JPS)* [5] and *Approval standards for manufacturing and selling OTC Kampo products (previous title: Approval Standards for OTC Kampo products)* [6].

Crude herbs used in Kampo medicine that are not regulated by the JP should follow the standards for *Non-JPS*. The latest version was launched in 2015 and includes 83 crude herbs.

Approval standards is the official specification used for controlling OTC Kampo medicines sold in Japan, which particularly gives requirements for the range for the quantity of each crude material used in OTC Kampo medicines, and the standard claims for indications and contraindications. In total, 294 OTC standard formulae preparations are regulated. Of all standard formulae, 148 formulae have been covered by Japanese national health insurance since 1967 and, therefore, apart from OTC uses, they could also be prescribed by statutory-regulated medical practitioners. These 148 formulae are also called "ethical" Kampo formulae in Japan, where "ethical" means it is for prescription use. The difference between OTC formulae and prescription formulae is that prescription formulae contain the highest amount of extract from regulated crude drug materials, while OTC formulae usually contain more than 50% of extract [7].

The English names for ingredients of these 148 formulae are available online as *Standards of Reporting Kampo Products* from the National Institutes of Biomedical Innovation, Health and Nutrition (NIBIOHN) [8]. This database includes a JP monograph for each formula, if it is a JP formula, and details of package inserts from manufacturers. Ingredients of each Kampo formula mentioned in this chapter are presented in Table 24.1. However, only JP English common names are available from NIBIOHN. One English common name may represent several species. It is beyond the scope of this chapter to list all potential species. Further information on species names can be found in JP17 [4].

The registration scheme for Kampo medicines under *Approval standards* is simpler, and is based on the long-term history of clinical experience, and common clinical reviews provided by an official expert committee. In contrast, newly developed herbal medicines, i.e., new Kampo formulae for prescription use and OTC Western herbal medicines, must follow strict data requirements including pre-clinical and clinical studies results [2].

For herbal products sold as health foods, the official registration scheme for health foods in Japan is called *Food with Health Claims*. In general, the scheme includes *Food for specified health uses* and *Food with nutrient function claims*, and *Food with Nutrient Function Claims* [9]. Still a large number of unlicensed herbal

Table 24.1 Names and listed ingredients for Kampo formulae discussed in this chapter

Japanese Alphabet	Japanese Characters	List of ingredients recorded in NIBIHN [8][a]	Source used by NIBIHN
Bofutsushousan	防風通聖散	Japanese angelica root, peony root, cnidium rhizome, gardenia fruit, forsythia fruit, mentha herb, ginger, schizonepeta spike, saposhnikovia root and rhizome, glehnia root and rhizome, ephedra herb, rhubarb, sodium sulfate, anhydrous sodium sulfate, atractylodes rhizome, platycodon root, scutellaria root, glycyrrhiza, gypsum, aluminum silicate hydrate with silicon dioxide	JP 17
Daisaikoto	大柴胡湯	Bupleurum root, pinellia tuber, scutellaria root, peony root, jujube, immature orange, ginger, rhubarb	JP 17
Hangeshashinto	半夏瀉心湯	Pinellia tuber, scutellaria root, processed ginger, ginger, ginseng, glycyrrhiza, jujube, coptis rhizome	JP 17
Hochuekkito	補中益気湯	Ginseng, atractylodes rhizome, *Atractylodes lancea* rhizome, astragalus root, Japanese angelica root, citrus unshiu peel, jujube bupleurum root, glycyrrhiza, ginger, processed ginger, cimicifuga rhizome	JP 17
Rikkunshito	六君子湯	Ginseng, atractylodes rhizome, atractylodes lancea rhizome, poria sclerotium, pinellia tuber, citrus unshiu peel, jujube, glycyrrhiza, ginger	JP 17
Saikokeishikankyoto	柴胡桂枝乾姜湯	Bupleurum root, cinnamon bark, trichosanthes root, scutellaria root, oyster shell, processed ginger, glycyrrhiza	Package insert
Saireito	柴苓湯	Bupleurum root, pinellia tuber, ginger, scutellaria root, jujube, ginseng, glycyrrhiza, alisma tuber, polyporus sclerotium, poria sclerotium, atractylodes rhizome, atractylodes lancea rhizome, cinnamon bark	JP 17
Seihaito	清肺湯	Japanese angelica root, ophiopogon root, poria sclerotium, scutellaria root, platycodon root, apricot kernel, gardenia fruit, mulberry bark, jujube, citrus unshiu peel, asparagus root, fritillaria bulb, glycyrrhiza, schisandra fruit, ginger, bamboo culm	Package insert
Shakuyakukanzoto	芍薬甘草湯	Peony root, glycyrrhiza	JP 17
Shosaikoto	小柴胡湯	Bupleurum root, pinellia tuber, ginger, scutellaria root, jujube, ginseng, glycyrrhiza	JP 17
Yokukansan	抑肝散	Japanese angelica root, uncaria hook, cnidium rhizome, atractylodes rhizome, atractylodes lancea rhizome, poria sclerotium, bupleurum root, glycyrrhiza	JP 17

NIBIHN the National Institutes of Biomedical Innovation, Health and Nutrition

[a] If it is a Japanese Pharmacopeoia (JP) formulae, the list of ingredients is presented as recorded in JP 17 in this table. JP does not specify any botanical species name, and only English common name is used to describe the ingredients

products, especially Western herbal products, on the market are beneath the borderline of pharmaceuticals to non-pharmaceuticals and are not able to meet any requirements for *Food with Health Claims*.

24.3 The Practice and Use of Herbal Medicine in Japan

As compared with some other Asian countries, where herbal medicines are mainstream medications, the number of herbal preparations available as a medicine in Japan is limited. According to *the Statistics of Production by Pharmaceutical Industry 2018* [10], the annual production value of Kampo medicine in 2018 was 179,453 million JPY (appx. 1.6 billion USD), a 13.2% increase from the previous year. Of these, 80.7% are used prescriptions. However, Kampo medicine production represents only around 2.6% of the total pharmaceutical industry (i.e., industry of pharmaceutical drugs and herbal medicines), although the number is slightly increasing over the past few years.

A proposal to formally incorporate Kampo education into the core curriculum model of Western medical schools in Japan started in 2001. At present, Kampo medicine is taught in all 80 medical schools in Japan as a core curriculum [11]. In most pharmacy schools, a Kampo course is provided in undergraduate courses and/or as an advance-learning program. Surveys conducted by the Japan Kampo Medicine Manufacturers Association (JKMA) showed that the number of medical practitioners who are currently prescribing Kampo medicine has increased from 83.5% in 2008 to 89% in 2011 [12]. Kampo products represent about 92.8% of the herbal product market in Japan [2].

A JKMA customer study shows that 54% of Japanese consumers have ever used prescription Kampo medicine, and 67% said they have been aware that Kampo medicine is sold as both prescription and OTC medicines [13]. A web-based JKMA survey [12] involving 627 medical practitioners in 2011 indicated that five commonly prescribed medical conditions are muscle cramp (44.1%), acute upper respiratory inflammation (40.1%), constipation (38.5%), general malaise and PMS (35.5%), and ileus (19.4%). In this study, 51.6% of practitioners stated that they prescribe Kampo medicine based only on conventional Western medical diagnoses. The top reasons for prescribing Kampo medicines were that pharmaceutical drugs were not effective (57%), patients' requests (42%), referring clinical evidence reported in the published literature or conference (34%) and limitations of Western medical treatment (31%). For reasons why Kampo medicine was not chosen, 46.4% indicated that the way of using Kampo medicine is difficult, 34.8% believed that Kampo medicine has insufficient clinical evidence, and 29% considered that Western medicine is enough for treatment. However, this study did not attempt to obtain practitioners' opinions on the safety of Kampo medicine.

24.4 Safety Issues Associated with Herbal Products in Japan

Although the traditional opinion that herbal medicines are safe and cause fewer adverse effects than conventional medicines still commonly exists, there is increasing awareness of herbal safety issues, especially after the announcement of MHLW that the Kampo formula Shosaikoto was associated with interstitial pneumonia in 1998 [14]. The first case was identified in 1989 for Shosaikoto [15], and the number of case reports subsequently increased, particularly in the 1990s [16–19]. Shosaikoto has been associated with 100 cases, including 10 deaths, among older, mostly male, patients [19].

Shosaikoto is a classical Kampo formula used for liver dysfunctions. The formula contains bupleurum, scutellaria root, pinellia tuber, ginger, jujube, ginseng, and glycyrrhiza. As well as being associated with interstitial pneumonia, it was also found to interact with interferon and cause death [19]. Other Kampo medicines suspected to cause interstitial pneumonia are Daisaikoto, Seihaito, and Saikokeishikankyoto, which mainly contain the suspected herb, bupleurum [19, 20]. However, other Kampo medicines that do not contain bupleurum, such as Hangesyashinto [21] and Rikkunshito [22], have also been reported to cause pneumonitis.

Several national studies of the adverse effects (AEs) associated with Kampo medicines have been conducted in Japan since 2010. A study exploring drug safety announcements from MHLW, conducted in 2012, found that between 30 Jul 2003 and 31 Mar 2012, of all reported prescription drugs (i.e., "ethical" medicines in Japan), there were 1862 (0.68%) reports of AEs associated with Kampo medicines. The most commonly reported ADRs were lung disorders (e.g., interstitial pneumonia), liver dysfunction, hypokalemia, pseudohyperaldosteronism, and rhabdomyolysis [23]. This was the first national study of its kind in Japan. However, due to the limitation of using spontaneous report data form the early 2010s, this study could only access safety announcements published by MHLW, which is a summary of spontaneous report data produced three times a year.

Another study, conducted using the Japanese Adverse Drug Event Report Database (JADER), managed by the Japanese regulatory authority the Pharmaceuticals and Medical Devices Agency (PMDA), collected the reports on 148 prescription Kampo medicines (from April 2004 to February 2013) [24]. This study found a total of 1958 reports of AEs associated with Kampo medicines. However, the proportion of Kampo reports of the total number of reports was not provided. Commonly reported AEs were liver disorders (34%; suspected Kampo medicines included Bofutsushousan, $n = 119$; Saireito, $n = 61$; Hangeshashinto, $n = 38$), lungs disorders (26%; suspected medicines included Saireito, $n = 51$; Bofutsushousan, $n = 48$; Shosaikoto, $n = 38$), and metabolic and nutritional disorders (9%; suspected Kampo medicines included Shakuyakukanzoto, $n = 103$; Yokukansan, $n = 28$; Hochuekkito, $n = 6$).

Apart from licensed herbal products, there is still a number of unlicensed herbal products, especially sold as Western herbs, on the market that are on the borderline of pharmaceuticals and non-pharmaceuticals and are not able to meet the requirement of *Food with Health Claims*. The pharmaceutical quality of these herbal products varies. For example, studies examining unlicensed dietary supplements containing black cohosh, chaste tree and ginkgo leaf found that 7/19 black cohosh products and 8/17 chaste tree products had used the wrong plant material as claimed, and that 2/17 chaste tree products and 5/10 ginkgo products failed to meet the disintegration test requirement for OTC medicines [1]. The daily doses of active ingredients in some of these dietary supplements are sometimes over 10 times that of products registered as OTC medicines in Japan. However, quality problems do not appear to exist in licensed OTC medicines containing the same herbal ingredients [1].

24.5 Safety Monitoring of Adverse Events Associated with Herbal Products in Japan

24.5.1 Spontaneous Adverse Events Monitoring Scheme

In Japan, licensed herbal medicines are under the same pharmacovigilance framework as pharmaceutical drugs. The spontaneous adverse events (AEs) monitoring scheme is operated by PMDA. This scheme first began among designated medical institutions in 1967; it was extended to include all healthcare professionals as reporters in 1997. In 2003, the scheme was stipulated in Pharmaceutical Affairs Law. Since 2012, the scheme has accepted patients' direct reporting of suspected adverse effects associated with medicinal products [25].

The Japanese spontaneous reporting scheme for adverse reactions covers all licensed medicines, including licensed herbal medicines, medical devices, regenerative medicine products, quasi-drugs, and cosmetics. Quasi-drugs refers to medication products that are considered to be pharmacologically active with respect to preventing and improving symptoms, but are not able to follow the regulations for medicines or cosmetics. However, this scheme does not aim to collect reports from health foods and unlicensed medicines, including unlicensed herbal medicines, which are under another reporting scheme (see below). The reporting time period should be within 15 days for reports involving death as an outcome, and within 30 days for other serious AEs. For serious AEs considered unexpected, reporting should be within 15 days.

The system has different reporting forms for each product category, i.e., medicines, medical devices, regenerative medicine products, and quasi-drugs/cosmetics. All forms are available to download from the PMDA website. Reporters are required to send the completed form by mail, fax, or email. There no electronic online reporting systems in Japan. Reporting items for medicines include patients' demographic information; medical and therapeutic history; information on AEs (e.g., type, seriousness, time periods, dechallenge and rechallenge information); details on suspected medicines (e.g., names, manufacturer, administration, reason for use, concomitant medicines); free text space for the reporter's opinion; and clinical laboratory test results. The reporting form is only in Japanese language; there is no separate reporting form for herbal medicines.

The drug dictionary used by PMDA for coding drug safety data is commonly known as Iyakuhinmei Data File (IDF), which combines codes for "category," "generic name," and "product name." The WHODrug Dictionary (WHODrug Global) [26] is currently recommended by PMDA for submitting new drug applications. At present, a WHODrug Cross Reference Tool Japan (WHODrug CRT Japan) is available via the Uppsala Monitoring Centre (UMC), which helps to convert Japanese IDF codes into WHODrug Global Codes [27]. Nevertheless, code translation is still a challenge for medicines that do not have standard English names, especially Kampo medicines.

In April 2004, PMDA formally introduced data mining methods for signal detection of reported AEs. The current principal approach adopted by PMDA is using the reporting odds ratio (ROR), while Bayesian confidence propagation neural network (BCPNN) and gamma Poisson shrinker (GPS) are applied as complementary approaches for signal detection. The reason for using ROR is that ROR is considered to have high sensitivity and to be relatively easy to apply [28]. Overall, signal detection processes and results are discussed by the signal detection team in PMDA before any action is taken. Medical Dictionary for Regulatory Activities Preferred Term or Lowest Level Term (MedDRA/PT or LLT) is used for coding reported AEs.

Current reporters under this scheme are marketing authorization holders (MAH), health professionals (HPs), and patients. The latest published summary of HP reporter groups shows that, in 2016, 72.6% of all HP reporters were pharmacists, 17.8% were medical practitioners, and 7.8% were nurses. In total, 88.6% were from hospitals, while 9.4% were from local pharmacies [29]. It is not clear which types of reporters are more likely to report AEs associated with herbal medicines. Although pharmacists were found to be the main reporter groups, a national survey of 3845 pharmacists in 2016 found that 23.1% of pharmacists "do not understand or do not know about the spontaneous reporting scheme" and 57.6% have no experience with reporting AEs. The main reasons for not reporting were that respondents only encounter "well-known AEs" and "mild side effects," and that the "causal relationship is not clear" [29].

The number of AE reports received annually is estimated to be around 30,000–40,000 for all types of products; with 580–770 reported weekly in 2012 [30]. Reports from MAH form the majority of the overall number of reports. For example, in 2012 the number of reports from MAH in Japan was 41,413, while the number of reports submitted by HPs was only 3304; in 2014, these numbers were 49,276 from MAH and 4782 from HP. In 2016, the number of reports increased to 55,817 from MAH and 4956 from HP [29]. Although there is an increasing trend in the number of reports submitted annually, the number of reports from HPs did not increase and remains low.

All reported cases under PMDA monitoring scheme from 1 Apr 2004 are available for online viewing using JADER [31]. Reports submitted before 1 April 2004 are available as archives. A study using JADER (from 2008 to 2014) [32] shows that, in total, 253,241 AEs were recorded in JADER, of which 0.46% (n = 1154) related to combination use of Kampo medicines and pharmaceutical drugs, and 1.0% (n = 2559) related to Kampo medicines only. Therefore, around 1.5% of reports of AEs are associated with Kampo medicines among all reports. However, the types of Kampo medicines (i.e., prescription use or OTC) were not clearly classified in this study. There are no published data to date on trends in numbers of reports only for herbal medicine using JADER. Little is known about the reports for newly registered licensed herbal medicines.

24.5.2 Reporting Suspected Adverse Reactions Associated with Health Foods and Unlicensed Medicines

In 2002, MHLW announced a reporting scheme for health foods and unlicensed medicines including herbal products sold as health foods or unlicensed medicines. Suspected AEs or other harms should be reported directly to local public health centers. Public health centers in Japan are official health institutes established in each local district of each city, and which provide direct health services to communities, such as prenatal care, health advice, and infectious disease control. The reporting scheme is open to all consumers. Local health centers collect the reports and send them to each prefecture government office; the reports are then forwarded to MHLW. Public health centers are expected to work with health professionals to provide professional advice for consumers [33]. The reporting form is available on the MHLW website and from each health center. Reporting items included consumer's details, product information, access/purchase route (e.g., pharmacy, drug store, internet), suspected adverse reaction(s) or medical complaints, and medical practitioner's opinions. The staff in public health centers are required to obtain detailed information from consumer reporters and to fill in the reporting form via consultation. At the time of writing, there has not been any analysis of the reported harms associated with herbal-containing health foods.

24.6 Safety Information for Herbal Products in Japan

At present, the safety profile of licensed herbal medicines is monitored by PMDA. Information on newly identified suspected adverse reactions associated with licensed herbal medicines is published in *Safety information of medicines and medical devices* by PMDA and available online [34].

To help ensure the safety of self-medication with herbal products, JKMA has recommended retail shops using an *Information sheet of OTC Kampo medicines for customers*, like a "confirmation slip," for consumers before purchasing OTC Kampo medicines. The slip includes questions on the user's medical history and current symptoms at the time of purchase [35]. Consumers are required to check all the answers in the slip and then the shop staff can decide whether or not this product is the correct medication for them. However, this confirmation slip is not compulsory. To date, "confirmation slips" have been developed for 39 commonly used OTC Kampo medicines.

Detailed information, including the formula name, English name in Japanese alphabet, ingredients, and other important information on package inserts of 148 Kampo prescription medicines, is published online in both Japanese and English by the National Institutes of Biomedical Innovation, Health and Nutrition (NIBIHN) as *Standards of Reporting Kampo Products* [8]. Therefore, further details of the Kampo medicines formulae mentioned in this chapter are available through NIBIHN website. In addition, there are several versions for translating Japanese character names into Japanese alphabet; for example, "shosaikoto" and "shosaikoutou" are the same formula with different spelling. The government has made a standard for alphabet spelling. Unfortunately, there is no English common name for Kampo formulae, only in Japanese alphabet.

For health foods, MHLW publishes reported suspected adverse reactions and harms online regularly [9]. Other safety information, and evidence of efficacy, is made available as *Information system on safety and effectiveness for health foods* by the National Institutes of Biomedical Innovation, Health and Nutrition [36].

24.7 Future Perspectives and Challenges

In conclusion, herbal products are available in Japan as medicinal products, registered health foods, and unlicensed products sold as health foods or unlicensed medicines. Licensed herbal medicines have to follow the same safety and quality procedures as for pharmaceutical products. Around 60% of registered Kampo formulae are available either as OTC products or as prescription medicines (and covered by National Health Insurance) under the same formula name. For OTC products, the quantity of ingredients and the price vary across different manufacturers. To achieve a better individualized treatment effectively, safely, and economically, patients are advised to consult a medical practitioner for Kampo prescriptions, rather than purchase products for self-medication.

Despite the long history of well-established herbal products registration scheme, several herbal ingredients are still not considered as medicines and are widely available as health foods. This raises safety issues, particularly with foreign unlicensed products that have been found to be adulterated with prescription drugs and sold via the Internet in Japan. This situation is likely to continue. Consumers are advised to consult health professionals when buying "foreign" health foods for self-medication, and to report any harms associated with these products to their local health center.

For licensed herbal medicines, safety problems due to poor quality ingredients or finished products are less common due to strict quality control processes in Japan.

Current quality problems mostly occur with herbal products considered as borderline, particularly new herbal products imported from overseas and sold over the internet. This is the major challenge for safety monitoring for herbal medicines in Japan at present. This situation may be improved by the implementation of the new JP and a revised version of *Approval standards for OTC Kampo products* in the future. In addition, modifying the *List of raw materials exclusively used as pharmaceuticals* to include more herbal ingredients into the national standards also is another important task for the future.

Several studies using JADER (the national database of spontaneous reports of suspected adverse reactions) have been conducted for Kampo medicines, as discussed above. However, these did not include reports of suspected adverse reactions associated with herbal products sold as health foods or unlicensed medicines. Although, MHLW frequently makes announcements and releases safety alerts relating to these products, there has been little investigation focusing only on herbal products reports among these heath foods and unlicensed medicines.

The PMDA publishes annual summaries on reporter statistics and has conducted several surveys exploring health professionals' views on reporting suspected adverse reactions. However, for most of these studies, results are available only as summaries under PMDA presentation files. Over the last 10 years, numbers of suspected adverse reactions reports submitted by healthcare professionals remain low in Japan. Further, no studies have examined the characteristics of reporters of adverse reactions associated with herbal medicines, or views on herbal-medicines' safety among health professionals in Japan. Future efforts should be focused on improving awareness of herbal safety through education, especially implementing pharmacovigilance content into official Kampo training courses and other advanced learning courses for healthcare professionals.

References

1. Hakamatsuka T (2014) About medicinal products from Western herb. Farumashia 50(10):978–982
2. Maekawa H, Nakamura T, Saito K (2014) Regulation of traditional herbal medicinal products in Japan. J Ethnopharmacol 158:511–515
3. Ministry of Health, Labour and Welfare (2019) The list of raw materials exclusively used as pharmaceuticals. https://www.mhlw.go.jp/file/06-Seisakujouhou-11130500-Shokuhinanzenbu/0000086062_1.pdf. Accessed 18 Feb 2020
4. Ministry of Health, Labour and Welfare (2016) Japanese Pharmacopoeia 17th Edition (with Suppl. 1 & 2). http://www.mhlw.go.jp/topics/bukyoku/iyaku/yakkyoku/english.html. Accessed 18 Feb 2020
5. Pharmaceutical and Food Safety Bureau (2018) The Japanese standards for non-pharmacopoeial crude drugs
6. Pharmaceutical and Food Safety Bureau (2017) Approval standards for manufacturing and selling OTC Kampo products
7. Nagata Y (2012) Essence on the guide to taking Kampo medicines. Kanpoiyakugakuzasshi 20(1):16

8. National Institutes of Biomedical Innovation, Health and Nutrition (2018) Standards of reporting Kampo products. http://mpdb.nibiohn.go.jp/stork/index.html
9. Ministry of Health, Labour and Welfare. About health foods. http://www.mhlw.go.jp/stf/seisakunitsuite/bunya/kenkou_iryou/shokuhin/hokenkinou/. Accessed 20 Feb 2020
10. Ministry of Health, Labour and Welfare (2019) The statistics of production by pharmaceutical industry 2018. https://www.mhlw.go.jp/topics/yakuji/2018/nenpo/
11. Hakamatsuka T (2011) Revision of Kampo extra formulae that recorded in Japanese pharmacopoeia and standard approval for OCT Kampo products. Farumashia 47:413–417
12. Japan Kampo Medicines Manufacturers Association (2011) Suvery of Kampo medicine for prescription uses. https://www.nikkankyo.org/serv/serv1.htm
13. Arai I, Sugiyama K (2009) Survey on the views of Kampo formulae. J Jpn Soc Orient Med 59:124
14. Ministry of Health, Labour and Welfare (1998) Pharmaceuticals and Medical Devices Safety Information No. 146. https://www.pmda.go.jp/safety/info-services/drugs/calling-attention/safety-info/0014.html#1
15. Tsukiyama K, Tasaka Y, Nakajima M et al (1989) A case of pneumonitis due to sho-saiko-to. Nihon Kyobu Shikkan Gakkai Zasshi 27(12):1556–1561
16. Takada N, Arai S, Kusuhara N et al (1993) A case of sho-saiko-to-induced pneumonitis, diagnosed by lymphocyte stimulation test using bronchoalveolar lavage fluid. Nihon Kyobu Shikkan Gakkai Zasshi 31(9):1163–1169
17. Hatakeyama S, Tachibana A, Morita M, Suzuki K, Okano H (1997) Five cases of pneumonitis induced by sho-saiko-to. Nihon Kyobu Shikkan Gakkai Zasshi 35(5):505–510
18. Kobashi Y, Nakajima M, Niki Y, Matsushima T (1997) A case of acute eosinophilic pneumonia due to Sho-saiko-to. Nihon Kyobu Shikkan Gakkai Zasshi 35(12):1372–1377
19. Suzuki H, Kumada H, Sato K et al (2000) Guidelines of Sho-saiko-to/Xiao-Chaihu-Tang treatment in patients with chronic hepatitis C. Wakaniyakugaku Zasshi 17(3):95–100
20. Mizushima Y, Oosaki R, Kobayashi M (1997) Clinical features of pneumonitis induced by herbal drugs. Phytother Res 11(4):295–298
21. Oketani N, Saito H, Ebe T (1996) Pneumonitis due to Hangeshashin-to. Nihon Kyobu Shikkan Gakkai Zasshi 34(9):983–988
22. Maruyama Y, Maruyama M, Takada T, Haraguchi M, Uno K (1994) A case of pneumonitis due to Rikkunshi-to. Nihon Kyobu Shikkan Gakkai Zasshi 32(1):84–89
23. Ishigawa S, Takuma H (2013) A signal identification methods for detecting side effect of Kampo prescription medicines. In: Conference proceedings of the 133th annual meeting of the Pharmaceutical Society of Japan 2013
24. Shimodaira H, Nozaki M, Kwon Y, Kamimura N, Kaiho F (2014) Analysis of adverse reaction in Kampo-medicines using JADER database of PMDA. Jpn J Drug Inf 16(1):16–22
25. Pharmaceutical and Medical Devices Agency (PMDA). http://www.info.pmda.go.jp/index.html. Accessed 18 Feb 2020
26. Uppsala Monitoring Centre (UMC) WHODrug Global. https://www.who-umc.org/whodrug/whodrug-portfolio/whodrug-global/. Accessed 6 Mar 2020
27. Uppsala Monitoring Centre (UMC) WHODrug Cross Reference Tool (CRT) Japan. https://www.who-umc.org/whodrug/whodrug-portfolio/whodrug-crt-japan/. Accessed 18 Feb 2020
28. Matsui K, Iimura Y (2014) Introduction and application of datamining method in safety monitoring. Farumashia 50(6):543–547
29. Ohmuta M (2017) Make your reports with Medi-navi: drug safety strategy and spontaneous reporting scheme for medicines and medical devices. In: The 50th academic conference of Japan Pharmaceutical Association, Tokyo
30. Matsui K (2013) Recent activities of drug safety monitoring strategy: a perspective from survey and analysis. Kurume biostatistics forum
31. Pharmaceuticals and Medical Devices Agency (PMDA). Japanese adverse drug event report database (JADER). http://www.info.pmda.go.jp/fukusayou/menu_fukusayou_new_attention.html. Accessed 18 Feb 2020

32. Arai I, Harada Y, Koda H, Tsutani K, Motoo Y (2020) Estimated incidence per population of adverse drug reactions to Kampo medicines from the Japanese adverse drug event report database (JADER). Tradit Kampo Med 7:3–16. https://doi.org/10.1002/tkm2.1234
33. Ministry of Health, Labour and Welfare (2002) Correspondence for preventing health harms from health foods and unlicensed medicines. https://www.mhlw.go.jp/houdou/2002/10/h1004-3.html. Accessed 18 Feb 2020
34. Pharmaceuticals and Medical Devices Agency (PMDA). Safety information of medicines and medical devices. https://www.pmda.go.jp/safety/info-services/drugs/calling-attention/safety-info/0043.html. Accessed 18 Feb 2020
35. Japan Kampo Medicines Manufacturers Association (JKMA). Information sheet of OTC Kampo medicines for customers. https://www.nikkankyo.org/seihin/seihin5.htm. Accessed 18 Feb 2020
36. National Institutes of Biomedical Innovation, Health and Nutrition (NIBIHN). Information system on safety and effectiveness for health foods. https://hfnet.nibiohn.go.jp. Accessed 18 Feb 2020

Chapter 25
Pharmacovigilance for Herbal Medicines in Iraq

Manal M. Younus and Inas R. Ibrahim

25.1 Introduction

Iraq is one of the 22 Arabic countries and the 59th largest country in the world by area; Iraq is located in South-West Asia, surrounded by Turkey from the north, Iran form the east, Syria and Jordan from the west, and Saudi Arabia and Kuwait from the south [1]. Geographically, Iraq mainly consists of desert; however, mountains are dominant in the north. Lands around the two major rivers (Tigris and Euphrates) that run from the north to the south are fertile plains and heavily populated. The country is rich in petroleum and natural resources. Around 99% of the country's inhabitants (total population = 38.34 million) are Muslims [2]. The official languages are Arabic and Kurdish. The country is a federal parliamentary republic consisting of 18 governorates (provinces), including one autonomous region (Kurdistan Region). Baghdad is the capital and the largest city [2].

Concerning the health system of the country, the Ministry of Health (MoH) offers health services at a subsidized cost to the population. Public health services are provided through MoH facilities, including public hospitals, primary healthcare centers, and tertiary health centers. All the public healthcare settings are supervised by Directorates of Health (DoHs), which are distributed throughout the 18 governorates (each governorate has one directorate, except Baghdad, which has three DoHs) [3].

M. M. Younus (✉)
Iraqi Pharmacovigilance Center, Directorate of Technical Affairs, Ministry of Health, Baghdad, Iraq

International Society of Pharmacovigilance (ISoP), London, UK

Council For International Organizations of Medical Sciences (CIOMS), Geneva, Switzerland

I. R. Ibrahim
College of Pharmacy, Uruk University, Baghdad, Iraq

J. Barnes (ed.), *Pharmacovigilance for Herbal and Traditional Medicines*,
https://doi.org/10.1007/978-3-031-07275-8_25

25.2 Epidemiology of Herbal Medicines in Iraq

The use of plant parts as therapy for different ailments extends back to the Babylonian era of old Iraq (the Mesopotamia), 60,000 years ago [4]. During that time, plants and animal products provided the fundamental ingredients of medical therapies. The diversity in the geographical area and climate of Iraq has resulted in a considerable wild flora [5].

Studies evaluating the use of traditional medicine by Iraqi society are not well documented in the scientific literature. However, some studies are available that provide insights into Iraqi traditional medicine practices concerning herbal medicines. Interviews with 75 herbalists in Iraq revealed that 53 species of plants were found at herbalists' shops for use in treating different diseases. These plants (described with their local name and scientific name) included babonage (*Chamaemelum nobile* (L.) All.), darceen (*Cinnamomum verum* J.Presl; synonym: *C. zeylanicum*), erksoos (*Glycyrrhiza glabra* L.), helba (*Trigonella foenum-graecum* L.), shaie kogarat (*Hibiscus subdariffa* L.), krenfel (*Syzygium aromaticum* (L.) Merr. & L.M.Perry), habit helwa (*Foeniculum vulgare* Mill.), yansoon (*Pimpinella anisum* L.), habit soda (*Nigella sativa* L.), erk haar (*Zingiber officinale* Roscoe), and kurkum (*Curcuma longa* L.). Participating herbalists claimed to have a comprehensive knowledge of phytotherapy, although none held a license for dispensing herbal products. Around 70% of practicing herbalists do not have a formal level of education. In addition, some of the plants sold in the domestic market had been imported and were stored for an indefinite period of time [6]. In the north of Iraq, interviews with 45 traditional healers revealed that 66 plant species were sold by healers at markets for the treatment of a variety of ailments. For example, plants commonly sold by these herbalists for the control of blood pressure were garlic (*Allium sativum* L.) preparations, cinnamon (*Cinnamomum cassia* (L.) J.Presl), rosella flowers (*Hibiscus sabdariffa* L.), flax leaves (*Linum usitatissimum* L.), wild chamomile (*Matricaria chamomilla* L.), and ginger (*Zingiber officinale* Roscoe). Additionally, inhabitants of this part of Iraq have a strong belief in traditional therapies as a way of maintaining and achieving good health [5]. The studies described above were limited in their scope, and there is a need to conduct more comprehensive studies to give a robust profile of herbal medicine use by Iraqi patients. Despite considerable access to modern medicine in Iraq, traditional medicines still attract substantial attention for historical, religious, and cultural reasons.

25.3 Regulation of Herbal and Traditional Medicines

The regulatory system for herbal medicines in Iraq consists of two units. The first unit involves regulations for crude herbal drugs; the second unit controls the formulated "natural" and botanical medicines and their distribution. A specific policy concerning herbal medicines in Iraq is not yet available.

25.3.1 Unit One: Regulatory System for Crude Herbal Drugs

This part could be considered as the national program for medicinal herbs in Iraq; it is the older part of the system and has passed many stages since its beginning, as described below:

1. In 1989: MoH established the Herbal Medicine Centre (HMC) as a unit connected to the therapeutic department. The HMC is now known as the Herbal Medicine Department (HMD) with the mandate of regulating crude medicinal herbs and of acting as a research center for medicinal herbs in Iraq [7].
2. In 1993: The inauguration of the first herbal clinic as part of the HMD to provide consultations and treatment for people willing to be treated with herbal products under a prescription from the center's physician and to be dispensed from the center's pharmacy by a pharmacist.
3. In 1997: The first legislation regarding medicinal herbs in Iraq was issued and published in the official journal of the Iraqi Ministry of Justice to regulate the selling of medicinal herbs, licensing of herbal shops and certifying of herbalists [8]. Currently, there are 391 registered herbalists in the HMD; all herbalists must pass a training program ranging from 3 to 9 months' duration, depending on the herbalist's educational background.
4. In 2000: A committee was established in the MoH to study and enroll 352 medicinal herbs endemic to Iraq in the forthcoming Iraqi herbal medicines pharmacopoeia.
5. In 2013: The subject of phytotherapy was included in the curriculum of medical schools in Iraq to increase awareness of herbal medicines among future clinicians. In addition, a herbal medicine coordinator was appointed in each of the 17 DoHs across the country to facilitate the work of HMD regarding increasing awareness of herbal medicines, as well as in certifying and licensing activities for herbalists in their respective governorates.
6. In 2019: The HMD in collaboration with the Directorate of Public Health at MoH issued a guideline of herbal medicines treatment for public health institutions in preparation of herbal clinics in the public healthcare settings in collaboration with the Directorate of Public Health at MoH. In addition, an update to the herbal legislation was made to include requirements for more details regarding the background of the herbalist, criteria for shops intended for use in herbalists' practice, record keeping and renewal of licensing on an annual basis, and to strengthen inspection of premises.

Despite the release of the herbal remedy Act in 1997, it took around 16 years to implement this law in practice. This delay resulted in the challenges that face the HMD today. For instance, only 8% of the total number of registered herbalists have renewed their license within the required timeframe, which extended from 2014 to 2017. This added to the shortage of data records in the HMD system, and many herbal shops appeared in the marketplace without a formal registration. A specialized laboratory unit for detection of active constituents of herbs is also still lacking in Iraq. The updated legislation in late 2019 empowered the HMD to overcome these challenges. Furthermore, a training program was developed by HMD to prepare pharmacists from both the private and public sectors to be specialists in the evaluation and control of herbal remedies.

25.3.2 Unit Two: Regulatory System for Manufactured Natural and Botanical Medicines

The components of this system are the same as the regulations for conventional medicines. It was implemented in the year 2000 to regulate imported herbal products. Approval of these products is based on the decision of a national committee of experts, and products are regulated as either plant products or supplements (food, herbal or sport and weight reduction) [7]. Accordingly, a national comprehensive list of approved products is developed as a reference. The required information for approval includes: efficacy and safety studies; method of extraction, and from which part of the plant the extraction is made; pharmacological activity studies (if available); compositions and concentrations of the ingredients in the finished product. In addition, the method of analysis of herbs included in the formula should be presented during approval and registration and any other studies to support the product.

After approval, registration of the product and the manufacturing site is mandatory at the registration department of Directorate of Technical Affairs (DoTA). This is followed by a series of tests by the national quality control labs, then obtaining the import authorization for the product to be imported and distributed in the Iraqi pharmaceutical market. The import of these products occurs exclusively via the marketing authorization holder/manufacturer representative (drug scientific bureau). At present, there are 853 approved products, of which 289 are available in the market as herbal supplements.

It is important to note that crude herbs are available and purchased by the lay public as over-the-counter products in community pharmacies. They are also available at herbalists' shops, supermarkets, and old "folk markets," whereas the manufactured natural and botanical medicines are restricted to sale from pharmacies and are subject to the same regulations as for conventional medicines; many are regulated to be dispensed only under medical supervision.

25.4 Pharmacovigilance for Herbal Medicines

The Iraqi Pharmacovigilance Centre (IPC) was established in 2010 under the Directorate of Technical Affairs (DoTA), one of the main MoH directorates, with defined roles and responsibilities; soon after, the national spontaneous reporting system began to operate. In the same year, Iraq achieved full membership of the

World Health Organization (WHO) Programme for International Drug Monitoring (PIDM) and became the 102nd member. The vision of the IPC is to create a safe medicinal environment, with a mission of contributing to ensuring patient safety through monitoring, evaluation and prevention of AEs and medication errors. The system is effective in both public and private sectors, but it is stronger in the public sector with clear roles and responsibilities and infrastructure. The IPC is government funded, and currently has ten full-time staff [9].

In the public sector, a regional center (RC) in each of the 17 DoHs around the country was established in 2012. The IPC is currently connected through the regional centers to around 260 remote hospitals where a hospital safety responsible is aligned with each of them. In addition, the IPC is connected with around 135 primary healthcare districts and, through them, to around 1017 main primary healthcare centers. Also, the IPC is connected to 144 tertiary health centers and 314 public health clinics across the country. Through these connections many adverse drug reaction (ADR) reports are received by the IPC.

There is no specific pharmacovigilance system for herbal medicines in Iraq: monitoring the safety and effectiveness of herbal products is embedded within the national spontaneous reporting system. The system depends on healthcare professionals for reporting suspected ADRs; however, public and patients' reports are also accepted.

A guideline on detecting and reporting on ADRs for health professionals was issued in 2012, and is still current [9]. A paper-based individual case safety report (ICSR) form has been available to all health professionals since 2010 in both Arabic and English languages and can be used to report suspected ADRs associated with herbal medicines, as well as for conventional medicines. In addition, an electronic reporting system was adopted as a joint project between the IPC and the Uppsala Monitoring Centre (UMC); this system provides an electronic form for reporting, in English language only, to all healthcare professionals [10].

As a member of the WHO PIDM, Iraq has a complete electronic data management system (VigiFlow®) at the central and regional levels, and also has access to VigiLyze® (the UMC's online search and analysis tool for use with VigiBase® data); both these services are provided and maintained by UMC on behalf of WHO.

25.5 Reporting Scenario and Reports Handling

The entry of information from submitted ADR reports occurs at the RCs level using Medical Dictionary for Regulatory Activities (MedDRA) terminologies to code ADRs/AEs and WHODrug dictionary to code medicines and herbal products. There are substantial challenges in coding poorly defined herbal and traditional medicines, and these are illustrated here by an example concerning coding a traditional product "Sagwa." This product does not have a specific content, rather different composition(s) from one area to another in Iraq, and from sample to sample. It was agreed to add "Sagwa" to the WHO-drug dictionary with the active ingredients "herbal NOS"; "mineral NOS" (NOS: not otherwise specified). However, previous reports concerning "Sagwa" have not been coded in a uniform way: other codes that have been used to code "Sagwa" include: "herbal extract NOS"; "herbal NOS w/ minerals NOS"; and "traditional medicine." In addition, different local names were recognized for "Sagwa" products available in different areas, so different trade names were used when entering the data.

All the ADR reports received are reviewed by the IPC assessors and subjected to quality control where data validation, completeness, timelines, accuracy, consistency, and integrity are assessed before committing the reports to the WHO-UMC international database. The received case reports are classified according to the seriousness of the condition(s)/reaction(s) in each case. Cases with death as an outcome are reported by health institutions to the RCs within 3 days of them being notified of the death; the RCs send the report to the NC within 3 days. Cases describing other serious condition(s)/reaction(s) are sent to the RCs within 7 days, then to the NC, also within 7 days. Reports of non-serious conditions/reactions should be sent to the RCs within 1 month and reach the NC within 6 months.

Reports of ADRs associated with herbal medicines constitute around 6% of the total number of reports in the IPC database as of April 2020. More than 570 reports of ADRs associated with herbal medicines have been stored in the database; the reports represent 1089 ADRs involving five different categories of herbal products. The reports involving "Sagwa" represent 95% of the total number of herbal reports in the database. The most frequently reported ADRs are related to gastrointestinal, general, and metabolism and nutrition disorders. Summary information on cases associated with herbal and traditional medicines reported to IPC from the year 2014 to 2020 are presented in Table 25.1.

Pharmacoepidemiological studies exploring the type and frequency of AEs associated with the use of herbal medicines in Iraq are scarce. In a cross-sectional study that explored the use of herbs for blood pressure control in 400 patients with hypertension, AEs were reported by the study participants and some were perceived to be related to the use of the herbs tested [11]. AEs reported were skin rash (1.7%, $n = 15$), headache (4.9%, $n = 13$), sleep disturbance (4.6%, $n = 12$), abdominal pain (4.6%, $n = 12$), diarrhea (3.8%, $n = 10$), skin deformity (3.1%, $n = 8$), frequent urination (1.9%, $n = 5$), and flatulence (1.5%, $n = 4$).

25.6 Signal Management for Herbal Medicines

Although the number of ADR reports involving herbal medicine is limited, scanning of the received reports is undertaken periodically by the IPC and the ICSRs are checked for any possible signal(s). A search for signals was specifically undertaken in relation to one of the most widely used unlicensed traditional remedies used in self-care for children with gastroenteritis—the so-called "Sagwa" mentioned earlier. The method used for signal detection was qualitative and a case-control study was used to confirm the signals. The WHO-UMC standardized case causality

Table 25.1 Characteristics of the reported cases to IPC from the year 2014 to 2020

Characteristics		Number (*n*)	Percentage (%)
Year	2020	81	14.0
	2019	202	34.9
	2018	192	33.2
	2017	94	16.2
	2016	8	1.4
	2015	0	0.0
	2014	2	0.3
Age of patient	0–27 days	17	2.9
	28 days–23 months	520	89.8
	2–11 years	36	6.2
	18–44 years	1	0.2
	45–64 years	5	0.9
Gender	Female	252	43.5
	Male	291	50.3
Drug (WHO Drug)	AI: Herbal NOS	130	22.5
	AI: Herbal NOS: Mineral NOS	412	71.2
	AI: Rhamnus cathartica: Rhamnus urshiana: Senna spp.	23	4.0
	AI: Herbal extract NOS	8	1.4
	AI: Zingiber officinale	3	0.5
	AI: Hedera Helix	1	0.2
	AI: Herbal anti-obesity preparations (AB Slim)	1	0.2
	AI: Traditional medicine	1	0.2
Reaction (MEdDRA)	SOC: Cardiac disorders	3	0.5
	SOC: Gastrointestinal disorders	444	76.7
	SOC: General disorders and administration site conditions	119	20.6
	SOC: Hepatobiliary disorders	1	0.2
	SOC: Immune system disorders	1	0.2
	SOC: Infections and infestations	15	2.6
	SOC: Injury, poising, procedure complications	5	0.9
	SOC: Investigations	8	1.4
	SOC: Metabolism and nutrition disorders	277	47.8
	SOC: Nervous system disorders	32	5.5
	SOC: Psychiatric disorders	9	1.6
	SOC: Renal and urinary disorders	8	1.4
	SOC: Respiratory, thoracic, and mediastinal disorders	43	7.4
	SOC: Skin and subcutaneous tissue disorders	2	0.3
	SOC: Vascular disorders	5	0.9

(continued)

Table 25.1 (continued)

Characteristics		Number (*n*)	Percentage (%)
Top Reported preferred terms (MedDRA)	PT: Vomiting	309	53.45
	PT: Diarrhea	244	42.1
	PT: Dehydration	227	39.2
	PT: Pyrexia	113	19.5
	PT: Poor feeding infant	77	13.3
	PT: Frequent bowel movements	41	7.1
	PT: Dyspnea	22	3.8
	PT: Cough	20	3.5
	PT: Lethargy	19	3.3
	PT: Nausea	17	2.9
Seriousness	Yes	464	80.1
	No	113	19.5
Seriousness Criteria	Death	12	1.8
	Life threating	440	76.0
	Caused/prolonged hospitalization	16	2.8
	Disability/incapacitating	1	0.2
	Other	6	1.0
Qualification of the reporter	Physician	1	0.2
	Pharmacist	575	99.3
	Other health professional	1	0.2
	Unknown	2	0.3

assessment tool was used to assess the reported cases with respect to the likelihood of a causal relationship. There has been an increase in the number of ICSRs of suspected AEs associated with "Sagwa" throughout the years; one reason for this could be the effect of increased awareness among health professionals regarding the reporting of AEs associated with use of "Sagwa," and the early call for reporting sent by the IPC. The call announcement involved all healthcare providers in the public sector: these individuals were asked to report any suspected cases of "Sagwa" poisoning and further reports received were very helpful in strengthening the signals associated with "Sagwa." Another reason for the increase in the number of reports could be an actual increase in the number of individuals experiencing AEs following exposure to "Sagwa." As part of the call, reporters were asked to send all relevant information, including laboratory test results, as well as completing the four mandatory fields (suspected patient, suspected herbal medicines, AE(s) or reaction(s), identified reporter) and to state the geographical area from which the affected cases were originated. Accordingly, a map of the most affected areas was produced, and found to be rural areas around north, south, east, and west of Baghdad, and nearby provinces, including Diyala and Salahaldeen.

"Sagwa" as a traditional herbal remedy is widely used based on the recommendations of respected family members, particularly grandmothers. In an attempt to confirm (or refute) the signals concerning "Sagwa," a case-control study was performed in the pediatrics, obstetrics, and gynecology Teaching Hospital in Diyala, as one of the most widely affected areas in Iraq. Around 200 children aged from 1 day to 5 years with severe and complicated gastroenteritis attending the hospital were included in the study for 6 months. One hundred patients with gastroenteritis and not taking "Sagwa" were the study controls and another 100 patients with gastroenteritis and taking "Sagwa" were the cases. AEs were more frequent in the "Sagwa" group compared to the controls. The confirmed clinical presentations and complications significantly (p-value <0.05) associated with "Sagwa" AEs were oligurea and anuria ($p = 0.0001$), acidosis ($p = 0.001$), arrhythmias ($p = 0.001$), central nervous system effects ($p = 0.0001$), vomiting ($p = 0.001$), and dehydration ($p = 0.001$). During the study period, more than 60% of the cases were improved on discharge and about 10% died [12]. More studies are required to confirm all the signals associated with "Sagwa" toxicity.

Analysis of the collected "Sagwa" samples revealed different compositions with different sample sources using atomic absorption spectroscopy (AAS) to measure the concentration of various elements in the samples. It was found that samples contained high concentrations of lead, mercury, cadmium, and arsenic exceeding safe limits set out in the WHO and European Pharmacopeia. These levels exceed the acceptable permissible levels of heavy metals. Other components that were found included: animal skin (mice); urchin (meat, skin); bentonite; undeclared conventional drugs; animal components (stool, urine, blood); cumin; coriander; digitalis; and many others [13].

The final results of the signal assessment are not yet published; however, the results of signal detection were made available to the Directorate of Public Health at the MoH, which is responsible for running different health-related programs, including the national tuberculosis program, national immunization program, mother and child programs and public health educational programs and so on. Risk minimization measures are ongoing, and no impact measurement is yet available.

25.7 Awareness of Herbal Medicines Reporting

To promote awareness of AEs reporting for herbal medicines, training activities were developed and conducted among healthcare providers in different institutions of health. These activities were the result of joint work with the National Centre for Training and Human Resources and the Herbal Medicines Department in the DoTA, both affiliated to the Iraqi MoH. Currently, 0.5 continuous medical education (CME) points are granted to each reporter to incentivize reporting (the total number of CME points required ranges from 20 to 40 depending on an individual's level of seniority and scientific degree). Educational programs have been delivered among undergraduate students of medical and health sciences colleges throughout the country to increase awareness of ADR reporting for herbal medicines.

In Iraq, like the rest of the world, very few reports concerning ADRs associated with use of herbal medicines are received; as long as reporting by herbal practitioners is not mandatory, the use of only spontaneous reporting systems as the main source of ICSRs information for herbal medicines, and while patients do not declare their use of herbal medicines and their healthcare professionals do not ask them about it, the situation will continue. In summary, achieving a successful pharmacovigilance system for herbal medicines in Iraq remains a challenging issue that requires substantial resources and efforts to resolve.

References

1. Khadduri M, Woods JE, Blake GH, Kennedy H, Chambers RL (2020) Iraq, country of southwestern Asia. https://www.britannica.com/place/Iraq. Accessed 10 Apr 2020
2. Central Statistical Organization (2012) Numbers and indicators of Iraq. Central Statistical Organization of Iraq, Ministry of Planning. http://cosit.gov.iq/en/74-others-e/434-publications-en. Accessed 10 Apr 2020
3. Ibrahim IR, Hassali MA (2014) The practice of pharmacy in Iraq. Health Med J 8(11):1231–1233
4. Alkhazrajy LA, Habib HA, Alsudani ZI (2014) Perception of Iraqi primary health care providers about herbs in practice. Eur J Bot 1(1):27–52
5. Ahmed HA (2016) Ethnopharmacobotanical study on the medicinal plants used by herbalists in Sulaymaniyah Province, Kurdistan, Iraq. J Ethnobiol Ethnomed 12:8. https://doi.org/10.1186/s13002-016-0081-3
6. Al-douri NA (2014) Some important medicinal plants in Iraq. Int J Adv Herbal Altern Med 2(1):10–20
7. Al-Jumaili AA (2020) Iraq pharmaceutical country profile. Prepared by the World Health Organization in collaboration with the Ministry of Health. https://moh.gov.iq/upload/upfile/ar/1375.pdf. Accessed 25 Sept 2020
8. World Health Organization (2005) National policy on traditional Medicine and regulation of herbal medicines: report of WHO global survey. https://apps.who.int/iris/bitstream/handle/10665/43229/9241593237.pdf;jsessionid=B757BC2327969E758676373E0F3D0D34?sequence=1. Accessed 25 Mar 2020
9. Alwan MH (2012) Guidelines for detecting & reporting adverse drug reactions in Iraq: individual case safety reports. Iraqi Pharmacovigilance Center. https://www.who-umc.org/media/1078/iraq.pdf. Accessed 25 Sept 2020
10. Iraqi Ministry of Health. Directorate of technical affairs. http://www.tec-moh.com/. Accessed 2 Apr 2020
11. Ibrahim IR, Hassali MA, Saleem F, Al Tukmagi HF, Dawood OT (2017) Use of complementary and alternative medicines: a cross-sectional study among hypertensive patients in Iraq. J Pharm Health Serv Res 9:59–65. https://doi.org/10.1111/jphs.12209
12. Shukur AM, Alezzi JI, Kadhim TJ (2019) Morbidity and mortality associated with community used herbal (sagwa) use in children with acute gastroenteritis in Diyala governorate. Diyala J Med 17(2):92–106. https://doi.org/10.26505/DJM.17024690527
13. Younus MM, Ali NK, Abdulgader HR (2017) Traditional medicine in Iraq, is it still a source of serious adverse events? A documentation of serious adverse events of sagwa among infants in Iraqi pediatric hospitals. Iraqi N Med J 3(2):112–114

Chapter 26
Pharmacovigilance for Herbal and Traditional Medicines in the Sultanate of Oman

Hussain Al Ramimmy, Shirly Varughess, and Nawal Al Alawi

26.1 Introduction

Like any other countries belonging to the eastern world, Oman too has a history of using traditional and herbal remedies in their treatment armamentarium [1]. This is so with the fact that the country is rich with a blend of unique biodiversity and is made up of desert rocky plains, sand and mountainous areas, as well as coastal plains. Therefore, Oman is a habitat for more than 1200 species of documented plants (3 globally threatened), 509 species of marine plants [2]. Located on the southeastern fringe of the Arabian Peninsula, Oman has been at the center of human and plant migration. Oman enjoys a rich mix of cultural influences originating from Asia, Persia, Africa, and south Eastern Europe [3]. As with many ancient cultures, much of the traditional knowledge pertaining to use of herbal remedies is passed from generations to the next orally only and there is a dearth of authentic scientific literature documentation. However, there are organizations, such as the Ethnobotany research group, Oman Animal and Plant Genetic resource Center (OAPGRC) [4], formed with a mission to promote the recognition, sustainable exploitation, and valuation of the genetic biodiversity inherent in Oman's animals, plants, and microorganisms, as natural heritage resources. A review [5] of 33 medicinal plants routinely used in folk medicine practice in Oman identified 22 plant families and 18 traditional treatment groups. Most of the plants were being used as infusions, pastes, or inhalations.

Traditional treatments are sought in Oman for various reasons. Lack of evidence-based scientific data on their safety and/or efficacy does not deter Omanis from seeking treatment from traditional healers. There are examples showing that, when applied in the treatment of ocular diseases, traditional medicines and healing practices may cause more harm than benefit for the patient [6].

H. Al Ramimmy (✉) · S. Varughess · N. A. Alawi
Department of Pharmacovigilance & Drug Information, Ministry of Health, Muscat, Oman

J. Barnes (ed.), *Pharmacovigilance for Herbal and Traditional Medicines*,
https://doi.org/10.1007/978-3-031-07275-8_26

Safety is a fundamental principle in the provision of herbal medicines and herbal products for healthcare, the critical component of quality control. Recognizing the importance of this, the World Health Organization (WHO) published the WHO Guidelines on safety monitoring of herbal medicines in pharmacovigilance systems in 2004 [7]. The guidelines were developed with the view that, within current pharmacovigilance systems, monitoring of the safety of medicines should be enhanced. The inclusion of herbal medicines in pharmacovigilance systems is becoming increasingly important given the extent of use of herbal medicinal products globally. Among consumers, there is a widespread misconception that "natural" always means "safe," and a common belief that remedies from natural origins are harmless and carry no risk. However, some medicinal plants are inherently toxic. Further, as with all medicines, herbal medicines are expected to have adverse effects.

In Oman, the Directorate General of Pharmaceutical Affairs and Dug Control (DGPA&DC) [8] introduced its herbal section under the Department of Drug Control in 2001. Due to the increased demand for herbal products and traditional formulations and availability of non-registered products in commercial establishments, the DGPA&DC decided to provide definite rules and regulations for the registration of these preparations within a legal framework.

The adverse drug reaction (ADR) monitoring system was implemented in Oman in 1993, but awareness of ADR reporting among healthcare professionals was initially low; this has improved more recently with advances in technology and social media. Originally, pharmacovigilance (PV) activities in the DGPA&DC were under the Department of Drug Control as a section. However, in 2015, a new milestone in the structure of DGPA&DC was reached, with the introduction of PV as a department—the Department of Pharmacovigilance & Drug information (DPV&DI)—and which implemented PV for herbal, traditional medicines and health products as one of its sections.

26.2 Regulatory Framework for Herbal Medicines in Oman

Herbal medicines, by definition, is the use of plants, plant parts, their water or solvent extracts, essential oils, gums, resins, or other forms of advanced products made from plant parts to treat, cure, or prevent a disease in humans. However, the herbal and traditional medicines that fall under the regulation as part of drug control in Oman are restricted to finished pharmaceutical products of herbal origin [9].

The presence of counterfeit, poor-quality, and adulterated herbal medicines worldwide has led to the necessity for the regulation of herbal medicines in Oman. Recognizing this, the Ministry of Health (MoH) Oman, under the auspices of DGPA&DC, started its initial steps, which involved the issue of import notes for herbal and traditional medicines in the year 1998. Prior to this, the importation of this group of medicines was not regulated and such products were available in supermarkets and other stores, which do not come under the purview of drug control. This step was followed by formulating a framework for the formal regulation of the import, sale, and use of herbal medicines in Oman. By the year 2008, the rules, regulations, and criteria for the registration of "herbaceutical" companies and traditional and complementary herbal medicine preparations were developed. During the initial framework, companies were asked to submit lists of products intended for import, with all documents and other requirements as per the registration criteria. With this, a database was prepared for all products available in the country.

In 2015, as part of the restructuring of DGPA&DC, three sections were dedicated to handling herbal drug regulations in Oman under the DGPA&DC; these sections were the Analysis of Herbal Medicines and Health Products, Registration of Herbal Medicines and Health Products, and Pharmacovigilance of Herbal Medicines & Health Products.

The herbal medicines registration requirements [9], as stipulated by the Department of Drug Control (DDC), state that to obtain registration of products companies should submit the following documentation:

- Certificate of pharmaceutical product/free sale certificate
- Scientific report
- Declaration letter indicating that the product is free from steroids, sex hormones, pesticides, insect debris, heavy metals, aflatoxins, etc.
- Declaration letter indicating that the product is free from adulteration with synthetic materials
- Clinical trials (i.e., summary of published literature on clinical trials)
- Other documents as per application form
- Company registration certificate

The pre-requisites for registration of herbal product companies are:

- Specified application form
- Manufacturing license certificate
- Good Manufacturing Practice (GMP) certificate
- Site master file
- Other documents as per the application form

However, the regulations for herbal medicines apply only to finished products in the form of oral solid dose forms or liquid oral preparations.

26.3 Pharmacovigilance for Herbal Medicines

The DPV&DI under the DGPA&DC is the authority responsible for pharmacovigilance in Oman and acts as the National Pharmacovigilance Centre (NPVC). Oman became a full member of the WHO Programme for International Drug Monitoring in 1995. The DPV&DI comprises three sections: PV for Human Medicines; PV for Herbal Medicines; Central Drug Information.

The scope of "herbovigilance" in Oman was initiated with a view to addressing issues relating to patient care, public health, risk-benefit assessment, and risk communication. The specific aims of the section included:

- To improve patient care and safety in relation to the use of herbal medicines and all medical and paramedical interventions.
- To improve public health and safety with the use of herbal medicines.
- To contribute to the assessment of benefit, harm, effectiveness, and risk of herbal medicines.
- To promote understanding, education, and clinical training in PV and its effective communication to the public.

Although the section for herbovigilance has been functioning since 2015, ADR reports for herbal medicines received at the NPC are very few, compared with the number of reports received for pharmaceutical products (conventional medicines). The low number of reports for herbal medicines could be because these products are not distributed through the public health system by the government, and most ADR reports are received from the government sector.

26.3.1 Safety/Risk Communication Strategies for Herbal Medicines and Other "Natural Health" Products

In Oman, safety/risk communication strategies for herbal medicines and other pharmaceutical products are according to the standard operating procedures adopted by the DGPA&DC for any medicines available in the country. If any counterfeit medicines, or other finished products that are labeled as being of herbal origin but are adulterated with chemical ingredients, are detected in the country, or notified from international sources, these are published as circulars on the MOH website and other social media platforms for wider circulation. The Drug Information section of DPV&DI is responsible for scanning all relevant competent authorities' websites for any new updates or additional safety issues or drug-drug interactions/food-drug interactions associated with any herbal products available. These updates are circulated to all healthcare providers who are dealing with such products.

26.3.2 International Collaboration in Pharmacovigilance for Herbal Medicines

The DPV&DI, in collaboration with the WHO office in Oman, held a workshop on "Regulation of Herbal Medicines in Oman: Challenges and Solutions," which involved stakeholders for the herbovigilance system. An outcome of the workshop was that a plan of action in line with the WHO Traditional Medicines Strategy, 2014–2023 was proposed (Box 26.1).

Box 26.1 Proposed Plan of Action for Herbovigilance

Short-term plan of action:

1. Begin collecting case reports in relation with herbal product. Review the reporting form to build a national database from which signals could be detected.
2. Sensitize stakeholders to the reporting.
3. Questionnaire survey about knowledge, perceptions, and practices of Herbovigilance in different health establishments.
4. Availability of reporting form (Facilitate access to reporting form).
5. Insertion of the theme of Herbovigilance into the programs of medical, pharmaceutical, and scientist days.

Long-term plan of action:

1. Collaboration with universities and faculties for thesis work on Herbovigilance.
2. Intensive monitoring programs by conducting studies in these establishments.
3. Identify how T &CM information is communicated through practitioners, product advertising, practices, and the media.

As well as with WHO, the PV for Herbal Medicines section has active communication and collaboration with other Gulf Cooperation Council (GCC) states, the Middle East and North Africa (MENA) members, the International Society of Pharmacovigilance (ISoP), and Special Interest Group (SIG) on herbal and traditional medicines [10], which was formed in 2017.

References

1. Ghazanfar SA, Al-Al-Sabahi AM (1993) Medicinal plants of Northern and Central Oman (Arabia). Econ Bot 47:89–98. https://doi.org/10.1007/BF02862209
2. Biological of Biodiversity. Environmental Authority, Sultanate of Oman. https://www.meca.gov.om/en/the-authority/authority-mandates/nature-conservation/biological-diversity/?csrt=14627527364481671471. Accessed 8 Mar 2021
3. Lupton D, Al Jabri T, Al Hinai A, Al Hatmi S (2012) People and plants in the Sultanate of Oman: ethnobany at Oman botanic garden. BGjournal 9(2):26–29
4. Al Saadi NA, Al-Khafaji HCA, Al-Saifi RH (2017) The Oman animal and plant genetic resources center's (OAPGRC) science café: a success story. Open Sci J 2(4). https://doi.org/10.23954/osj.v2i4.870
5. Divakar MC, Al-Siyabi A, Varghese SS, Rubaie MA (2016) The practice of ethnomedicine in the northern and southern provinces of Oman. Oman Med J 31(4):245–252. https://doi.org/10.5001/omj.2016.49

6. Shenoy R, Bialasiewicz A, Khandekar R, Al Barwani B, Al Belushi H (2009) Traditional medicine in Oman: its role in ophthalmology. Middle East Afr J Ophthalmol 16(2):92–96. https://doi.org/10.4103/0974-9233.53869
7. WHO Guidelines on safety monitoring of herbal medicines in pharmacovigilance systems. http://apps.who.int/medicinedocs/documents/s7148e/s7148e.pdf. Accessed 8 Mar 2021
8. Directorate General of Pharmaceutical Affairs and Drug Control. https://www.moh.gov.om/en/web/dgpadc/about-us. Accessed 9 Mar 2021
9. Registration of herbal companies and product. Ministry of Health, Oman. https://www.moh.gov.om/en_US/web/dgpadc/-2. Accessed 9 Mar 2021
10. Barnes J (2020) The International Society of Pharmacovigilance (ISoP) Special Interest Group on Herbal and Traditional Medicines: towards progress in pharmacovigilance for herbal and traditional medicines and other "natural health" products. Drug Saf 43(7):619–622. https://doi.org/10.1007/s40264-020-00937-0

Chapter 27
Pharmacovigilance for Herbal Medicines in Sudan

Ammar Abbas

27.1 Introduction

Sudan is a federal state in North East Africa encompassing a geographically strategic area of 1.87 million square kilometres between the Sahara desert to the north and west, the Red Sea and the Ethiopian plateau to the east [1], the rich savanna and equatorial marshlands to the south and south west and poor savanna to the west reflecting a rich variation in vegetation and medicinal plants. Sudan has a population of 42.8 million (2019) [2] and is equally ethnically rich in diversity as the country lies in the intersection between North Africa with its Mediterranean influence, the Middle East and Sub-Saharan Africa and has been subject to continuous migrations. Historically, Sudan is a land of very old civilisations, such as the Kingdom of Kush, which flourished at the same time as Ancient Egypt and shared many common traditions, religions and cultural traditional and complementary medicinal practices continuing to this present day. Arab migration into Sudan started around the beginning of the second millennium [3] adding new traditional medicinal practices and Islamic cultural influences [4]. Consequently, Sudanese herbal and traditional medicine is very diverse and is widely practised as access to conventional healthcare is variable across the country with an average of 2.5 physicians per 10,000 population [5].

In Sudan, 90% of the rural population depends mainly on traditional medicine for its healthcare, since admission to hospitals and obtaining modern synthetic medicines are limited and a high percentage of the population is nomads [6]. The prevalence of use of herbal medicines varies slightly in urban areas, but is still considered high [7]. Sudanese medicinal plants are associated with a wide spectrum of

A. Abbas (✉)
Countess of Chester Hospital NHS Foundation Trust, Chester, UK

Pharmacovigilance Committee—National Medicines & Poisons Board, Khartoum, Sudan
e-mail: ammar.abbas@nhs.net

J. Barnes (ed.), *Pharmacovigilance for Herbal and Traditional Medicines*,
https://doi.org/10.1007/978-3-031-07275-8_27

traditional medicinal uses, including antimicrobial indications, gastrointestinal disorders, malaria, diabetes, rheumatic pain, respiratory system disorders, jaundice, urinary system inflammations, wounds and, possibly, anticancer uses. In some cases, findings of pharmacological studies are in agreement with traditional uses; numerous bioactive compounds have been isolated from Sudanese medicinal plants, including flavonoids, saponins, alkaloids, steroids, terpenes, tannins, fatty acids and essential oils [6].

Commonly used medicinal plants in Sudan amongst many others include *Vachellia nilotica* (L.) P.J.H.Hurter & Mabb. (synonym: *Acacia nilotica* (L.) Willd. ex Delile (commonly known as 'garad')), *Adansonia digitata* L. (commonly known as 'tabaldi' or 'gongolaise'), *Hibiscus sabdariffa* L. (commonly known as 'karkadi' or 'karkade'), *Albizia amara* (Roxb.) Boivin (commonly known as 'arrada') and *Grewia tenax* (Forssk.) Fiori (commonly known as 'guddeim') [6].

27.2 Adverse Reactions Associated with Sudanese Herbal and Traditional Medicines

Literature describing adverse drug reactions associated with Sudanese herbal medicines is limited to a few individual case reports. Examples of these, for which causality has not been established, include:

- A case of per-apical abscess and burning ulcers (chemically induced) in a 40-year-old female patient developing a few weeks after short-term use of clove oil (Eugenol oil) obtained over the counter to treat dental pain and applied excessively [8].
- A case of intracerebral haemorrhage developing in a 78-year-old female who had been taking warfarin for pulmonary embolism for the past 3 months and which was controlled with an international normalised ratio (INR) of 2.5; this increased to 10, resulting in bleeding, after a few days of concurrent daily intake of anise tea (*Pimpinella anisum* L.) for indigestion [9].
- Drug-herb interactions:

Hibiscus sabdariffa L. flower juice (Malvaceae) has a potential interaction with the angiotensin-converting enzyme inhibitor (ACEI) lisinopril. A case report described an emergency admission due to profound hypotension following use of *H. sabdariffa* flower juice in a 70-year-old male patient taking lisinopril 10 mg daily [10]. Many effects of hibiscus are attributed to the strong antioxidant actions of the extracts and individual chemical constituents, particularly the anthocyanins. In a study, 10 mg lisinopril was given to 75 Nigerian subjects once daily for 4 weeks to determine the effect on renal function and the renin-angiotensin-aldosterone system. *H. sabdariffa* was associated with increased urine volume and creatinine clearance compared to placebo and lisinopril. A greater effect on systolic and diastolic blood pressure was reported for *H. sabdariffa* relative to lisinopril. *H. sabdariffa* and lisinopril both reduced plasma aldosterone, but there was no effect on serum angiotensin-converting enzyme or plasma renin activity compared to placebo [11]. Another documented interaction of *H. sabdariffa* is with chloroquine, a drug used in the treatment or prevention of malaria. Oral co-administration of chloroquine with hibiscus in healthy males resulted in a reduction in the area under the curve (AUC) and C of chloroquine leading to a potential reduction in antimalarial efficacy [12].

The extent of dependence on herbal remedies in Sudan can also have an indirect negative impact on treatment outcomes for some diseases, even if adverse drug reactions (ADRs) associated with use of herbal remedies are not experienced. For example, Sudan suffers from a high burden of mycetoma (an infectious disease affecting subcutaneous tissues) in some areas, and patients with mycetoma frequently present with advanced disease, partly because of delays in seeking conventional treatment whilst trying herbal medicines as they are accessible, cheap and available [13]. Additionally, field studies have demonstrated prevalent use of herbal remedies during the COVID-19 pandemic in Sudan [14].

27.3 Regulation of Herbal and Traditional Medicines in Sudan

In Sudan, the regulation of all medicinal products, including herbal and traditional medicines, is conducted by the National Medicines and Poisons Board (NMPB), which sits under the Federal Ministry of Health (FMOH). The NMPB mandates that all herbal medicines marketed as finished products (imported or locally manufactured) should be registered and subject to licensing procedures as human medicines [15]. However, a majority of herbal remedies commonly used in Sudan is not finished products: they may, therefore, come under the category of traditional medicines, or food supplements, and are not regulated. These preparations may be used within the diet, or may be recommended by herbalists and other traditional-medicine practitioners. The current traditional and complementary medicine (T&CM) policy of Sudan, published in 2016 [16], is based on the World Health Organization (WHO) Traditional Medicines Strategy 2014–2023 [17]. There are currently no formal professional training requirements or professional licensing procedures for herbalists and other traditional-medicine practitioners in Sudan, and previous attempts to regulate this sector within the state of Khartoum have not been successful (personal communication with Abdelrahman D (9 June 2021)). There has been recent legislation introduced within Khartoum state to regulate herbalist shops in terms of premises, clear labelling of products with ingredients, and restricting the sale of herbal products containing more than two ingredients; however, legislation fell short of licensing specific premises as herbal pharmacies regulated by law and overseen by registered health professionals (personal communication with Mustafa B (14 June 2021)).

27.4 Pharmacovigilance for Herbal and Traditional Medicines in Sudan

The current 2018 NMPB Guidelines for Detecting and Reporting Adverse Drug Reactions (ADRs) [18] encourage healthcare professionals to report all suspected ADRs associated with herbal and 'complementary health products'. However, as this policy is mainly aimed at registered conventional healthcare practitioners, such as doctors, pharmacists and nurses, a significant gap in capturing important safety data from the herbal and traditional medicine sector currently exists.

Sudan has been a full member of the WHO Program for International Drug Monitoring, managed by the Uppsala Monitoring Centre (UMC), since 2008. The main mode of monitoring the safety of medicinal products in Sudan is via a spontaneous reporting system overseen by the NMPB and which facilitates paper, online and telephone reporting of ADRs by healthcare professionals and patients/members of the public. However, under-reporting of ADRs is a major problem; before 2018, only 24 reports in total were submitted by the Sudan spontaneous reporting system to the Vigibase database of the UMC. After a substantial ADR reporting campaign was launched in 2018, 103 reports were submitted in 2018 [19]. A search of the database at NMPB revealed only one report relating to a registered herbal drug product and none relating to any herbal remedies or traditional medicines not currently subject to registration/licensing requirements (personal communication with Mohammed M (13 June 2021)). A potential technical limiting factor, alongside the perceived lack of awareness of the importance and mechanisms of reporting ADRs, and possibly the misperception that herbal medicines are safe, is that the ADR reporting form is not intuitive to facilitate the easy inclusion of essential product characteristics relating to unlicensed herbal products. These herbal entities often have variations in pharmaceutical quality, unknown methods of extraction from crude herbal raw materials, in addition to potentially multiple additional ingredients as well as the main herbal remedy being reported (personal communication with Mohammed M (13 June 2021)). These descriptive challenges are not unique and have been encountered previously by other spontaneous reporting systems, such as the UK Yellow Card Scheme [20]. A recent multinational published survey into the current status of the spontaneous reporting and classification/coding system for herbal and traditional medicine in pharmacovigilance included Sudan amongst respondents; due to the relative lack of reports, the response stated that traditional and herbal medicines were not included in the spontaneous reporting system of the NMPB without specifically clarifying that licensed registered herbal products are included [21].

27.5 Conclusion

The relative scarcity of pharmacovigilance data is inconsistent with the extent of use of herbal medicines in Sudan as described above. This warrants the urgent review of current pharmacovigilance strategies to address the under-reporting of ADRs to the

NMPB from across Sudan in general and with a specific focus on ADRs associated with herbal medicines given the extent of use in Sudan and the recorded unpublished stories from the public.

References

1. The World Bank (2021) Sudan at-a-glance. https://www.worldbank.org/en/country/sudan. Accessed 19 Sept 2021
2. The World Bank (2021) Sudan data. https://data.worldbank.org/country/sudan. Accessed 19 Sept 2021
3. Hasan YF (1967) Main aspects of the Arab migration to the Sudan. Arabica 14(Fasc. 1):14–31
4. Metz HC (1992) Sudan: a country study. Federal Research Division, Library of Congress, Washington
5. WHO Regional Office for the Eastern Mediterranean (2017) Sudan health profile 2015. https://extranet.who.int/iris/restricted/bitstream/handle/10665/254895/EMROPUB_2017_EN_19610.pdf;jsessionid=08B54951B8AB929361F566CB7B28B8E2?sequence=1. Accessed 19 Sept 2021
6. Karar MGE, Kuhnert N (2017) Herbal drugs from Sudan: traditional uses and phytoconstituents. Pharmacognosy Rev 11(22):83–103. https://doi.org/10.4103/phrev.phrev_15_15
7. Awad AI, Eltayeb IB, Capps PA (2006) Self-medication practices in Khartoum State, Sudan. Eur J Clin Pharmacol 62(4):317–324. https://doi.org/10.1007/s00228-006-0107-1
8. Alaa M (2016) Clove oil and dental pain. Sudan J Rational Use Med 11:6
9. Randa A (2016) Anise may cause serious bleeding when taken concurrently with warfarin. Sudan J Rational Use Med 11:7
10. Sarra R (2016) Karkadi serious interaction. Sudan J Rational Use Med 11:8
11. Herbal Medicines. www.medicinescomplete.com. Accessed 19 Sept 2021
12. Mikyung K, Yeonju W, Chang-ho H (2021) Current status of the spontaneous reporting and classification/coding system for herbal and traditional medicine in pharmacovigilance. Integr Med Res 10(1):100467. https://doi.org/10.1016/j.imr.2020.100467
13. Kunna E, Yamamoto T, Fahal A (2021) The use of traditional medicines among mycetoma patients. Trans R Soc Trop Med Hyg 115(4):297–306. https://doi.org/10.1093/trstmh/traa135
14. Mohamed AA, Humaida MA, Saeed AA (2020) Sudanese experience of herbal formulas used during covid-19 infection. Univers J Pharm Res 5(5):39–42
15. National Medicines and Poisons Board. www.nmpb.gov.sd. Accessed 19 Sept 2021
16. Traditional and complementary (T&CM) policy of Sudan (2016) Sudan's Federal Ministry of Health. http://sho.gov.sd/controller/knowledge_hub.php?mid=110&sm_id=132&lid=1. Accessed 19 Sept 2021
17. World Health Organisation (2013) The WHO traditional medicine strategy 2014–2023. https://www.who.int/publications/i/item/9789241506096. Accessed 19 Sept 2021
18. The current (2018) NMPB Guidelines for Detecting and Reporting Adverse Drug Reactions (ADRs). http://sho.gov.sd/controller/knowledge_hub.php?sm_id=132&mid=110&lid=1. Accessed 19 Sept 2021
19. Fadoul KM, Abbas A (2019) Positive reaction to Sudanese reporting campaign. Uppsala Reports. https://www.uppsalareports.org/articles/positive-reaction-to-sudanese-reporting-campaign/. Accessed 19 Sept 2021
20. Barnes J (2003) Pharmacovigilance of herbal medicines: a UK perspective. Drug Saf 26(12):829–851. https://doi.org/10.2165/00002018-200326120-00001
21. Izzo AA (2012) Interactions between herbs and conventional drugs: overview of the clinical data. Med Princ Pract 21(5):404–428. https://doi.org/10.1159/000334488

Epilogue

It is the 15th of June 2031, early in the morning. We are in a bustling hospital in one of the crowded and rather run-down areas of the city. In the pharmacy situated on the ground floor, next to the emergency department, the clinical pharmacist in charge is making herself a cup of tea before starting her working day. As she carefully strains the tea leaves, her eyes fall on a book left on the table. Recognising it, a smile lights up her face, and her thoughts wander back to the day she ordered the book, nearly a decade ago.

Having started her work in the hospital as a junior pharmacist a couple of years previously, she had impressed her manager by showing both dedication and a creative spirit, and soon was offered more challenging tasks, which she took on with great zest. When asked if she would lead a new project on 'pharmacovigilance for safer use of medicines and safer patients', her answer was an immediate 'yes!' From her own encounters she was familiar with adverse drug reactions, interactions and medication errors as sources of serious, sometimes fatal, harm to patients, and had for some time been concerned at the lack of a systematic follow-up of patients' medications. Now it was her opportunity to bring about real change.

Being a conscientious person, she wanted to make sure that she knew all about the latest thinking in pharmacovigilance: the science, the tools and the processes. As she assembled her reading list, she found to her surprise that there was relatively little published on the specific challenges involved in monitoring the safety of herbal products—and yet so many of the patients she met in her daily work relied heavily on the local traditional plant-based medicines as the primary remedies for many of their common illnesses. Contrary to the widespread belief that 'natural' means 'safe', it was long known among healthcare workers that herbal medicines, as well as conventional medicines, could cause adverse reactions. And she knew from her own experience that drug interactions involving herbals could be a problem, but one that often went undetected, unless you specifically asked the patient if they had taken any herbal medicines.

J. Barnes (ed.), *Pharmacovigilance for Herbal and Traditional Medicines*,
https://doi.org/10.1007/978-3-031-07275-8

After some searching, she had found the book she needed—the book you are now reading—and she was delighted when her manager agreed that she could buy it. She was confident that she now would acquire all the background knowledge needed to do a good job and was so looking forward to her first meeting with the project team.

¤¤¤

Lost in her reminiscences, her tea has gone almost cold. She still has a few minutes before the rush starts. As she fetches more hot water from the kettle, an idea springs to mind: she will write a paper about the progress made since they started the pharmacovigilance project 10 years ago. One of the main achievements, which she would like to describe, is how the project team managed to change the mindset of hospital leaders and clinical staff to move away from a reporting system where the focus was on the prescription, not on the patient. At the first team meeting, she had argued that pharmacovigilance should be seen as a tool for systematic learning and improvement, not only identifying problems with medicines. 'We need to move beyond simply recording that something went wrong, to be able to find out *why* it happened and *how* we can prevent it happening again', she contended. 'Good pharmacovigilance does not create problems, it solves them'. 'To succeed', she continued, 'we need to build our system based on a deep understanding of the culture, experience, values and attitudes of everyone involved'.

Initially, there were concerns about the cost, and a good deal of scepticism about the value of the new approach—'we already have a reporting system in place' and 'I don't get this patient involvement stuff' were among the comments raised. 'How can we call something a patient-focused system if we don't ask our patients what their key concerns and priorities are?' she had then asked. 'And we can't ensure a rational use of medicines', she asserted, 'if we don't understand the problems and issues facing hospital workers in their daily practice'.

It took some convincing, but once the project began, it soon became apparent that the time spent on discussions with patients and hospital staff, including health professionals and ward cleaners, was paying off. After 2 years, the team could report a reduction in both incidence and seriousness of patient harm caused by adverse reactions and interactions, and the number of mistakes and incidents with medications had also substantially decreased. It was a day of joy when it was decided to make pharmacovigilance a permanent function, with a reasonable budget for its continued operation.

With this support from the hospital management, the team was able to continue the developments: better prescription support, with automatic and manual check points throughout the prescription chain; reduced medicine waste, thanks to an individual dosing system for each patient; more, and more relevant, information from patients who have control of their complete medical health records on smart cards, and can grant access to their data for screening and studies; faster detection of potential new adverse reactions by continuous screening of electronic medical records; fewer interactions, since all medicinal and health products used, including herbal medicines, dietary/food supplements, and conventional non-prescription medicines, are recorded and included in the analyses; increasing competence and motivation of staff through an imaginative mix of classroom and digital education and training; and an ongoing dialogue with patients using a monitored chat group. An unexpected positive effect noted by the Human Resources team is that it has been able to recruit more competent staff, thanks to the hospital's growing reputation as a good place to work. Even the people in the finance department are content: overall costs have not increased, and the innovative research programme has attracted more funding.

And, most important, as a direct result of these efforts, many patients' lives have been saved and unnecessary suffering avoided. This, she thought, as she washed her cup and put it away, is what really counts.

Marie Edwards Lindquist
Uppsala Monitoring Centre
Göteborg, Sweden

GPSR Compliance

The European Union's (EU) General Product Safety Regulation (GPSR) is a set of rules that requires consumer products to be safe and our obligations to ensure this.

If you have any concerns about our products, you can contact us on ProductSafety@springernature.com

In case Publisher is established outside the EU, the EU authorized representative is:

Springer Nature Customer Service Center GmbH
Europaplatz 3
69115 Heidelberg, Germany

Batch number: 10732031

Printed by Printforce, the Netherlands